Nuh Bilgin
Hanifi Copur
Cemal Balci

TBM Excavation in Difficult Ground Conditions

Ernst & Sohn
A Wiley Brand

TBM Excavation in Difficult Ground Conditions
Case Studies from Turkey

Nuh Bilgin
Hanifi Copur
Cemal Balci

Ernst & Sohn
A Wiley Brand

Prof. Dr. Nuh Bilgin
Prof. Dr. Hanifi Copur
Prof. Dr. Cemal Balci

Istanbul Technical University
Faculty of Mines, Mining Engineering Department
34469 Maslak/Istanbul
Turkey

Cover: Methane Explosion in the Pressure Chamber of a Tunnel Boring Machine
Photo: Bilgin/Copur

Library of Congress Card No.:
applied for

British Library Cataloguing-in-Publication Data
A catalogue record for this book is available from the British Library.

Bibliographic information published by
the Deutsche Nationalbibliothek
The Deutsche Nationalbibliothek lists this publication in the Deutsche Nationalbibliografie;
detailed bibliographic data are available on the Internet at <http://dnb.d-nb.de>.

© 2016 Wilhelm Ernst & Sohn, Verlag für Architektur und technische
Wissenschaften GmbH & Co. KG, Rotherstraße 21, 10245 Berlin, Germany

Coverdesign: Sonja Frank, Berlin
Production Management: pp030 – Produktionsbüro Heike Praetor, Berlin
Typesetting: Reemers Publishing Services GmbH, Krefeld
Printing and Binding:

Printed in the Federal Republic of Germany.
Printed on acid-free paper.

Print ISBN: 978-3-433-03150-6
ePDF ISBN: 978-3-433-60722-0
ePub ISBN: 978-3-433-60720-6
eMobi ISBN: 978-3-433-60721-3
oBook ISBN: 978-3-433-60719-0

This book is dedicated to our lovely wives
Ayfer Bilgin, Nurten Copur and Nurgul Balci
and our beloved children
Damlanur Bilgin, Serkan and Busra Copur and Cem Eren Balci

Preface

The use of tunnel boring machine (TBM) tunneling has increased considerably in the past ten years in Turkey. It is planned to excavate 200 km of tunnels in the near future in Istanbul alone, and 100 km of tunnels in other parts of Turkey. Thirty new TBMs are predicted to start working in Istanbul during 2017.

The geology of Turkey is complex, and the country is in a tectonically active region; on a broad scale, the tectonics of the region are controlled by the collision of the Arabian Plate and the Eurasian Plate. The Anatolian block is being squeezed to the west. The block is bounded to the north by the North Anatolian Fault and to the south-east by the East Anatolian Fault. The effects of these faults are seen clearly on the performance of TBMs used in these regions.

This book is written with the intention of sharing the tunneling experiences gained during several years in difficult ground and complex geology. The methane explosion in an earth pressure balance (EPB) TBM chamber, the clogging of a TBM, the need to change disc cutters to chisel cutters, the need to change CCS-type discs cutters to V-type disc cutters, excessive disc cutter consumption, the optimum selection of TBM type in complex geology, magmatic inclusions or 'dykes', the effect of blocky ground on TBM performance, the mechanism of rock rupture in front of TBMs, TBM face collapses and blockages, the effect of opening ratio in EPB-TBMs in fractured rock, squeezing of the TBM or jamming of the cutterhead, probe drilling and the use of umbrella arching ahead of TBMs are discussed within this book.

We hope that the experiences shared in this book may help project designers and practicing engineers dealing with TBM drivages in complex geology in different parts of the world.

Istanbul, June 2016 Nuh Bilgin
Hanifi Copur
Cemal Balci

Acknowledgements

The contents of this book were discussed at some of the World Tunneling Congresses organized by the ITA, and some of the data has been published in different technical journals such as Tunneling and Underground Space Technology, Rock Mechanics and Rock Engineering and International Journal of Rock Mechanics and Mining Sciences. However, the topics in this book include more data and it has been analyzed more comprehensively. We are grateful to the organizers of the World Tunneling Congresses and cited journal authorities who gave us the chance to discuss the subject in advance.

The following contractors and state organizations have given us the opportunity to access field data and have helped in analyzing tunneling performances. Without their generous help it would have been impossible to write this book.

Contracting companies:
Aga Enerji
Celikler-Akad JV
Dogus
E-Berk
Eferay-Silahtaroglu JV
Fugro-Sial
Gulermak
Gulermak-Kolin-Kalyon JV
Intekar-Biskon JV
Kolin
Mosmetrostoy
NAS-YSE JV
NTF
Ozka-Kalyon Construction JV
Sargin Construction
Statkraft
STFA
Unal Akpinar
Yapi Merkezi
Yertas

Government organizations:
Istanbul Metropolitan Municipality (IBB)
State Hydraulic Works (DSI)
State Railway Organization (TCDD)
Istanbul Water and Sewerage Administration (ISKI)

Our thanks are also due to Istanbul Technical University authorities and Mining Engineering Department Staff, especially to Assoc. Prof. Dr. Deniz Tumac, and PhD students Emre Avunduk, Aydin Shaterpour Mamaghani and Ugur Ates. Finally we must thank also to Prof. Dr. Ergin Arioglu who has always encouraged us to write papers, articles and books since our school days.

About the Authors

Nuh Bilgin graduated from the Mining Engineering Department of Istanbul Technical University, and completed his PhD studies at the University of Newcastle, UK, in 1977 in "Mechanical cutting characteristics of some medium and high strength rocks". He joined the Mining Engineering Department of Istanbul Technical University in 1978. He was appointed as a full-time professor in 1989. He spent one year in Colorado School of Mines, USA, and one year in Witwatersrand University, South Africa, as visiting professor. He has published more than one hundred papers on mechanized mining and tunneling technologies, and is currently working in ITU and is chairman of Turkish Tunneling Society.

Hanifi Copur graduated from the Mining Engineering Department of Istanbul Technical University, and completed his PhD on rock cutting mechanics and mechanical excavation of mines and tunnels at the Colorado School of Mines in 1999. He worked as a research engineer at the Earth Mechanics Institute of the Colorado School Mines between 1995 and 1999, and has been working as an academician in the Mining Engineering Department of the Istanbul Technical University since 1999, and currently works in the same department. He has written many national and international scientific and industrial research reports and publications on mechanical mining and tunneling. He is a board member of the Turkish Tunneling Society.

Cemal Balci graduated from the Mining Engineering Department of Istanbul Technical University, and in 2004, received his PhD after completing his thesis on "Comparison of small and full-scale rock cutting tests to select mechanized excavation machines" at the Mining Engineering Department of Istanbul Technical University, Mine Mechanization and Technology Division. He worked with the research group of the Earth Mechanics Institute, Colorado School of Mines, between 2001 and 2004 as a researcher. He has published numerous papers on mechanized mining and tunneling technologies and has worked on many national and international implementation and research projects. He is currently working as full-time professor in ITU and is a board member of the Turkish Tunneling Society.

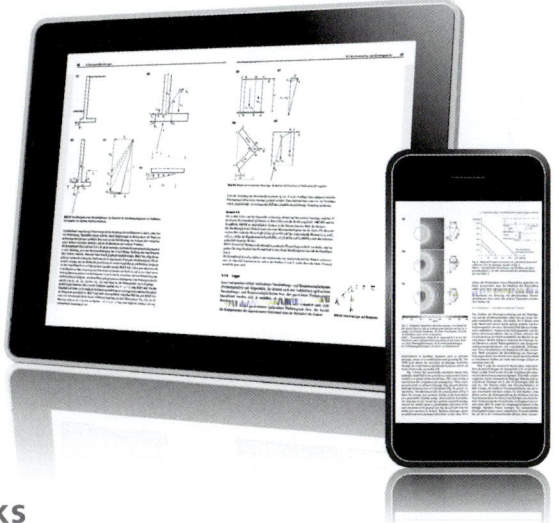

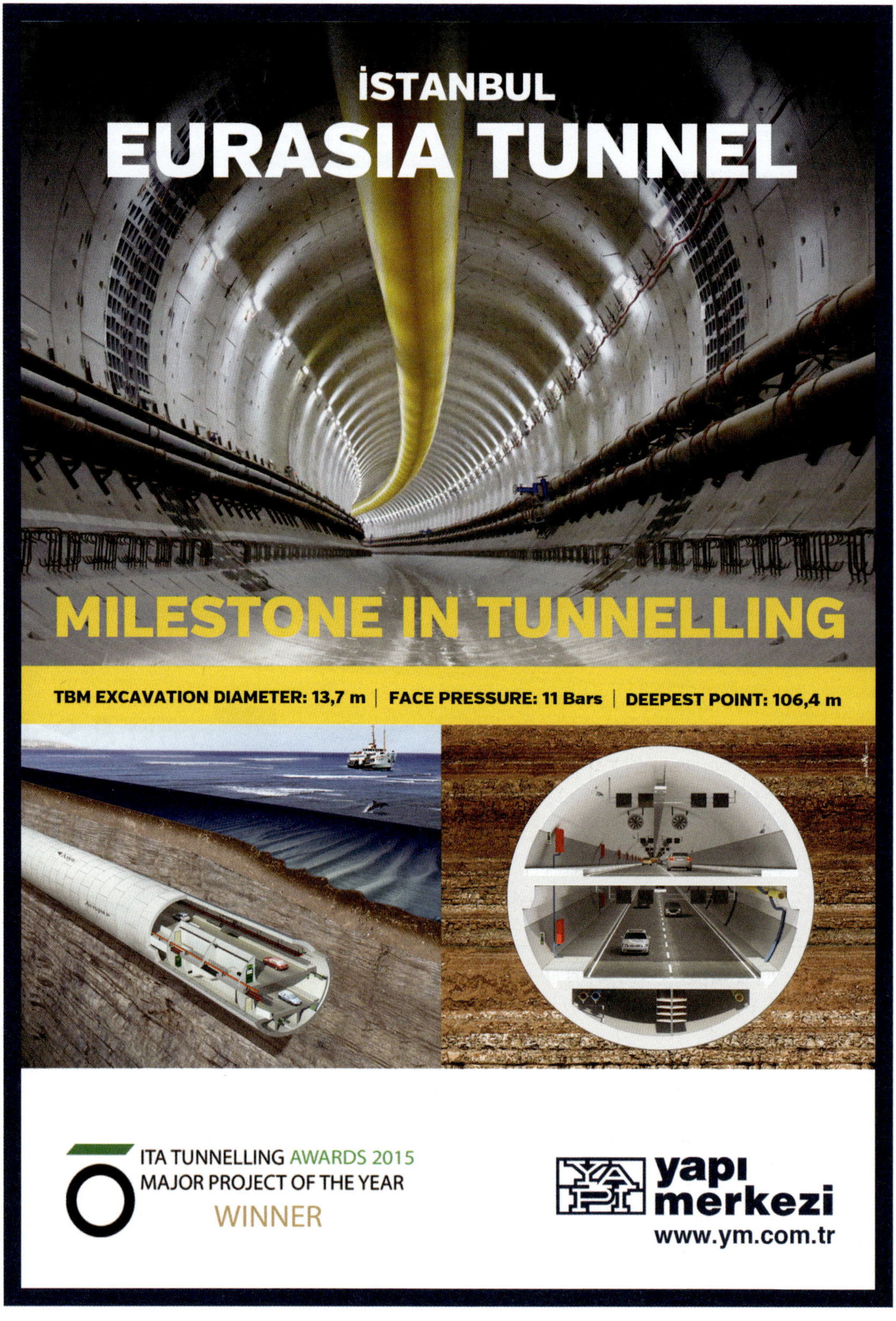

İSTANBUL
EURASIA TUNNEL

MILESTONE IN TUNNELLING

TBM EXCAVATION DIAMETER: 13,7 m | FACE PRESSURE: 11 Bars | DEEPEST POINT: 106,4 m

ITA TUNNELLING AWARDS 2015
MAJOR PROJECT OF THE YEAR
WINNER

yapı
merkezi
www.ym.com.tr

Table of contents

Preface VII

Acknowledgements IX

About the Authors XI

1   **Introduction** 1

2   **Geology of Turkey and Istanbul, expected problems, some**
         cuttability characteristics of the rocks 3
2.1   Introduction 3
2.2   Geology of Turkey 3
2.3   Geology of Istanbul 4
2.4   TBM performance in different projects in Istanbul 6
2.5   Description of geological formations in Istanbul, physical and
        mechanical properties 8
2.5.1   The stratigraphy of Istanbul and description of geologic formations 8
2.5.2   Typical frequency of RQD and GSI in the geological formations
        in Istanbul 16
2.5.3   Physical and mechanical properties of rocks taken from different
        geological formations 28
2.5.4   Schmidt hammer tests carried out in station tunnels in Uskudar–
        Umraniye metro line 37
2.6   Full-scale linear rock cutting tests with disc cutters in rock
        samples collected from different projects in Istanbul 40
2.6.1   Description of laboratory full-scale linear rock cutting test 41
2.6.2   Testing methodology and test results 43
2.6.3   Comparison of laboratory full-scale linear rock cutting test results
        with in-situ cutter performance – the effect of rock discontinuities 52
2.7   Conclusions 68

References 70

3   **Difficult ground conditions dictating selection of TBM type**
         in Istanbul 73
3.1   Introduction 73
3.2   Case study of open TBM in complex geology (1989), in
        Baltalimani tunnel: Why open type TBM failed 73
3.2.1   Collapse between chainage 0+920 and 0+935 km 78
3.2.2   Collapse between chainage 0+965 and 0+982 km 79
3.2.3   Collapse between chainage 1+148 and 1+155 km 80
3.2.4   Collapse between chainage 1+220 and 1+235 km 80
3.3   Double shield TBM in the Istanbul–Moda collector tunnel, 1989/90 80
3.4   Double shield TBM working without precast segment, difficulties
        in difficult ground: Tuzla-Dragos tunnel in Istanbul 80
3.5   Difficulties in using slurry TBMs in complicated geology,
        Marmaray tunnel project 84

3.6   Difficulties in single-shield TBM working in open mode in
         complex geology: An example from Kadikoy–Kartal metro tunnel 86
3.7   Eurasia tunnel excavated by a large diameter slurry TBM 88
3.8   Conclusions 88

References 90

4   Difficult ground conditions affecting performance of EPB-TBMs 91
4.1   Introduction 91
4.2   Factors affecting performance of EPB-TBMs 92
4.3   Performance prediction of EPM TBMs in difficult ground conditions 93
4.3.1   A model to predict the performance of EPB-TBMs
         in difficult ground conditions 94
4.3.2   Estimation of optimum specific energy from full-scale laboratory
         cutting experiments 95
4.3.3   Estimation of optimum field specific energy 95
4.3.4   Estimation of machine utilization time 98
4.3.5   Numerical example to estimate daily advance rate of an EPB-TBM 99
4.3.6   Comparison of predicted and realized EPB-TBM performance values 100
4.3.7   Verification and modification of the model for silty-clay and sand
         in the Mahmutbey–Mecidiyekoy metro tunnels 103
4.4   Conclusions 106

References 107

5   Selection of cutter type for difficult ground conditions 109
5.1   Introduction 109
5.2   Comparative studies of different type of cutters for Tuzla–Dragos
         tunnel in Istanbul – test procedure and results 109
5.2.1   Efficiency of chisel cutters as against disc cutters 111
5.3   The inefficient use of tungsten carbide studded disc cutters in the
         Marmaray–Istanbul project 114
5.4   Conclusions 115

References 116

6   Effects of North and East Anatolian Faults on TBM performances 117
6.1   Introduction 117
6.2   Kargi tunnel 117
6.3   Gerede tunnel 118
6.4   Dogancay energy tunnel 120
6.5   Nurdagi railway tunnel 122
6.6   Uluabat energy tunnel 123
6.7   Tunnels excavated by drill and blast methods 126
6.7.1   Ayas Tunnel: the most difficult tunnel in Turkey affected by North
         Anatolian Fault 126
6.7.2   Bolu tunnel 127
6.8   Conclusions 127

References 128

**7   Effect of blocky ground on TBM performance and the
        mechanism of rock rupture** 129
7.1   Introduction 129
7.2   Mechanism of rock rupture and face collapse in front of the TBM
        in the Kozyatagi–Kadikoy metro tunnels in Istanbul 131
7.2.1   Kozyatagi and Kadikoy metro tunnels and problems related to
        blocky ground 131
7.2.2   Mechanism of rock rupture and face collapse in front of the TBM 134
7.2.3   Other factors affecting the efficiency of tunnel excavation on the
        Kozyatagi–Kadikoy metro line 138
7.3   Conclusions 140

References 141

**8   Effects of transition zones, dykes, fault zones and rock
        discontinuities on TBM performance** 143
8.1   Introduction 143
8.2   Beykoz sewerage tunnel 143
8.2.1   Description of the project 143
8.2.2   Geology of the area 144
8.2.3   Description of the TBM 146
8.2.4   Effect of rock formation on chip formation and machine utilization time 147
8.2.5   Effect of dykes on tunnel face collapse and TBM blockage 149
8.2.6   Effect of transition zones on TBM performance 150
8.2.7   Effect of fault zones on TBM performance 152
8.3   Kartal–Kadikoy metro tunnels, methodology of understanding
        critical zones 153
8.3.1   Geology and physical and mechanical properties of rocks 154
8.3.2   Mechanism of face collapse and TBM blockages 155
8.3.3   Change of TBM performance in problematic areas 161
8.4   Conclusions 163

References 165

9   Squeezing grounds and their effects on TBM performance 167
9.1   Introduction 167
9.2   Basic works carried out on squeezing ground 167
9.3   Uluabat tunnel 168
9.3.1   Description of the project 168
9.3.2   Geology of the project area 169
9.3.3   Description of the TBM used and the general performance 171
9.3.4   Effect of TBM waiting time on squeezing 176
9.3.5   Effect of bentonite application on TBM squeezing 178
9.3.6   Conclusions 179
9.4   Kargi Tunnel 179
9.4.1   Squeezing of the cutterhead and related problems 179

9.4.2   Effect of Q values on squeezing of TBM 184
9.4.3   Discussions and conclusions on TBM swelling in Kargi project 185

References 186

10   Clogging of the TBM cutterhead 189
10.1   Introduction 189
10.2   What is clogging of a TBM cutterhead and what are the clogging
          materials? 189
10.3   Testing clogging effects of the ground 189
10.4   Mitigation programs to eliminate clogging 191
10.5   Clogging of TBMs in Turkish projects 192
10.5.1   Suruc Project 192
10.5.2   Selimpasa sewerage tunnel in Istanbul 196
10.5.3   Zeytinburnu Ayvalidere-2 wastewater tunnel project 201
10.6   Conclusions 207

References 208

11   Effect of high strength rocks on TBM performance 211
11.1   Introduction 211
11.2   Beykoz sewerage tunnel, replacing CCS disc cutters with V-type
          disc cutters to overcome undesirable limits of penetration for a
          maximum limit of TBM thrust 211
11.3   Nurdagi tunnel, full-scale cutting tests to obtain optimum TBM
          design parameters in very high strength and abrasive rock formation 213
11.4   Beylerbeyi–Kucuksu wastewater tunnel, TBM performance in
          high strength rock formation 216
11.5   Tuzla-Akfirat wastewater tunnel, TBM performance in
          high strength rocks 221
11.6   Conclusions 222

References 223

12   Effect of high abrasivity on TBM performance 225
12.1   Introduction 225
12.2   Determination of the abrasivity 229
12.3   Empirical prediction methods for disc cutter consumption 232
12.3.1   Colorado School of Mines (CSM) model for CCS type 17-inch
          single-disc cutters 232
12.3.2   Norwegian Institute of Technology (NTNU) model 233
12.3.3   Maidl et al. (2008) model for CCS type 17-inch single-disc cutters 238
12.3.4   Frenzel (2011) model for CCS type 17-inch single-disc cutters 238
12.3.5   Gumus et al. (2016) model for CCS type 12-inch monoblock
          double-disc cutters of an EPB-TBM 238
12.4   Examples of cutter consumptions on TBMs in Turkey 239
12.4.1   Tuzla-Akfirat wastewater project in Istanbul 240
12.4.2   Yamanli II HEPP project in Adana 250
12.4.3   Beykoz wastewater project in Istanbul 260

12.4.4   Buyukcekmece wastewater tunnel in Istanbul 262
12.4.5   Uskudar–Umraniye–Cekmekoy–Sancaktepe metro tunnel in Istanbul 264
12.5   Conclusions 269

References 271

13   Effect of methane and other gases on TBM performance 273
13.1   Properties of methane 273
13.2   Selimpasa wastewater tunnel, methane explosion in the pressure
          chamber of an EPB-TBM 275
13.2.1   Introduction to the Selimpasa wastewater project 275
13.2.2   Occurrence and causes of methane explosion in the Selimpasa
          wastewater tunnel 278
13.2.3   Consequences of methane explosion in the Selimpasa wastewater tunnel   279
13.2.4   Precautions against methane and excavation performance in the
          Selimpasa wastewater tunnel 280
13.3   Gas flaming in the Silvan irrigation tunnel 284
13.4   More gas-related accident examples for mechanized tunneling 287
13.5   Conclusions 290

References 291

14   Probe drilling ahead of TBMs in difficult ground conditions 293
14.1   Introduction 293
14.2   General information on probe drilling and previous experiences in
          different countries 293
14.3   Melen water tunnel excavated under the Bosphorus in Istanbul 295
14.4   Methodology of predicting weak zones ahead in the Melen water tunnel   297
14.4.1   Data analysis and results 300
14.5   Kargi energy tunnel 308
14.5.1   General information on the Kargi project 308
14.5.2   Probe drilling operations 309
14.5.3   Analysis of probe drilling data in the Kargi project 309
14.6   Conclusions 316

References 318

15   Application of umbrella arch in the Kargi project 321
15.1   Introduction 321
15.2   General concept of umbrella arch and worldwide application 321
15.3   Methodology of using umbrella arch in the Kargi project 322
15.4   Criteria used for umbrella arch in the Kargi project and the results .. 325
15.5   Conclusions 330

References 331

16   Index 333

1   Introduction

A man who carries a cat by the tail learns something he can learn in no other way.

~Mark Twain

This book is written with the intention of sharing the experiences gained in difficult ground conditions with TBMs in Turkey.

Turkey is in a tectonically active region; at a large scale, the tectonics of the region are controlled by the collision of the Arabian Plate and the Eurasian Plate. The Anatolian block is being squeezed to the west. The block is bounded to the north by the North Anatolian Fault and to the south-east by the East Anatolian Fault. The effects of North and East Anatolian Faults on TBM performances in Kargi energy tunnel, Dogancay energy tunnel, Nurdagi railway tunnel and Uluabat energy tunnels are explained in detail giving the causes, effects and precautions to be taken in order to eliminate the problems created by two large sets of faults. Some information is also given about the most difficult tunnels (Ayas and Bolu) that have ever been excavated by drill and blast method.

We believe that the Selimpasa and Silvan tunnels also provide unique experience since one suffered a methane explosion in the EPB chamber and the other hit a natural gas reservoir completely destroying a TBM and its related accessories.

The clogging of a TBM, as is encountered in clay-containing ground, has extensive consequences for the construction process and can severely affect the performance of the machine, increasing the torque, thrust and specific energy and lowering the advance rates with the extra cleaning efforts needed. Chapter 10 is written with the intention of clarifying the subject by giving three examples of tunneling projects in Turkey: Suruc Tunnel plus Selimpasa and Zeytinburnu Ayvalidere, two wastewater tunnels that were studied in detail in this respect. Experimental studies performed in the soil conditioning laboratory indicated that regular application of foam selected by the contractor was adequate to solve the sticking and clogging problems in Selimpasa, while an anti-clay agent different from the one selected by the contractor was suggested for Zeytinburnu. The representatives in the two cases applied the laboratory results in the field. The field measurements validated the experimental studies and the net advance rates of the EPB-TBMs increased at least 1.3 to 1.5 times and the stoppages due to clogging problems were reduced to normal ranges.

One of the most difficult tunnels ever opened in Istanbul was Beykoz sewerage tunnel, which encountered a complex geology. The need to change disc cutters to chisel cutters, CCS-type discs cutters to V-type disc cutters, excessive disc cutter consumption and TBM squeezing problems were also experienced in this tunnel.

Istanbul has a very complex geology, and in the near future the majority of TBM tunneling projects of Turkey are planned to be carried out in this fast-growing city. Bearing in mind this reality, the main objective of Chapter 3 is to show how the optimum selection of TBM type for Istanbul, has gradually changed from open type TBM (Baltalimani Tunnel), to double shield TBM (Moda-Tuzla Tunnel), to slurry type TBM

TBM Excavation in Difficult Ground Conditions. Case Studies from Turkey. First Edition. Nuh Bilgin, Hanifi Copur, Cemal Balci.
© 2016 Ernst & Sohn GmbH & Co. KG. Published 2016 by Ernst & Sohn GmbH & Co. KG.

(Marmaray tunnels) and finally to EPB-TBMs over the past 25 years. This gradually progressing selection based on the complex geology of Istanbul is a typical example to the concept of 'learning costs'. A model of the performance prediction of EPB-TBMs is also given based on experiences and data collected in several metro tunnels as Uskudar–Cekmekoy and Mahmutbey–Mecidiyekoy metro tunnels.

As already explained, Turkey is widely affected by two major fault systems, the North Anatolian and East Anatolian Faults. These two fault systems and magmatic inclusions 'dykes', fracture the host rock creating problematic blocky ground for TBM excavations. This problem is explained in Chapter 7 which is aimed to explain the effect of blocky ground on TBM performance and the mechanism of rock rupture in front of the TBM. Typical examples are given from Kozyatagi–Kadikoy Metro tunnels.

The causes and effects of TBM blockages are explained for Kadikoy–Kozyatagi metro tunnels. Eleven different TBM face collapses and blockages which have occurred in very complex geology within the Kadikoy–Kozyatagi Metro tunnels are analyzed considering TBM parameters such as opening ratio, working modes and geological parameters. It is determined that the TBM excavation parameters fluctuate while approaching the collapse regions, and these parameters show an increasing or decreasing trend in-site 'during collapse' region and it is concluded that this trend is a good indicator of face collapses, which will serve as a guide to foresee critical areas in front of TBM.

Squeezing of TBM or jamming the cutterhead is a nightmare for tunnel engineers, since it affects machine utilization time and realization of the project scheduled time. The salvation (rescue) of a jammed cutterhead can considerably reduce the mean advance rate. This problem was studied for Kargi, Uluabat and Dogancay tunnels, where the causes and effects of TBM squeezing are discussed with respect to remedial works needed for these three tunnels.

Cutter consumption is one of the most important cost items in mechanized tunneling due to replacement costs, cutting efficiency (penetration rate reduction with worn tools), and also man-hours spent on replacement. Yamanli II HEPP Tunnel, Buyukcekmece wastewater tunnel, Beykoz sewerage tunnel and Uskudar–Umraniye–Cekmekoy–Sancaktepe Metro Tunnels are detailed in this respect in Chapter 12.

Probe drilling ahead of a TBM is a time-consuming and tedious operation. If it is not interpreted correctly, it can give misleading results in complex geology. The research study summarized in this book shows that for correct interpretation of the drilling data, muck from the excavated area should be collected continuously for petrographic identification and strength tests. Two typical examples are Melen water tunnel and the Kargi Project. The experience gained in the umbrella arch in front of the TBM in the Kargi Project is also shared within this book.

2 Geology of Turkey and Istanbul, expected problems, some cuttability characteristics of the rocks

2.1 Introduction

Recent studies have revealed that tunnels were excavated under hundreds of Neolithic settlements all over Europe, and the fact that so many tunnels have survived 12,000 years indicates that the original networks must have been huge, from Scotland to Turkey. Some experts believe that the network was a way of protecting man from predators, while others believe that some of the linked tunnels were used like motorways are today, for people to travel safely regardless of wars or violence or even weather above ground [1]. There are several underground cities from Roman Imperial Times, and even older in Cappadocia, Nevsehir in Turkey, the historical underground cities are linked with a network of tunnels. Tunneling activities in Turkey have being carried out for centuries since then.

There are currently 1,700 hydropower projects, with more than 800 tunnels being excavated. Turkey is a mountainous country necessitating continuous road tunneling activities, all over the country. The national target up to 2023 is 330 km of road and highway tunnels and 78 km of railway tunnels. The need for metro tunnels in big cities is increasing, reflecting the growing population; the length of metro lines in Istanbul was 141 km in 2013 and it is expected that this will reach to 400 km by 2019. The length of utility tunnels to be built in the near future in Istanbul is expected to be around 85 km. Altogether, it is predicted that Turkey will invest more than 35 billion USD in tunneling projects in the near future, and the majority of tunnels will be driven using tunnel boring machines (TBM). The anticipated problems are directly related to the complexity of the geology of Turkey, including frequent face collapses, squeezing of TBM, high water inflow and excessive cutter wear. In this book a brief summary of the geology of Turkey, and Istanbul in particular, will first be given, describing the main geological formations with physical, mechanical and cuttability characteristics. It is hoped that the information given in this chapter will help to make rational decisions in designing and executing the tunneling projects.

2.2 Geology of Turkey

Turkey's varied landscapes are the product of a wide variety of tectonic processes that have shaped Anatolia over millions of years and which continue today, as evidenced by frequent earthquakes and occasional volcanic eruptions. Turkey's terrain is structurally complex. Nearly 85% of the land is at an elevation of more than 450 m, the median altitude of the country being 1,128 m. It is hard to explain the complexity of the geology of Turkey within the limited length of this book, but the following paragraph taken from a well-known paper published by Okay [2] explains the complexity of the geology.

"Geologically Turkey consists of a mosaic of several terranes, which were amalgamated during the Alpide orogeny. The relics of the oceans, which once separated these terranes, are widespread through the Anatolia; they are represented by ophiolite and

TBM Excavation in Difficult Ground Conditions. Case Studies from Turkey. First Edition. Nuh Bilgin, Hanifi Copur, Cemal Balci.
© 2016 Ernst & Sohn GmbH & Co. KG. Published 2016 by Ernst & Sohn GmbH & Co. KG.

accretionary complexes. The three terranes, which make up the Pontides, namely the Strandja, Istanbul and Sakarya terranes, have Laurasian affinities. These Pontic terranes bear evidence for Variscan and Cimmeride orogenies. Their Palaeozoic and Mesozoic evolutions are quite different from the Anatolide-Taurides. The Pontides and the Anatolide-Taurides evolved independently during the Phanerozoic and they were first brought together in the Tertiary. In contrast to the Pontic terranes, the Anatolide-Tauride terrane has not been affected by the Variscan and Cimmeride deformation and metamorphism but was strongly shaped by the Alpide orogeny. It was part of the Arabian Platform and hence Gondwana until the Triassic and was reassembled with the Arabian Platform in the Miocene. The Anatolide-Tauride terrane is subdivided into several zones mainly on the basis of type and age of Alpide metamorphism. The southeast Anatolia forms the northernmost extension of the Arabian Platform and shares many common stratigraphic features with the Anatolide-Tauride terrane. The final amalgamation of the terranes in the Oligo-Miocene ushered a new tectonic era characterized by continental sedimentation, calc-alkaline magmatism, extension and strike-slip faulting. Most of the present active structures, such as the North Anatolian Fault, and most of the present landscape are a result of this neotectonic phase."

As Bozkurt and Mittwede [3] stated, Anatolia forms a superb laboratory for the study of subduction, ophiolite obduction, continent-continent collision, metamorphism, the relationship between lithospheric deformation and magmatism, fold and thrust belts, suture zones, active strike-slip faulting, active normal faulting and associated basin formation.

All the complexity of the geology as explained above makes it really difficult to excavate tunnels in some parts of Turkey and this book is aimed at summarizing the possible difficulties that may occur during TBM excavation, with possible solutions given based on past experience.

2.3   Geology of Istanbul

Palaeozoic, Mesozoic and Cenozoic formations are recognized in Istanbul as seen in Figure 2.1 [4].

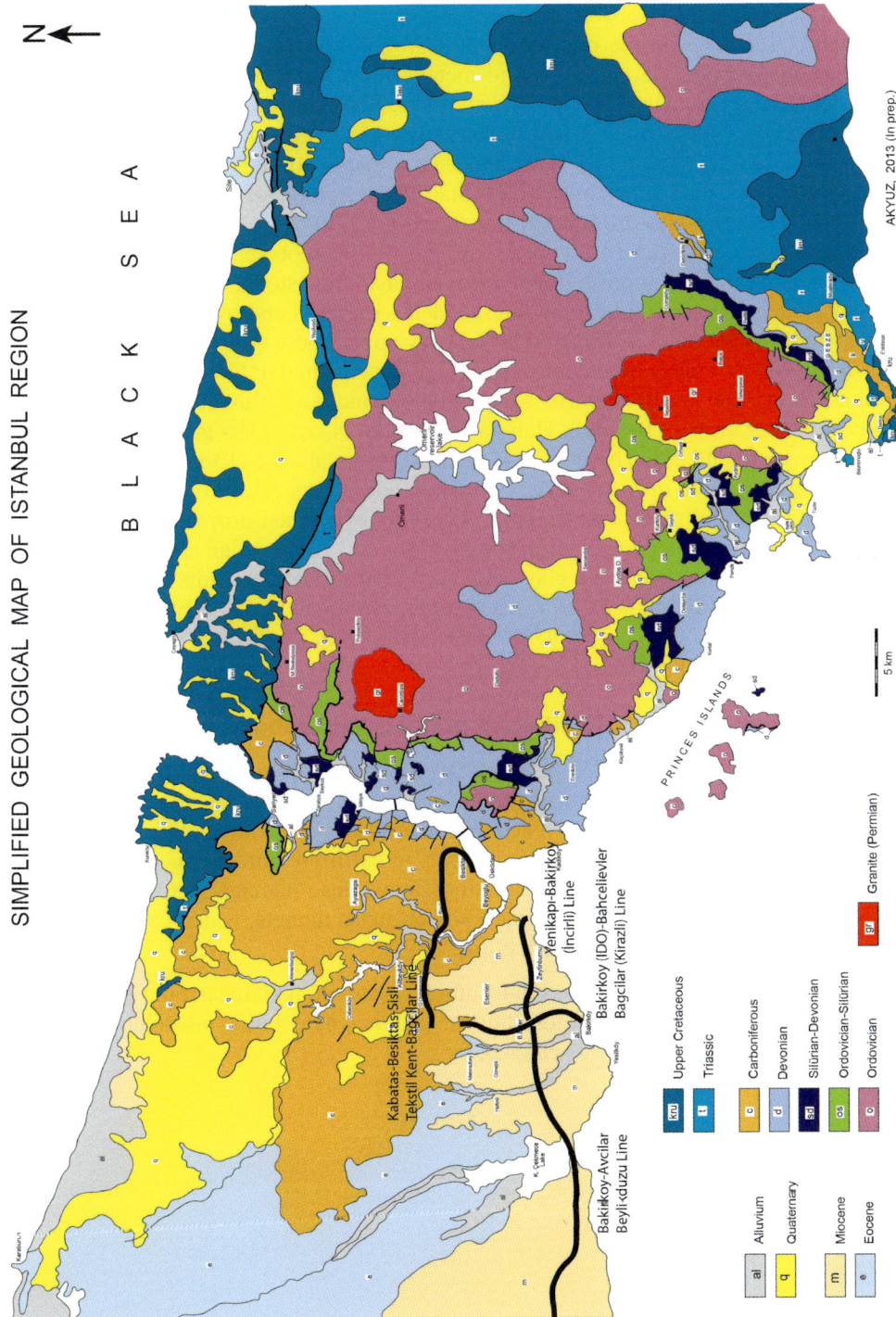

Figure 2.1 The geology of Istanbul [4].

The oldest rocks are Palaeozoic rocks consisting of quartz, quartz arenite and arcose. From the Ordovician to the middle of the Carboniferous there is a concordant rock sequence outcropping several thousand meters thick. This sequence mainly consists of variable facies of clastic and carbonate rocks. Some of the main characteristics of the Palaeozoic sequence are horizontal and vertical transitions, alternations of different rocks and lenticular structures. Palaeozoic rocks cover large areas of Istanbul. Rocks of Ordovician, Silurian and Devonian age outcrop mostly on the Asian side, and Carboniferous rocks are located mostly on the European side. Andesitic and diabasic dykes are also present. Triassic formations are represented by conglomerate, sandstone, dolomite, dolomitic limestone and clayey limestone, and Cretaceous units by sandstone, shale, and limestone interbedded with lavas and pyroclastic rocks. The Mesozoic rocks, in turn, are covered by Tertiary units. These units are generally fossiliferous limestone, clayey limestone or marl, and uncemented or loosely cemented sand, silt and clay. Halic and Bosphorus sediments, old and new alluvium and artificial fill cover the other units in Istanbul. Halic and Bosphorus sediments generally consist of black clay, sand and silt. The thickness of these materials is 5–13 m on the Marmara Sea coasts, and 60–70 m on the Halic (Golden Horn) coasts. Old and new alluvium can be seen in the Baltalimani, Istinye and Bebek Valleys on the European side, and close to the Kurbagali, Maltepe-Cevizli and Buyuk Rivers on the Asian side. They consist of uncemented gravel, sand and clay which originate from other units in the area. The thickness of the alluvium is 8–20 m, and artificial fill 2–40 m thick can be seen on the Historical Peninsula (Eminonu). This consists of gravel, sand, silt, clay, rubble, brick and tile fragments [5].

2.4   TBM performance in different projects in Istanbul

Some typical examples of TBM performance in different projects are given in Tables 2.1, 2.2 and 2.3. As seen from these tables main daily advance rates vary in different geologic formations. The main expected problems are frequent changing of the strata, dykes, ancient water wells etc. Sometimes in earth pressure balance (EPB) TBM applications, excessive ground deformations may cause damage to the surrounding buildings as experienced in the Otogar–Esenler metro tunnels, which caused an extra cost of 35.6 million USD of the project [4].

For an efficient tunneling operation in Istanbul it is essential to understand the behavior of TBMs in different geological formations. That is why detailed information of each stratum will be given within this chapter.

Table 2.1 Some of the completed metro tunnels [4].

Line	Kozyatagi–Kadikoy	Otogar–Kirazli	Basaksehir–Ikitelli
Geology	Kartal formation	Gungoren formation	Gungoren formation
TBM Type	2 Herrenknecht open mode + EPB	Herrenknecht +Lovat EPB	2 Lovat (EPB)
Cutterhead diameter, m	6.57	6.5	6.5
Total power, kW	2,000	1,622	2,100
Cutterhead torque, kNm	5,200 at 1.6 rpm 1,515 at 5.5 rpm	4,450 at 2.5 rpm 4,350 at 1.95 rpm	6,600 at 1.7 rpm 4,400 at 2.1 rpm
Max. thrust, kN	42,575	—	—
Best daily advance, m	19.6	25	29
Mean daily advance, excluding main stoppages, m	7.2; 7.7	11.3; 11.1	—

Table 2.2 Some of recent TBM applications in sewerage projects in Istanbul [4].

Line	Beykoz–Kavacik	Ambarli	Baltalimani Sariyer
Geology	Gozdag–Dolayoba–Kartal formation	Sediments-Mudstone Formation	Dolayoba–Tuzla–Trakya formation
TBM	Robbins	Herrenknecht	Herrenknecht
Diameter, m	3.175	4.6	2.915
Total power, kW	400	—	620
Torque, kNm	527 at 4.3 rpm 254 at 8.5 rpm	3,117	725
Max. thrust, kN	9,950	16,625	3,750
Best daily advance, m	23	28	24.8
Mean daily advance, incl. main stoppages, m	7.0	13.7	11.5

Table 2.3 Some of tunnels completed within Marmaray Project [4].

Line	TBM type	Thrust (kN)	Torque (kNm)	Cutting power (kW)	Mean advance rate (m/day)
H-1	EPB CAT	6,250	2,045	2,100	—
H-2	EPB CAT	6,250	2,045	2,100	13.5 m
H-3	Slurry, Hitachi-Zosen	7,500	4,767–10,582	2,000	5.1 m
H-4	Slurry, Hitachi-Zosen	7,500	4,767–10,582	2,000	7.50 m
H-5	Slurry, Hitachi-Zosen	7,500	4,767–10,582	2,000	4.6 m
H-6	Slurry, Hitachi-Zosen	7,500	4,767–10,582	2,000	3.9 m

H-1 and H-2 are excavated Miocene aged Bakirkoy and Gungoren formation. Marl, consolidated clay and sand lenses, H-4-6 re-excavated in Trakya formation interbedded sandstone, marl, siltstone and diabase dykes

2.5   Description of geological formations in Istanbul, physical and mechanical properties

2.5.1   The stratigraphy of Istanbul and description of geologic formations

The stratigraphy of Istanbul is summarized in Figure 2.2 [6].

Some of the geological formations are described below from old to young units [7].

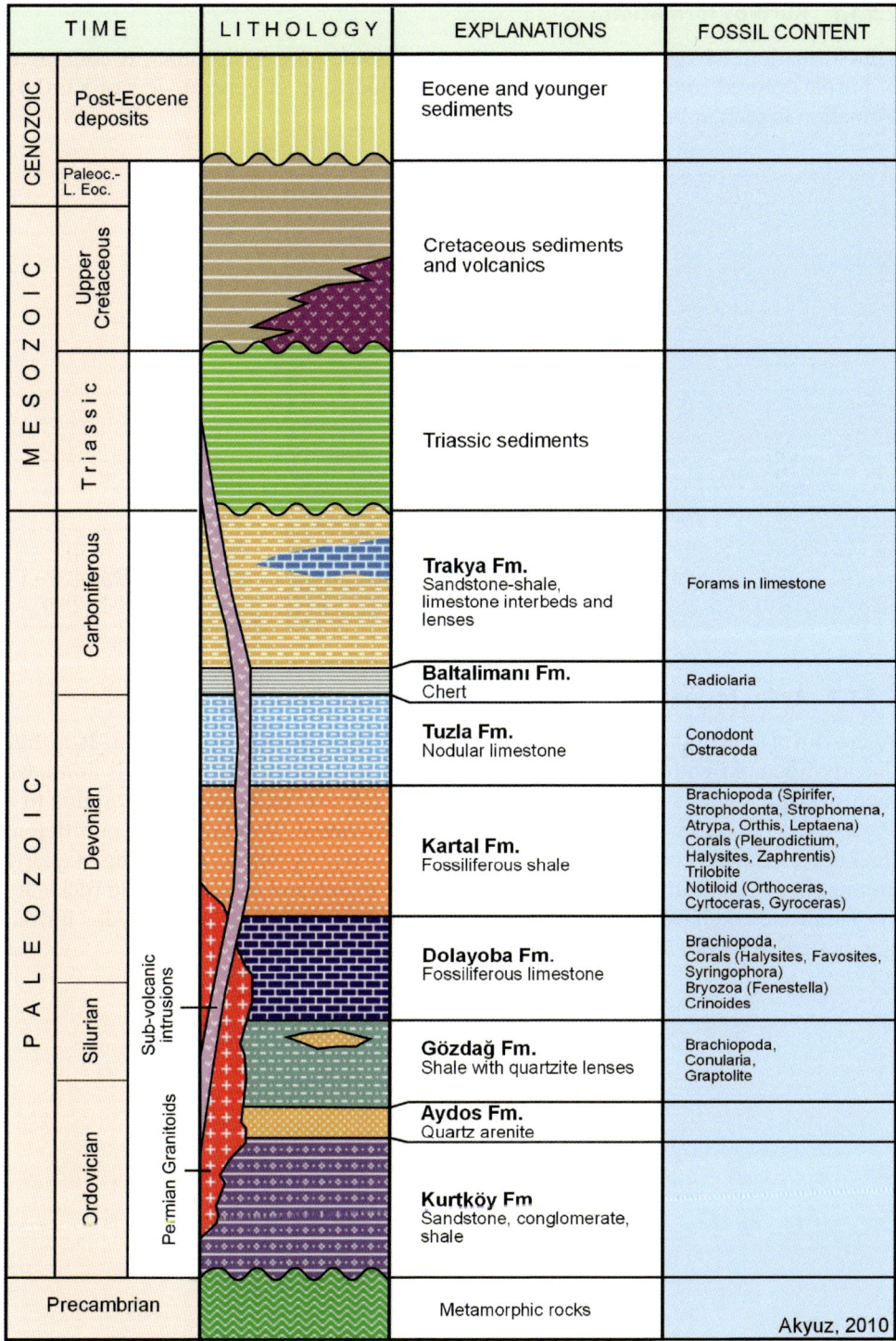

TIME		LITHOLOGY	EXPLANATIONS	FOSSIL CONTENT
CENOZOIC	Post-Eocene deposits		Eocene and younger sediments	
CENOZOIC	Paleoc.-L. Eoc.		Cretaceous sediments and volcanics	
MESOZOIC	Upper Cretaceous		Cretaceous sediments and volcanics	
MESOZOIC	Triassic		Triassic sediments	
PALEOZOIC	Carboniferous		**Trakya Fm.** Sandstone-shale, limestone interbeds and lenses	Forams in limestone
PALEOZOIC	Carboniferous		**Baltalimanı Fm.** Chert	Radiolaria
PALEOZOIC	Devonian		**Tuzla Fm.** Nodular limestone	Conodont Ostracoda
PALEOZOIC	Devonian		**Kartal Fm.** Fossiliferous shale	Brachiopoda (Spirifer, Strophodonta, Strophomena, Atrypa, Orthis, Leptaena) Corals (Pleurodictium, Halysites, Zaphrentis) Trilobite Notiloid (Orthoceras, Cyrtoceras, Gyroceras)
PALEOZOIC	Silurian		**Dolayoba Fm.** Fossiliferous limestone	Brachiopoda, Corals (Halysites, Favosites, Syringophora) Bryozoa (Fenestella) Crinoides
PALEOZOIC	Silurian		**Gözdağ Fm.** Shale with quartzite lenses	Brachiopoda, Conularia, Graptolite
PALEOZOIC	Ordovician		**Aydos Fm.** Quartz arenite	
PALEOZOIC	Ordovician		**Kurtköy Fm** Sandstone, conglomerate, shale	
Precambrian			Metamorphic rocks	Akyuz, 2010

Sub-volcanic intrusions
Permian Granitoids

Figure 2.2 Stratigraphy of Istanbul [6].

2.5.1.1 Kurtkoy formation

This formation, having a thickness of 150 m in some parts of Istanbul, is composed of purple-colored conglomerate, sandstone and mudstone. A general view of Kurtkoy formation is seen in Figure 2.3 [6].

Figure 2.3 Shale and sandstone within Kurtkoy formation [6].

2.5.1.2 Aydos formation

Aydos formation is situated in the upper level of the Kurtkoy formation. It is frequently fractured with rough surfaces filled with clay. The rocks within this unit have high strength values in most places not weathered; however, in some places, due to a shear zone, the rocks are highly fractured and weathered. It is laminated in different thicknesses, having colors ranging between white and pink-purple, consisting of quartz arenite, highly rich in feldspar with interbedded shales and siltstones. The rocks are very abrasive and have a range of high strength values. Typical view of Aydos formation is seen in Figure 2.4 [6].

Figure 2.4 Typical view of Aydos formation [6].

2.5.1.3 Gozdag formation

Gozdag formation is found in the upper level of Kurtkoy formation, having a thickness of 700 m in some places, consists of laminated shale layers, and in upper levels shales having medium thickness sandstone layers. Typical view of Gozdag formation is given in Figure 2.5 [6].

Figure 2.5 Typical view of Gozdag formation [6].

2.5.1.4 Doloyaba formation

This starts from the bottom with dark blue-gray limestone and later interbedded laminated mudstone–shale–limestone. In general, it is fractured, containing karstic cavities in some places.

A general view of Doloyaba formation is seen in Figure 2.6 [6].

Figure 2.6 Typical view of Doloyoba formation [6].

2.5.1.5 Kartal formation

This consists of fossilized shale and limestone, sometimes interbedded siltstone and sandstone, and it is widely found on the Asian side of Istanbul. Shale bands generally contain feldspars, mica and quartz minerals in sand grain size. Shale bands are highly fractured and have low strength characteristics, although limestone bands are relatively less fractured and have high strength characteristics. A general view of Kartal formation is seen in Figure 2.7 [6].

Figure 2.7 Typical view of Kartal formation [6].

2.5.1.6 Tuzla formation

Limestone is found with very thin laminated mudstone with a thickness of 50 m in most places. Typical view of Tuzla formation is seen in Figure 2.8. Kurtkoy, Aydos, Gozdag are Ordovician aged, Doloyoba is Silurian aged, Kartal and Tuzla formations are, Devonian aged [6].

Figure 2.8 Typical view of Tuzla formation [6].

2.5.1.7 Baltalimani formation

This is carboniferous aged and consists of shales with cherts and phosphate inclusions. It is very thinly laminated. Typical view of Baltalimani formation is seen in Figure 2.9 [6].

Figure 2.9 Typical view of Baltalimani formation with andesite dykes (with kind permission of Akyuz, S.).

2.5.1.8 Trakya formation

This is also carboniferous aged and is found widely all over Istanbul. It contains of laminated interbedded siltstone, mudstone and sandstone, frequently fractured. Fractures are filled with clay, calcite and silts. It is very abrasive and fractured in some places. Typical view of Trakya formation is seen in Figure 2.10 [6].

Figure 2.10 Typical view of Trakya formation (with kind permission of Akyuz, S.).

2.5.1.9 Belgrad formation

This consists of clay, sand and silt and is up to 50 m thick; it is found occasionally in the metro lines. Typical view of Belgrad formation is seen in Figure 2.11. It should be emphasized that the geologic formations are frequently cut with dykes of different thickness, fracturing and weathering the main rock [6].

Figure 2.11 Typical view of Belgrad formation. (with kind permission of Akyuz, S.).

2.5.1.10 Kusdili, Belgrad, Cekmece–Gungoren–Cukurcesme, Bakirkoy Suleymaniye, Ikitelli geological formation units

The formations cited above are Quartenary, Pliocene and Miocene aged formations consisting mainly of sand, sandy clay, silty clay and stiff clay, sometimes interbedded with sandstone bands and limestone. The geological formations in Istanbul are summarized in Table 2.4. In most cases silty clay has natural water content of 20–30%, liquid limit of 4–60, plasticity limit of 12–20 and plasticity index of 16. Typical size distribution of sand in these formations is as shown in Figure 2.12. However, it should be remembered that the size distribution of silty sand is quite variable.

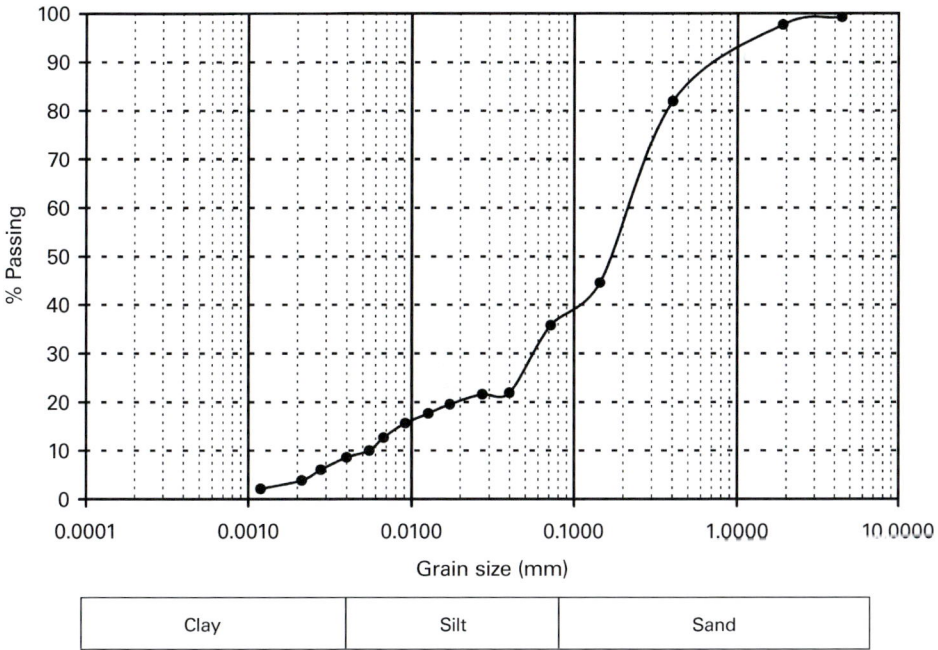

Figure 2.12 Typical grain size in silty sand in Gungoren–Cukurcesme formation.

Table 2.4 Summary of geological formations in Istanbul.

Geologic Age		Formation	Material
Cenozoic	Quartenary	Fillings	
		Alluvium Kusdili	Sand, clay
Cenozoic	Pliocene	Belgrad Cekmece–Gungoren–Cukurcesme	Uncemented clay, sand, silt and gravel, sandy clay, silty clay
		Bakirkoy	Clay, interbedded limestone
	Miocene	Suleymaniye	Mostly blue clay
		Ikitelli	Yellow sand, silt and gravel
Palaeozoic	Carboniferous	Trakya	Alternating sandstone, siltstone, claystone
		Baltalimani	Bedded chert, siliceous shale
	Devonian	Buyukada	Limestone, nodular limestone
		Kartal	Sandstone, limestone
	Silurian	Istinye	Limestone
		Dolayoba	Fossiliferous limestone
	Ordovician	Gozdag	Sandstone
		Aydos	Pink sandstone, conglomerate
		Kurtkoy	Sandstone, conglomerate, shale

2.5.2　Typical frequency of RQD and GSI in the geological formations in Istanbul

A statistical analysis was carried out on the drill cores obtained during side investigations for Kadikoy metro line, Phase 1 by Yuksel [8] and Phase 2 by Tuysuz [9]. RQD variations and GSI in different geological formations are given in Figures 2.13–2.33.

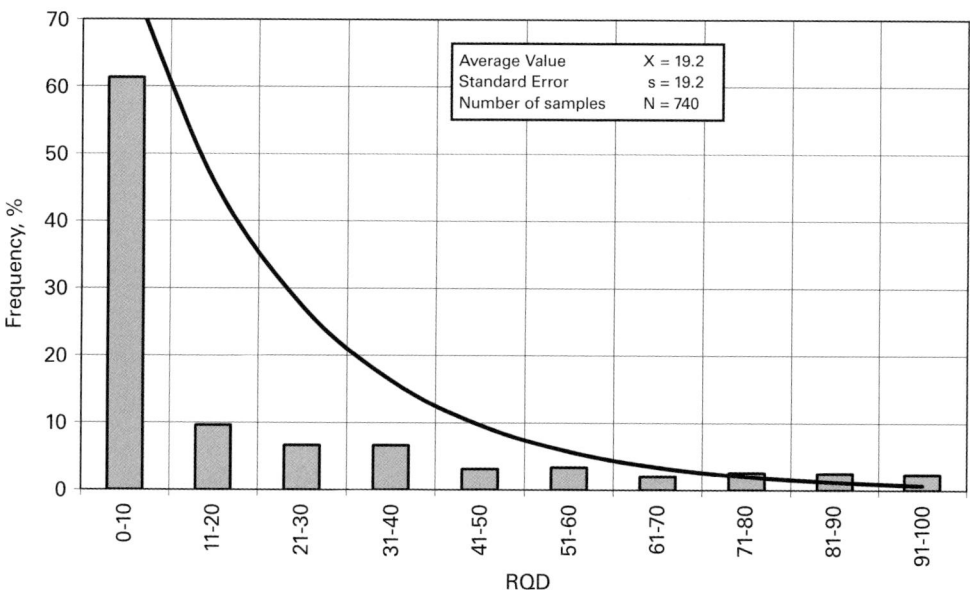

Figure 2.13 Variation of RQD in Trakya formation [8].

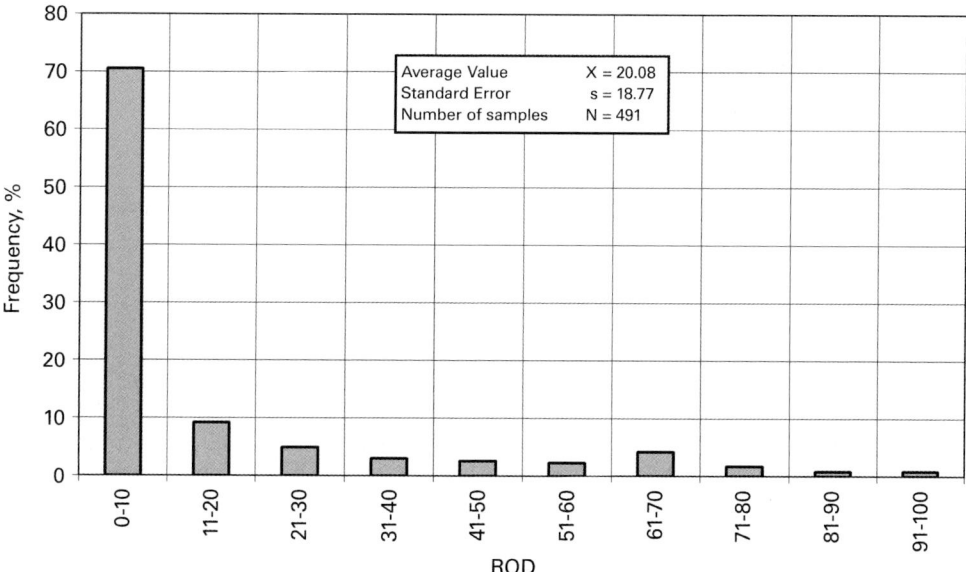

Figure 2.14 Variation of RQD in Trakya formation, [9].

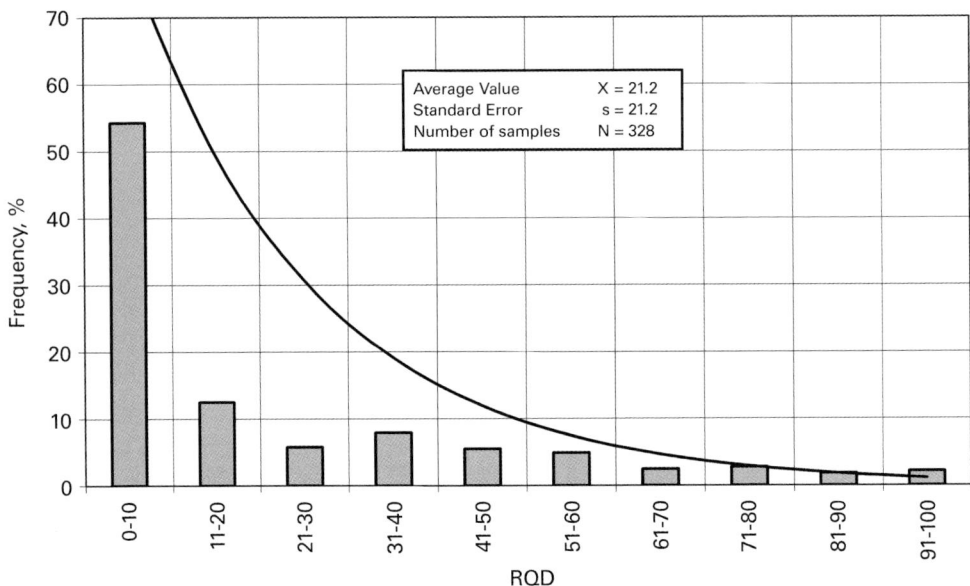

Figure 2.15 Variation of RQD in Kartal formation Zone A [8].

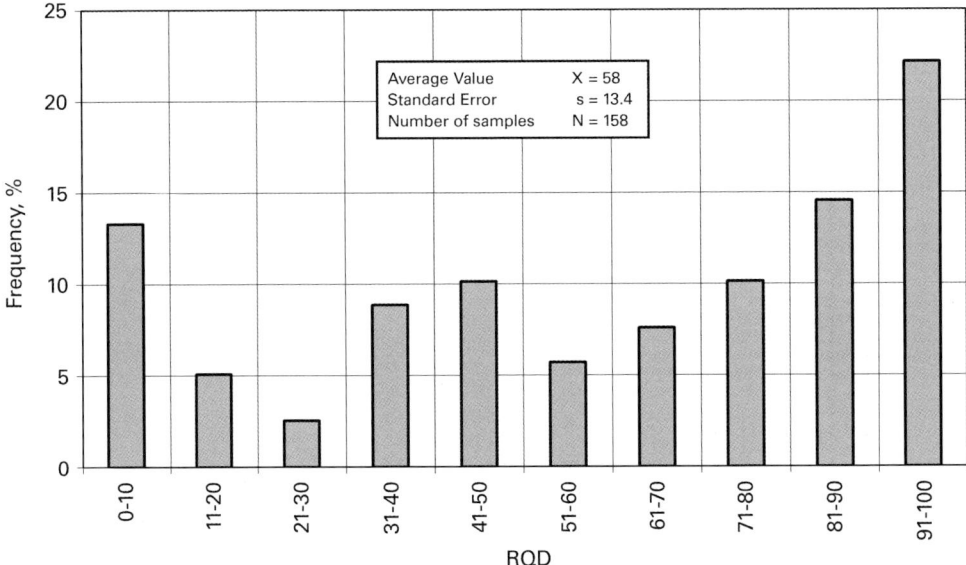

Figure 2.16 Variation of RQD in Kartal formation Zone B [8].

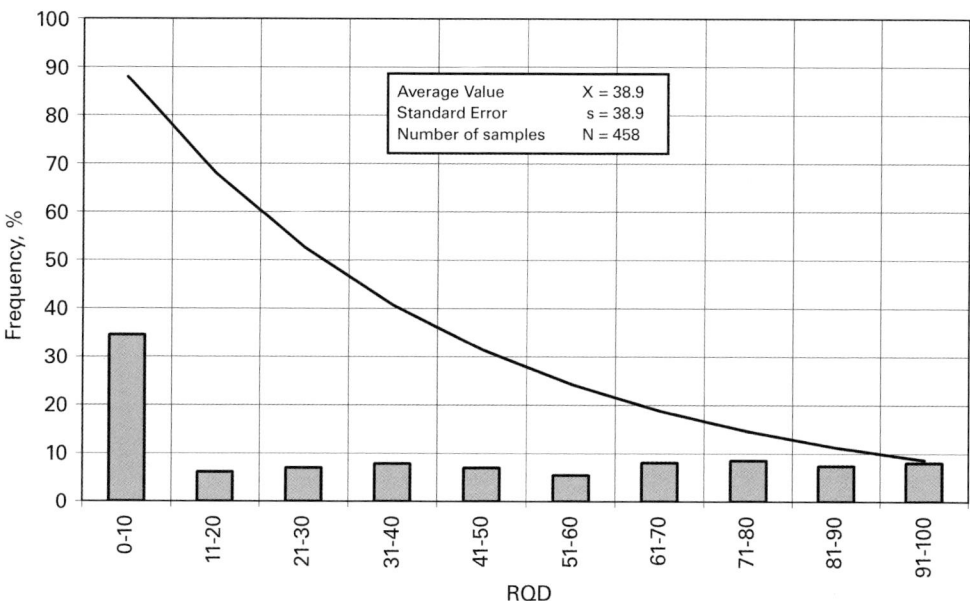

Figure 2.17 Variation of RQD in Kurtkoy formation RQD [8].

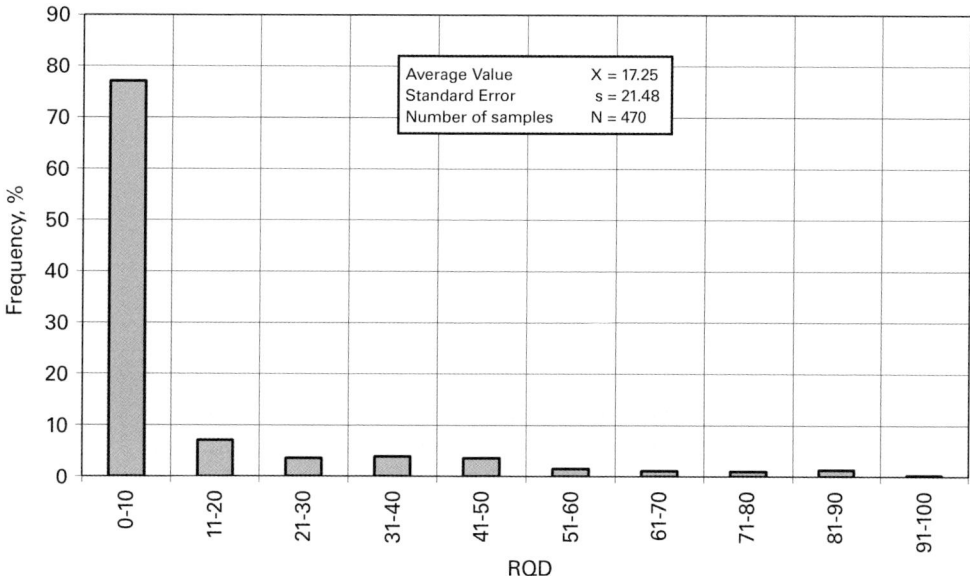

Figure 2.18 Variation of RQD in Kurtkoy formation, [9].

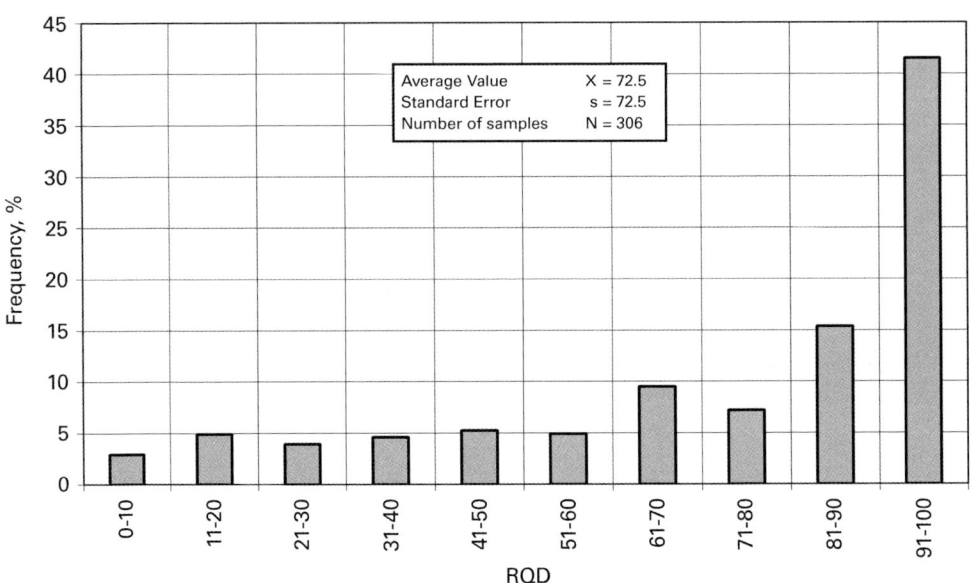

Figure 2.19 Variation of RQD in Dolayoba formation [8].

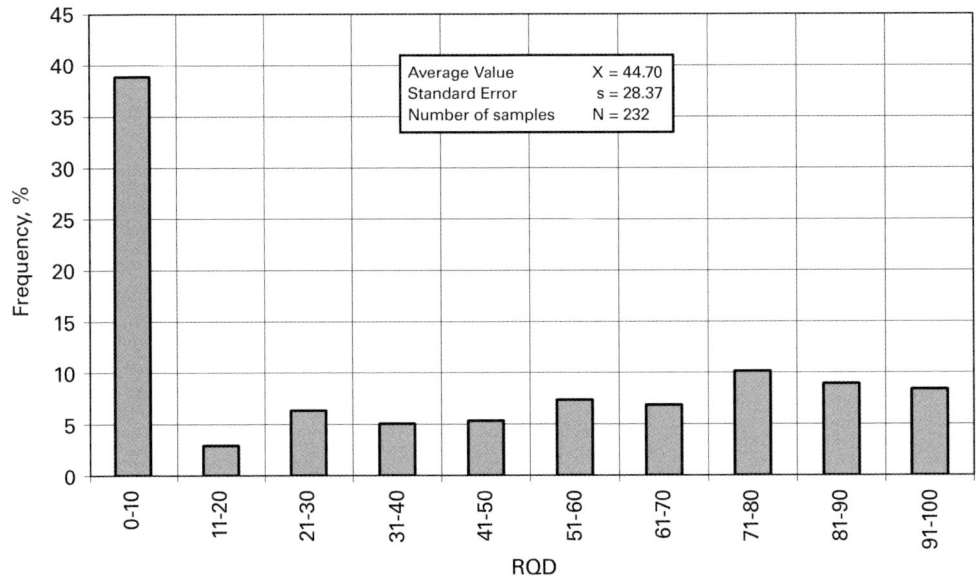

Figure 2.20 Variation of RQD in Doloyaba formation, [9].

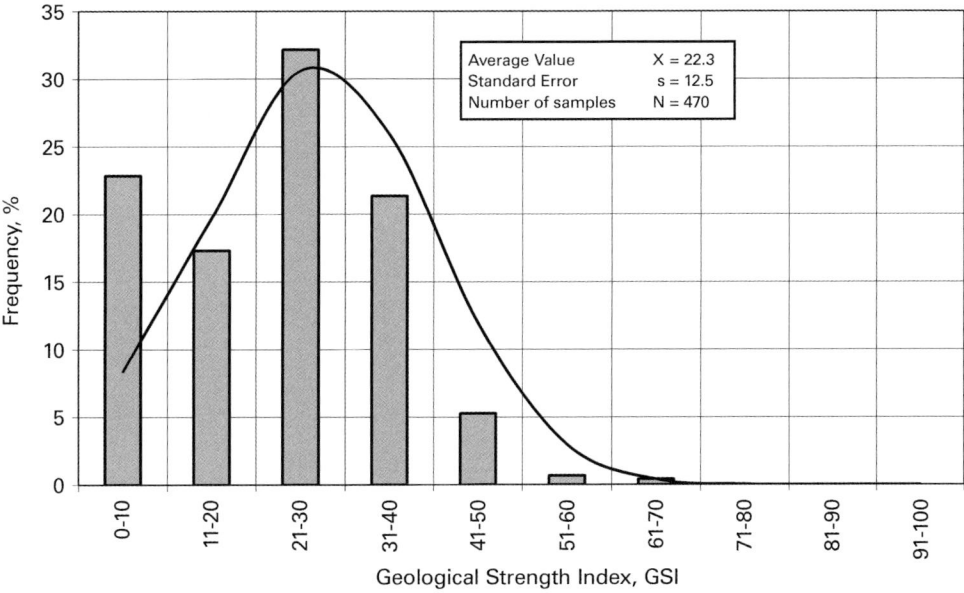

Figure 2.21 Variation of GSI in Trakya formation [8].

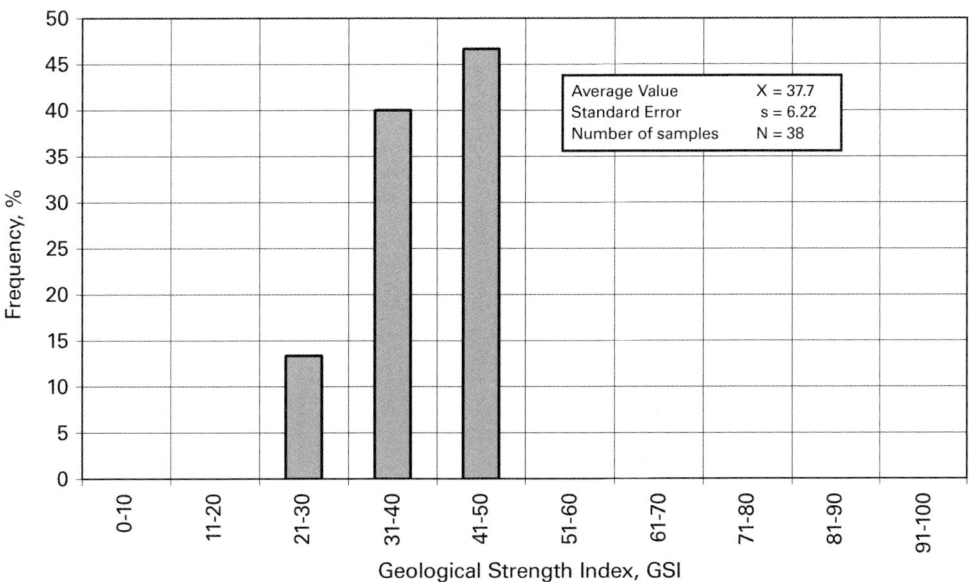

Figure 2.22 Variation of GSI in Trakya formation [9].

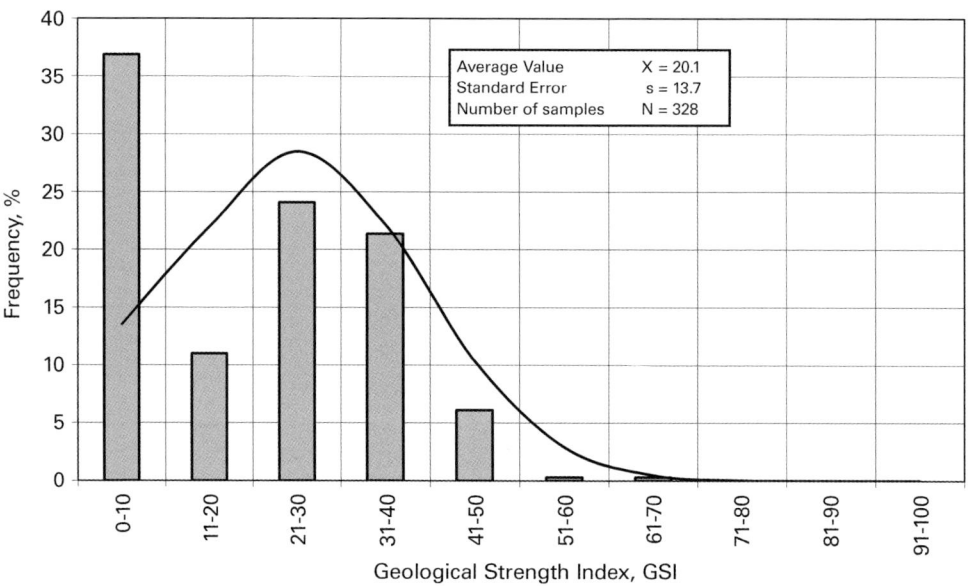

Figure 2.23 Variation of GSI in Kartal formation Zone A [8].

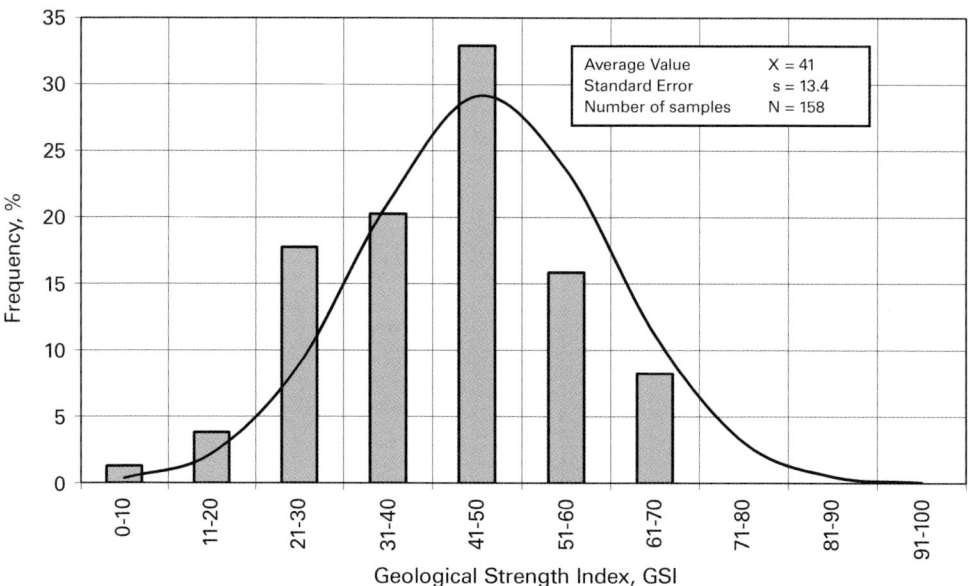

Figure 2.24 Variation of GSI in Kartal formation Zone B [8].

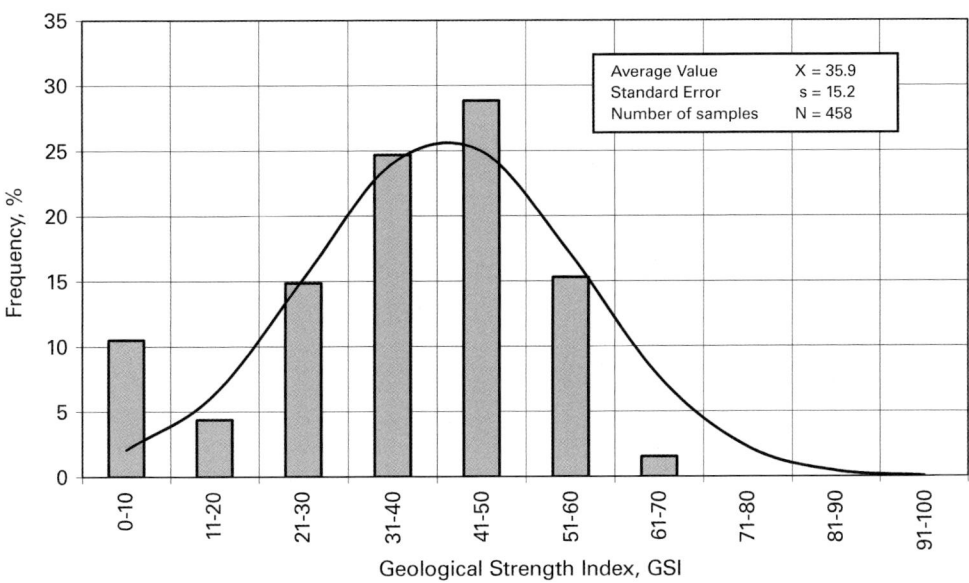

Figure 2.25 Variation of GSI in Kurtkoy formation [8].

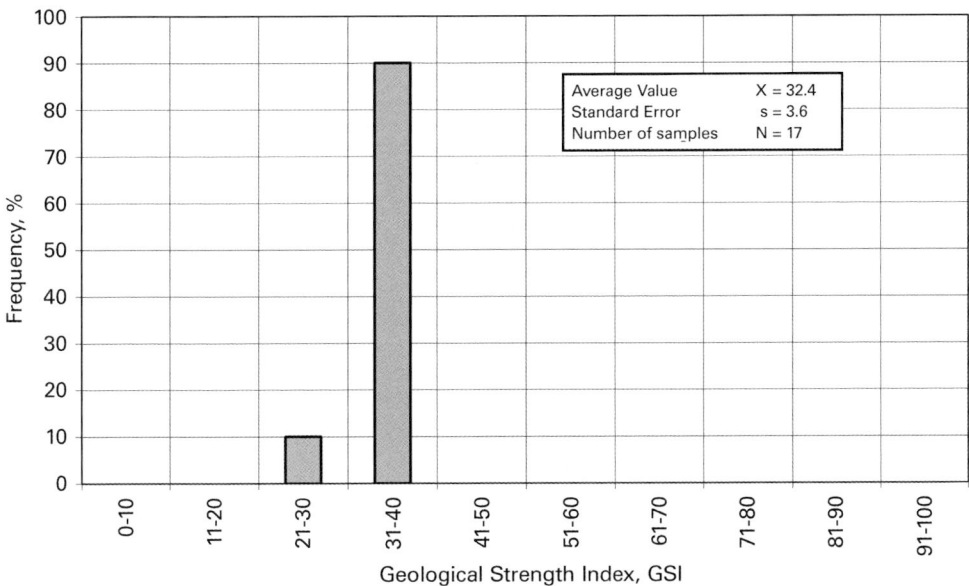

Figure 2.26 Variation of GSI in Kurtkoy formation [9].

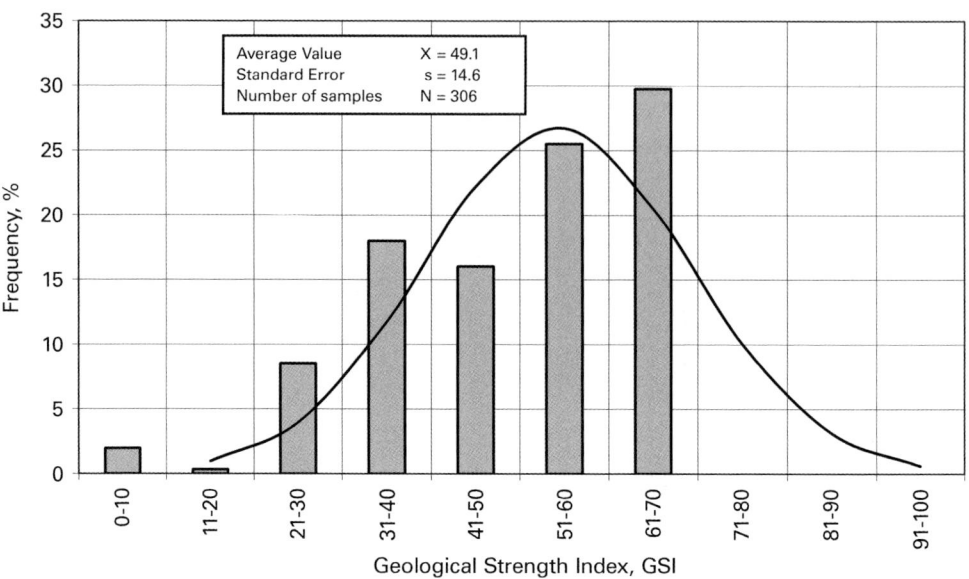

Figure 2.27 Variation of GSI in Kurtkoy formation [8].

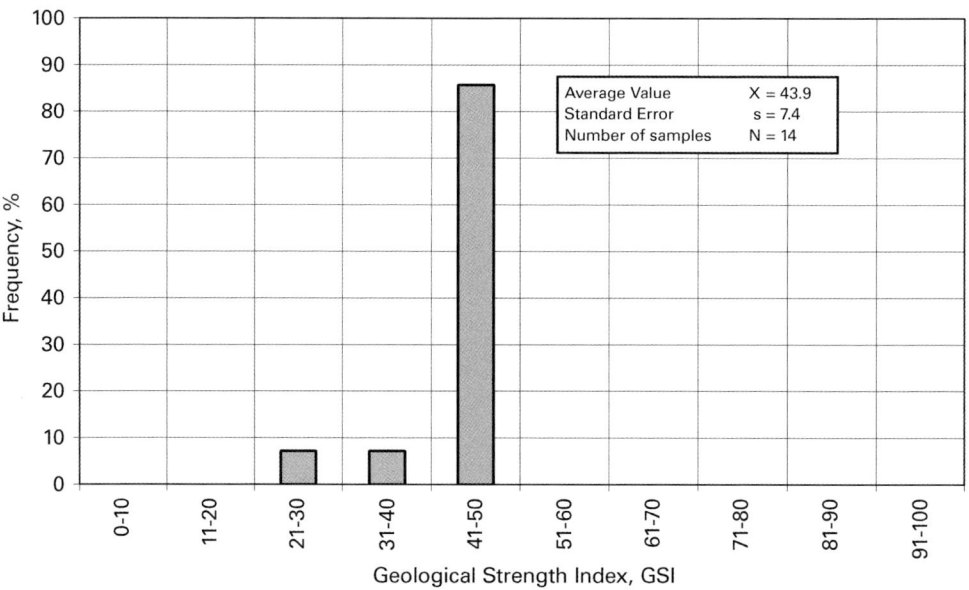

Figure 2.28 Variation of GSI in Doloyaba formation [9].

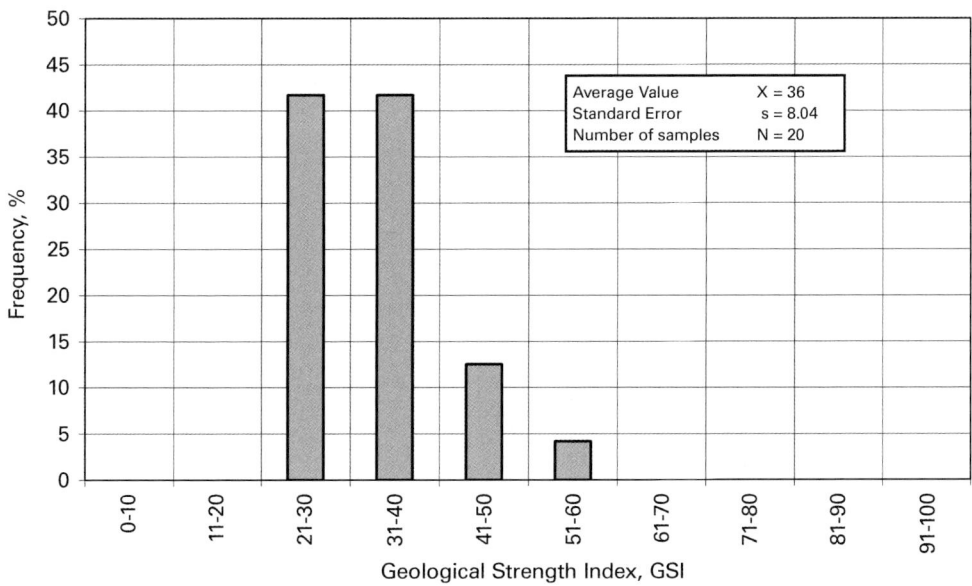

Figure 2.29 Variation of GSI in Gozdag formation [9].

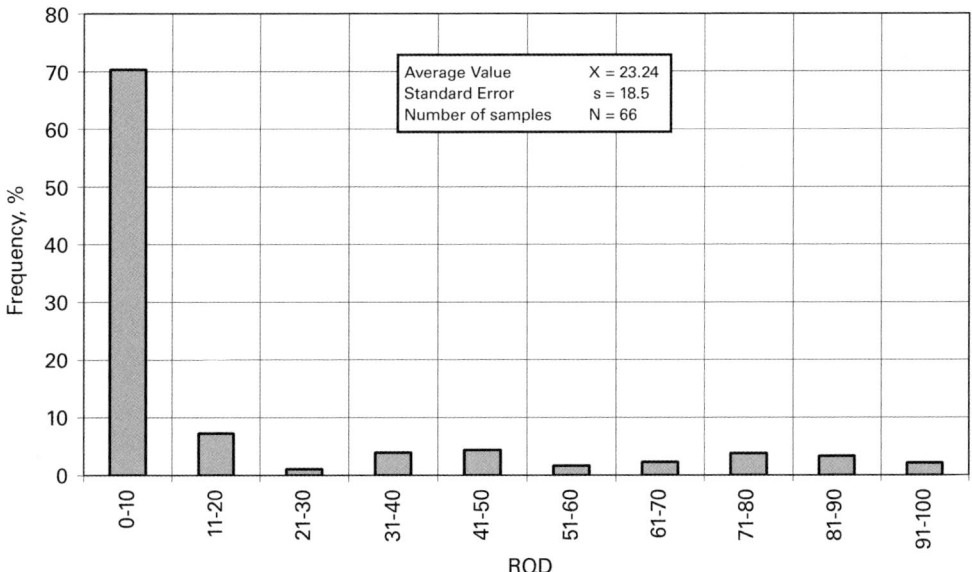

Figure 2.30 Variation of RQD in Tuzla formation (claystone shale) [9].

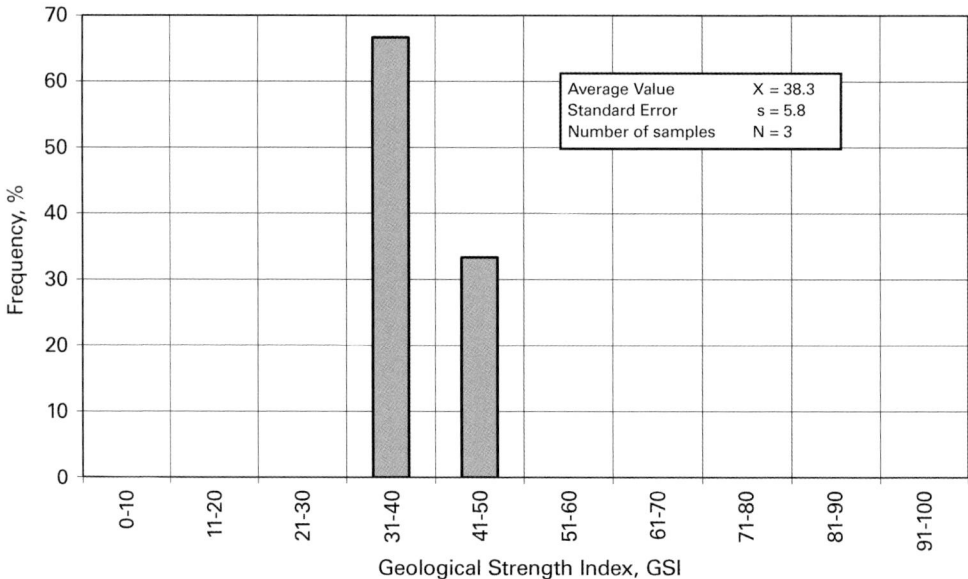

Figure 2.31 Variation of GSI in Tuzla formation (claystone-shale) [9].

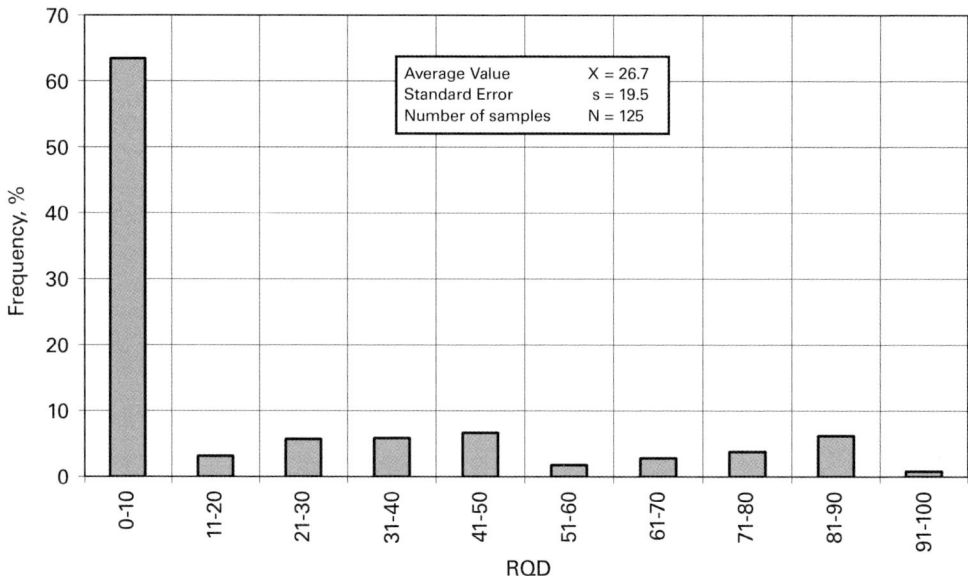

Figure 2.32 Variation of RQD in Tuzla formation (limestone) [9].

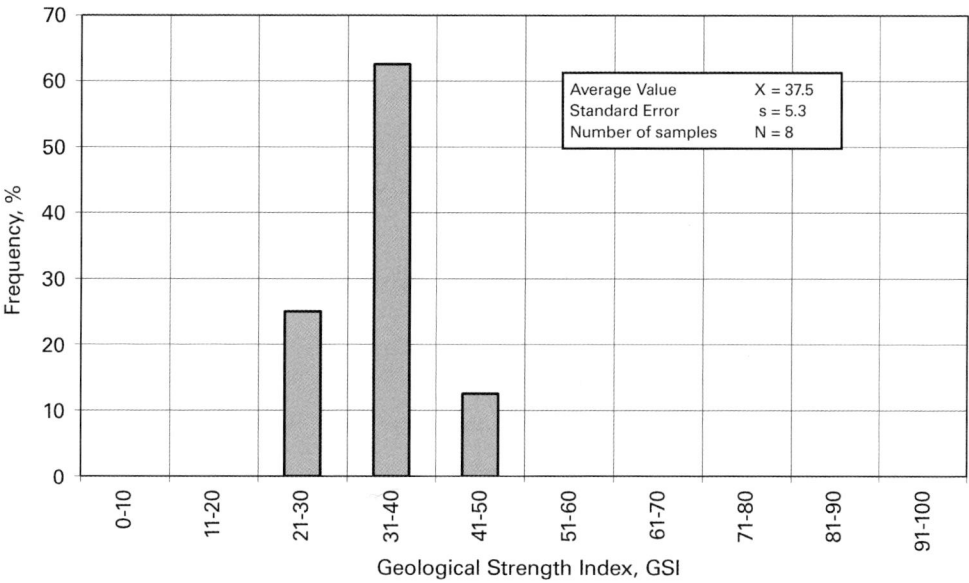

Figure 2.33 Variation of GSI in Tuzla formation (limestone) [9].

Geological strength index (GSI) for different drill holes is calculated according to Hoek–Brown criteria. The main advantage of this index is that in situ strength of the rock formation, cohesion, internal friction angle, elastic modulus and other geotechnical properties can be calculated empirically from this index value [7]. GSI shows a normal variation characteristic. According to the rate of structural property (SR) and the rate of surface discontinuity (SCR) as defined by Sonmez and Ulusay [10], Trakya and Kartal formation zone A can be classified as weak/fractured, Kurtkoy formation as blocky/medium fractured, Kartal and Dolayoba formations as blocky/good quality. As can be seen from these figures, Trakya formation and Kartal formation zone A may be classified as very weak rock, Kurtkoy formation as weak rock, Kartal formation zone B as middle quality rock and Doloyaba formation as good quality rock.

Laboratory test results realized for different geological formations are given in Tables 2.5–2.9. As noted in these tables, although the variance is low in rock density, Poisson ratio, the variance the other parameters are found to be high, especially in porosity. This may be due to different weathering characteristics of different samples.

The variance (V) in uniaxial compressive strength, elastic modulus and internal friction angle is around 50%. If the mean values of uniaxial compressive strength values are considered, Trakya, Kartal and Dolayoba formations may be classified as medium strength rocks, and Kurtkoy formation as a high strength formation. However, it should be emphasized that all the rocks show a wide range of variance in strength and in rock quality.

2.5.3 Physical and mechanical properties of rocks taken from different geological formations

Physical and mechanical properties of rocks taken from different geological formations are given in Tables 2.5–2.15.

Table 2.5 Physical and mechanical properties of rocks taken from Trakya formation [8].

Engineering properties	Symbol	Unit	Data number, n	Variation		Mean $\bar{X}$	Standard deviation sd	Variance V (%) (*)
				Min.	Max.			
Rock density	γ	(kN/m^3)	101	20.6	27.6	26.5	1	38
Porosity	n	(%)	37	0.37	27.5	4.85	6.30	130.0
Uni. compressive strength	σ_c	(MPa)	72	1.9	155.3	46.6	35.7	7.7
Tensile strength	σ_t	(MPa)	32	1.1	14.0	6.4	3.0	4.7
Elastic modulus	E	(MPa)	44	670	20,300	9,487.7	5,440.2	5.7
(Elast. mod./compre str.) ratio	E/σ_c	—	44	65:1	631:1	237:1	120:1	50.6
Poisson ratio	υ	—	31	0.18	0.42	0.27	0.05	18.5
Cohesion (tri axial)	c	(MPa)	8	4.0	32.5	13.9	9.3	6.7
Internal friction angle (tri axial)	Ø	(°)	8	27.4	51.7	37.9	9.2	24.3
Cherchar abrasivity index	CAI	—	18	0.50	2.50	1.37	0.59	43.1

Table 2.6 Physical and mechanical properties of rocks taken from Trakya formation [9].

Engineering properties	Symbol	Unit	Data number, n	Min.	Max.	Mean $\overline{X}$	Standard deviation, sd	Variance V(%) (*)
Rock qual. designation	RQD	(%)	16	0	36.6	10.5	11.9	113.5
Tock mass rating	RMR	(%)	16	33	65	48.8	8.2	16.9
Geological strength ind.	GSI	(%)	15	25	50	37.67	6.2	16.5
Internal friction angle	Ø	(°)	8	46.1	67.9	57.2	6.3	11.0
Cohesion	c	(MPa)	8	1.19	15	6.3	5.0	7.9
Compressive strength	σ_c	(MPa)	12	5.7	119.2	40.8	29.5	72.4
Elasticity modulus	E	(MPa)	18	450	29,800	6,393.9	6,776.1	106
Menard modulus	Em	(MPa)	10	20.7	266.8	224.1	74.2	33.1
Poisson ratio	υ	—	16	0.2	1.26	0.51	0.36	71.8
Permeability	p	(m/s)	13	1.1E-08	7.41E-06	2.51E-06	2.5E-06	97.9
Rock density	γ	(kN/m^3)	14	24.8	26.9	26.6	0.59	2.2
Porosity	n	(%)	7	0.3	7	3.9	2.8	70.7

Table 2.7 Physical and mechanical properties of the rocks taken from Kartal formation (Zone A) [8].

Engineering properties	Symbol	Unit	Data number, n	Variation		Mean values, $\overline{X}$	Standard deviation, sd	Variance V (%) (*)
				Min.	Max.			
Rock density	γ	(kN/m^3)	42	22.2	27.5	26.1	1.1	42
Porosity	n	(%)	19	0.95	10.51	4.24	2.81	66.2
Compressive strength.	σ_c	(MPa)	39	1.13	94.5	32.6	24.7	7.6

Engineering properties	Symbol	Unit	Data number, n	Variation		Mean values, $\overline{X}$	Standard deviation, sd	Variance V (%) [*]
				Min.	Max.			
Tensile strength	σ_t	(MPa)	20	1.8	9.1	4.8	2.2	4.6
Elasticity modulus	E	(MPa)	28	1,200	25,100	7,844	5,308	6.8
(Elasticity mod /Comp. str.) ratio	E/σ_c	—	28	96:1	753:1	273:1	132:1	48.4
Poisson ratio	υ	—	23	0.18	0.38	0.30	0.06	20.0
Cohesion (triaxial)	c	(MPa)	5	4.3	19.8	13.1	5.9	4.5
Internal friction angle (triaxial)	Ø	(°)	5	32.3	45.3	40.5	5.1	12.6
Cherchar abrasivity	CAI	—	13	0.50	2.30	0.94	0.52	55.3

Table 2.8 Physical and mechanical properties of the rock samples taken from Kartal formation (zone B) [8].

Engineering properties	Symbol	Unit	Data number n	Variation		Mean value, $\overline{X}$	Standard deviation sd	Variance V (%)
				Min.	Max.			
Rock density	γ	(kN/m^3)	55	23.5	28.5	26.9	0.9	33
Porosity	n	(%)	19	0.25	3.23	1.01	0.84	83.2
Compressive strength	σ_c	(MPa)	51	7.8	158.3	52.7	29.3	5.6
Tensile strength	σ_t	(MPa)	35	0.2	17.9	7.7	4.3	5.5
Elasticity modulus	E	(MPa)	24	4,500	18,800	10,838	3,813	3.5
(Elasticity mod. /cmp. str) ratio	E/σ_c	—	24	134:1	387:1	202:1	59:1	29.0
Poisson ratio	υ	—	23	0.21	0.38	0.30	0.05	16.7

Engineering properties	Symbol	Unit	Data number n	Variation		Mean value, $\overline{X}$	Standard deviation sd	Variance V (%)
				Min.	Max.			
Cohesion (triaxial)	c	(MPa)	6	5.4	28.0	15.1	9.7	6.4
Internal friction angle (triaxial)	Ø	(°)	6	40.0	54.2	45.4	5.0	11.0
Cherchar abrasivity index	CAI	—	26	0.50	3.50	1.79	0.68	38.0

Table 2.9 Physical and mechanical properties of the rock samples taken from Kurtköy formation [8].

Engineering properties	Symbol	Unit	Data number, n	Variation		Mean value, $\overline{X}$	Standard deviation, sd	Variation V (%)
				Min.	Max.			
Rock density	γ	(kN/m³)	55	23.5	28.5	26.9	0.9	33
Porosity	n	(%)	19	0.25	3.23	1.01	0.84	83.2
Compressive strength	σ_c	(MPa)	51	77.5	158.3	52.7	29.3	5.6
Tensile strength	σ_t	(MPa)	35	0.2	17.9	7.7	4.3	5.5
Elasticity modulus	E	(MPa)	24	4,500	18,800	10,838	3812	3.5
(Elasticity mod. / (comp.str.) ratio	E / σ_c	—	24	134:1	387:1	202:1	59:1	29.0
Poisson ratio	υ	—	23	0.21	0.38	0.30	0.05	16.7
Cohesion (triaxial)	c	(MPa)	6	5.4	28.0	15.8	9.7	6.4
Internal friction angle (triaxial)	Ø	(°)	6	40.0	54.2	45.4	5.0	11.0
Cherchar abrasivity	CAI	—	26	0.50	3.50	1.79	0.68	38.0

Table 2.10 Physical and mechanical properties of the rocks taken from Kurtkoy formation [9].

Engineering properties	Symbol	Unit	Data number, n	Min.	Max.	Mean	Standard deviation, sd	Variance V (%) (*)
Rock quality designation	RQD	(%)	14	0	54	17.7	16.5	93.1
Rock mass rating	RMR	(%)	19	29	62	39.3	9.3	23.8
Geological strength in.	GSI	(%)	17	25	40	32.4	3.6	11.1
Internal friction angle	Ø	(°)	7	8.7	68	50.3	19.3	38.4
Cohesion	c	(MPa)	6	1	6	3.9	1.9	0.5
Compressive strength	σ_c	(MPa)	21	4	98.7	24.5	22.4	93.6
Elasticity Modulus	E	(MPa)	17	640	7,800	3,833	2,351	61.3
Menard Modulus	Em	(MPa)	14	143	226	173	31.3	18.1
Poisson ratio	υ	—	5	0.24	0.37	0.3	0.05	17
Permeability	p	(m/s)	11	1.87E−08	1.2 E−0.6	3.67E−0.7	3.92E−07	166.7
Rock density	γ	(kN/m³)	15	26.3	27.1	26.7	0.36	1.34
Porosity	n	(%)	13	0.57	15.8	3.74	3.76	110.5

Table 2.11 Physical and mechanical properties of the rocks from Dolayoba formation [8].

Engineering properties	Symbol	Unit	Data number, n	Variation		Mean value, $\bar{X}$	Standard deviation, sd	Variation V (%)
				Min.	Max.			
Rock density	γ	(kN/m³)	48	23.2	27.6	27.0	0.7	26
Porosity	n	%	26	0.37	6.96	1.82	1.49	82.1
Compressive strength	σ_c	(MPa)	43	10.4	89.6	45.2	17.5	3.8
Tensile strength	σ_t	(MPa)	12	3.7	10.7	6.4	2.1	3.3

Engineering properties	Symbol	Unit	Data number, n	Variation		Mean value, $\overline{X}$	Standard deviation, sd	Variation V (%)
				Min.	Max.			
Elasticity modulus	E	(MPa)	37	6,215	54,851	18,118	12,140	6.7
(Elasticity mod. /comp str.) ratio	E / σ_c	—	37	186:1	909:1	422:1	226:1	53.7
Poisson ratio	υ	—	10	0.21	0.36	0.30	0.04	13.3
Cohesion (tri axial)	c	(MPa)	10	0.9	5.6	1.7	1.4	0.8
Internal friction angle (tri axial)	Ø	(°)	10	26.6	56.0	46.2	8.3	18.0
Cherchar abrasivity index	CAI	—	10	0.75	1.50	1.22	0.24	19.7

Table 2.12 Physical and mechanical properties of the rocks taken from Doloyaba formation [9].

Engineering properties	Symbol	Unit	Data number, n	Min.	Max.	Mean	Standard deviation sd	Variance V (%) (*)
Rock quality designation	RQD	(%)	12	3.5	81	41.3	24.8	60
Rock mass rating	RMR	(%)	10	27	61	45.2	13.8	30.6
Geological strength index	GSI	(%)	10	25	50	43.5	7.5	17.2
Internal friction angle	Ø	(°)	7	18.8	69	57	17.7	71.7
Cohesion	c	(MPa)	7	0.9	0.44	27.4	13.8	0.5
Compressive strength	σ_c	MPa	14	18.2	96.5	49.1	24	49
Elasticity modulus	E	MPa	11	330	14,900	6,306	4,710	74.7
Menard modulus	Em	MPa	8	182.2	325	261	55.7	21.3

Engineering properties	Symbol	Unit	Data number, n	Min.	Max.	Mean	Standard deviation sd	Variance V (%) (*)
Poisson ratio	υ	—	6	0.2	0.78	0.37	0.21	58.5
Rock density	γ	kN/ m³	10	25.9	27.3	27.1	0.454	1.66
Porosity	n	(%)	9	0.12	3.7	1.5	1.3	86.6

Table 2.13 Physical and mechanical characteristics of the rocks from Gozdag formation [9].

Engineering properties	Symbol	Unit	Data number	Min.	Max.	Mean	Standard deviation sd	Variance. V (%) (*)
Rock quality designation	RQD	(%)	10	2	64	36	18	52
Rock mass rating	RMR	(%)	20	25	65	43.2	12.6	29.2
Geological strength index	GSI	(%)	20	25	55	36	8.1	22.4
Internal friction angle	Ø	(°)	4	41	63	53	9	17.2
Cohesion	c	(MPa)	4	5	75.6	26.9	32.8	1.22
Compressive strength	σ_c	MPa	24	3.2	102	28.1	22.8	81.3
Elasticity modulus	E	MPa	20	1,600	198,000	25,289	48,398	191
Menard modulus	Em	MPa	9	37.6	292	205	103	50
Poisson ratio	υ	—	11	0.19	0.4	0.3	0.07	23.1
Rock density	γ	kN/ m³	10	22	27.2	26	1.52	5.83
Porosity	n	(%)	9	1.52	14.5	6	3.7	62.3

Table 2.14 Physical and mechanical characteristics of rocks from Tuzla formation (clay stone-shale) [9].

Engineering properties	Symbol	Unit	Data numb.	Min.	Max.	Mean	Standard deviation sd	Variance V (%) (*)
Rock quality designation	RQD	(%)	2	1.7	3.2	2.45	1.1	43.3
Rock mass rating	RMR	(%)	4	45	58	51.5	5.5	10.6
Geological strength index	GSI	(%)	3	35	45	38.3	5.8	15.1
Internal friction angle	Ø	(°)	3	47.5	50.5	49.3	1.5	3.4
Cohesion	c	(MPa)	3	1.49	4.06	2.68	1.29	0.5
Compressive strength	σ_c	(MPa)	2	15.4	66.1	40.8	35.9	88
Elasticity modulus	E	(MPa)	3	3,200	13,200	8,200	7,071	86.2
Menard modulus	Em	(MPa)	2	141.7	269.2	205.5	90.2	43.2
Poisson ratio	υ	—	2	0.16	0.19	0.18	0.02	12.1
Rock density	γ	(kN/m^3)	2	25.5	26.8	26.2	0.92	3.5
Porosity	n	(%)	3	0.76	10	3.9	5.6	136

Table 2.15 Rock physical and mechanical properties of the samples from Sultanbeyli formation [9].

Engineering properties	Symbol	Unit	Data number, n	Min.	Max.	Mean	Standard deviation sd	Variance. V (%) (*)
Rock quality designation	RQD	(%)	6	0	93	29.50	34.02	115.3
Rock mass rating	RMR	(%)		—	—	—	—	—
Geological strength index	GSI	(%)	—	—	—	—	—	—

Engineering properties	Symbol	Unit	Data number, n	Min.	Max.	Mean	Standard deviation sd	Variance. V (%) (*)
Internal friction angle	Ø	(°)	2	11.56	19.8	15.7	5.85	37.3
Cohesion	c	(MPa)	3	0.14	1.14	0.65	0.50	0.76
Compressive str.	σ_c	(MPa)	4	31	99.6	62.3	29.7	47.7
Elasticity modulus	E	(MPa)	1	14,440	14,440	14,440	—	—
Menard modulus	Em	(MPa)	6	37.6	42.2	38.4	1.9	4.9
Poisson ratio	υ	—	1	0.27	0.27	0.27	—	—
Density	γ	(kN/m³)	3	22	22	22	0	0
Porosity	n	(%)	0	0	0	—	—	—

$(*)\ V = \left(s/\ \overline{X}\right) \times 100, \quad \%$

2.5.4 Schmidt hammer tests carried out in station tunnels in Uskudar–Umraniye metro line

The Schmidt hammer test was originally developed for testing concrete, and it is also widely used to test in situ strength of rocks. The readings are largely affected by the geological discontinuities, and that is why it is prepared as an in situ rock strength testing device.

Within the study for this report an N-type Schmidt hammer was used. In situ tests were carried out at seven different station tunnels on the Uskudar–Umraniye–Cekmekoy Metro lines. Tunnel faces where Schmidt hammer tests were carried are seen in Figures 2.34–2.40 [13].

Figure 2.34 Tunnel face in Uskudar–Umraniye metro line, Ihlamurkuyu Station, Kurtkoy formation [13].

Figure 2.35 Tunnel face in Uskudar–Umraniye metro line, Shaft of Inkilap Station Area Kartal formation [13].

Figure 2.36 Tunnel face in Uskudar–Umraniye metro line, shaft of Carsi Station Kartal formation [13].

Figure 2.37 Tunnel face in Uskudar–Umraniye metro line Altunizade station Kartal formation [13].

Figure 2.38 Tunnel face in Uskudar–Umraniye metro line Libadiye (Dudullu) station Kartal formation [13].

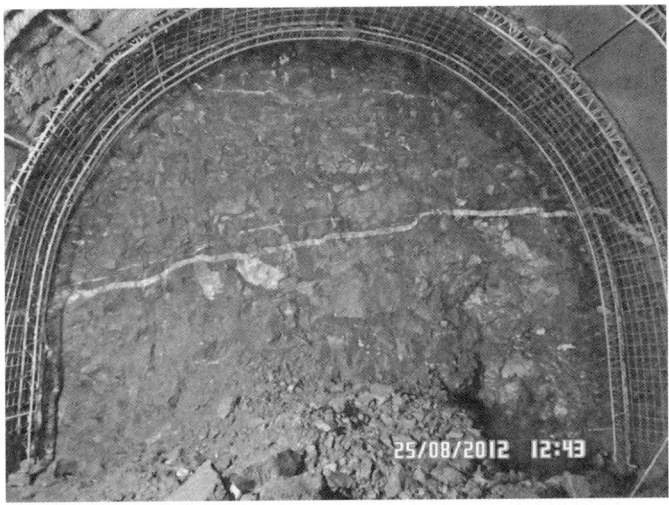

Figure 2.39 Tunnel face in Uskudar–Umraniye metro line, Umraniye Station, Kartal formation [13].

Figure 2.40 Tunnel face in Uskudar–Umraniye metro line, Kisikli Station, Aydos formation [13].

2.5.4.1 Comparison between point load and Schmidt hammer test results in Uskudar–Umraniye–Dudullu metro lines

Some rock samples for point load strength tests were collected from tunnel faces where Schmidt hammer tests were carried out. The test results are summarized in Table 2.16. Schmidt hammer tests are realized in 82 different points [13].

Table 2.16 In situ Schmidt hammer test values and point load test values [13]

The place where Schmidt hammer tests are realized	Geologic formation and Rock unit	In situ Schmidt hammer value in massif ± sd	In situ Schmidt hammer test value in fractured rock ± sd	Compressive strength calculated from Point load test MPa ± sd	Cerchar abrasivity value
Ihlamur shaft	Kurtkoy fm., Arcozic sandstone	44 ± 3	34.3 ±3	30 ± 2	3
Inkilap area	Kartal fm., limestone	52 ± 4	36 +4/−5	90 ± 3	—
Carsi	Kartal fm., andesite	57 ± 5	—	117 ± 12	5
Carsi	Kartal fm., mudstone	49 ± 1	40 ± 3	98 ± 3	—
Altunizade	Kartal fm., andesite	45 ± 8	27 ± 2	86 ± 17	4.5
Altunizade	Kartal fm., mudstone	45 ± 4	33 ± 4	54 ± 2	—
Libadiye	Kartal fm., limestone	64 ± 2	52 ± 2	120 ± 5	—
Umraniye	Kartal fm., limestone	64 ± 2	52 ± 2	139 ± 5	—
Kisikli	Aydos fm., quartzite	58 ± 1	—	—	5.5
Kisikli	Conglomerate	—	—	—	3.5

2.6 Full-scale linear rock cutting tests with disc cutters in rock samples collected from different projects in Istanbul

The rock cutting experiments are the most reliable and economic solution for selection, designing and predicting/optimizing performance of TBMs. Many types of testing devices have been developed for these purposes. Today, there are different rock cutting devices used in the laboratories of universities, research institutes and machine manufacturers. Since full-scale cutting equipment requires experienced personnel, and because it is usually not possible or too expensive to obtain large block samples before starting the excavation project, the possibility of developing a reliable portable or small-scale device using small samples and small cutting tools is today a basic task for researchers working on rock cutting mechanics, although other types of testing cannot be as reliable as full-scale testing devices. One such portable rock cutting device was developed in the mining engineering department of Istanbul Technical University. It is used for cutting small core samples or small block samples using a mini scale index disc cutter [11].

2.6.1   Description of laboratory full-scale linear rock cutting test

The full-scale linear cutting tests measure full-scale forces acting on a real life cutter of any type (e.g. single disc, conical, radial) while cutting a block of rock sample cast in a sample box. Full-scale testing minimizes the uncertainties of scaling and of any unusual rock cutting behavior not being reflected in its physical properties. Results of this test can be used as input for selection, designing and predicting/optimizing the costs and performance (excavation, production, cutting rate) of TBMs for feasibility purposes. This test, along with deterministic computer simulation, is accepted as the most reliable and economical method for these purposes.

The full-scale linear cutting machine features a large stiff reaction frame on which the cutter and load cell assembly are mounted. A block sample up to $0.6 \times 0.8 \times 1.0$ m in size is cast within a stiff (metal) sample box with a fast curing concrete at a certain dip angle or parallel or perpendicular to the bedding planes, in order to simulate the real cutting conditions of the deposit. A servo-controlled hydraulic cylinder moves the sample within the box through the cutter at a preset depth of cut (penetration), width of spacing (line spacing of tool cutting tracks) and constant velocity. Cutting velocity, depth of cut and line spacing of the tool can be adjusted by hydraulic cylinders as desired. A triaxial load cell located between the cutter and the frame monitors/measures/records the orthogonal forces (normal, drag-cutting-rolling and sideways forces) acting on the tool. After each cut, the rock box is moved sideways by a desired spacing to duplicate the action of the multiple cutters on the cutterhead of a mechanical excavator. After each cut, the cut length is measured; the cut materials are collected, weighted and sieved. Sieve analysis is then performed to measure the size gradation of the muck samples; this also gives information on the efficiency of the cutting. Each cut should be replicated at least three times; the results should be averaged and the cut surface should be photographed. A picture and a schematic drawing of the full-scale linear cutting machine found in the laboratories of the mining engineering department of Istanbul Technical University are presented in Figures 2.41 and 2.42, respectively.

Figure 2.41 Full-scale linear rock cutting machine in ITU Laboratories.

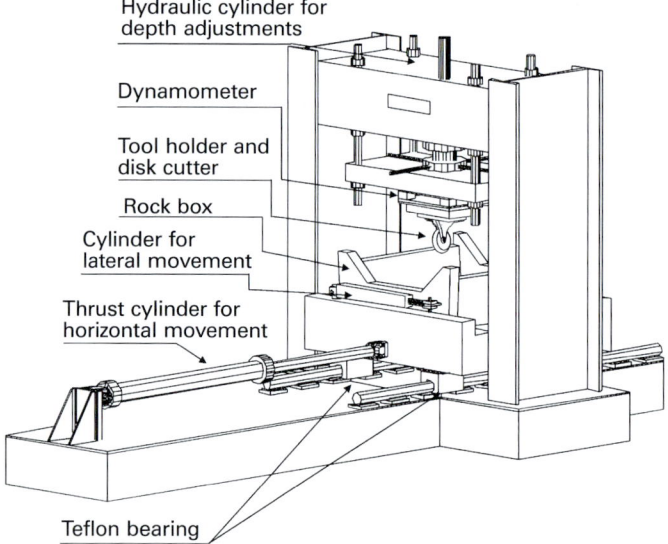

Hydraulic cylinder for depth adjustments

Dynamometer

Tool holder and disk cutter

Rock box

Cylinder for lateral movement

Thrust cylinder for horizontal movement

Teflon bearing

Figure 2.42 A schematic drawing of full-scale linear rock cutting machine.

A data acquisition system used with this testing system includes a dynamometer (load cell), A/D card, signal conditioning amplifier and personal computer. The data is recorded at required gain and sampling rate by commercial software. The data acquisition card should include at least four independent channels for monitoring/measuring/recording data from the dynamometer. Excitation voltage of the amplifier should be 10 V. Data sampling rate should be a minimum of 1,000 Hz. The recorded data can then be evaluated by using a custom-made spreadsheet macro program.

The force data is used as input for selection and design of a mechanical excavator, selection of a cutter, definition of optimum cutting geometry and prediction of performance.

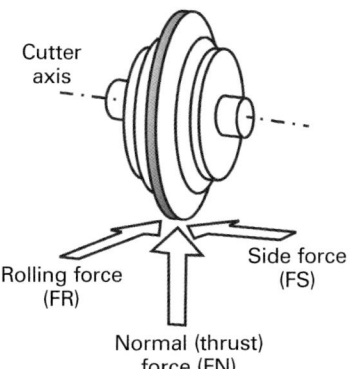

Cutter axis

Rolling force
(FR)

Side force
(FS)

Normal (thrust)
force (FN)

Figure 2.43 Three components of forces acting
on a disc cutter.

Figure 2.43 shows a schematic drawing of forces acting on a disc cutter. The normal force (the average force required to be applied perpendicular to the direction of rolling, to maintain the disc at the prescribed level of penetration) recorded by the LCM is used to calculate the thrust requirement of the machine. This is important to ensure that the machine is able to provide the necessary thrust, so that the cutters can effectively penetrate the rock. The rolling force (the average force required to be applied in the direction of cutting to cause the disc to roll at the prescribed level of penetration), recorded during LCM testing, is used to calculate the torque and power requirements for excavating the rock. The rolling force is directly related to the torque requirement of an excavator, and is also used to calculate the specific energy requirement. Specific energy is defined as the amount of energy required to excavate a unit volume of rock. Using the specific energy (kWh/m^3), achievable production rates are calculated for a machine with a known horsepower available at the cutterhead. The side force may be used along with normal and rolling forces to balance the cutterhead design [12].

2.6.2 Testing methodology and test results

Rock samples having minimum sizes of $1 \times 0.5 \times 0.7$ m were collected from the surface along the tunnel line of the Kartal–Kadikoy metro tunnels in Istanbul. Rock samples were subjected to full-scale laboratory cutting tests with different depths of cut and cutter spacings using a CCS disc cutter [14]. Cutter forces (i.e. thrust force and rolling force) and specific energy values in kWh/m^3 were recorded for each cut. It is believed that the results will serve as a guide for efficient selection and use of TBMs.

A constant cross-section disc (CCS) cutter having tip width of 1.2 cm and diameter of 13″ is used during cutting experiments. Cutter spacing is keep constant at 7 and 8 cm. Unrelieved cutting results are given in Tables 2.17–2.20, and relieved cutting results are tabulated in Tables 2.21–2.24. The variations of thrust and rolling forces with depth in unrelieved cutting mode are given in Figures 2.44–2.47; the variations of thrust and rolling forces with depth in relieved cutting mode are given in Figures 2.48–2.51 and the variations of specific energy with optimum s/d ratio are given in Figures 2.52–2.55.

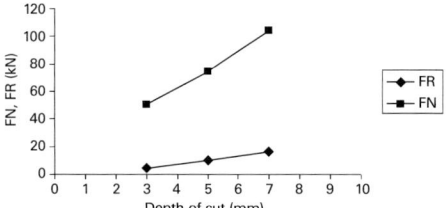

Figure 2.44 Disc cutter forces in Arcose in unrelieved mode [14].

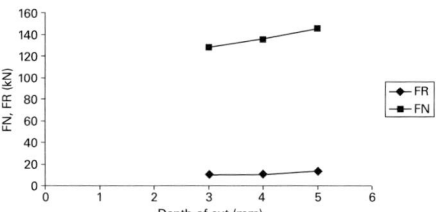

Figure 2.45 Disc cutter forces in Dolayoba limestone in unrelieved mode [14].

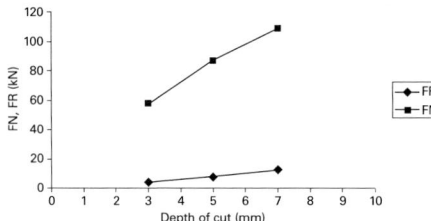

Figure 2.46 Disc cutter forces in Kartal limestone in unrelieved mode [14].

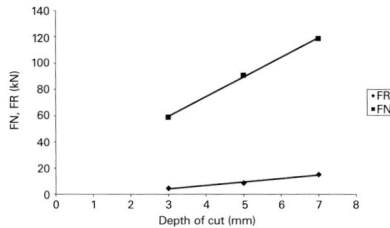

Figure 2.47 Disc cutter forces in Trakya siltstone in unrelieved mode [14].

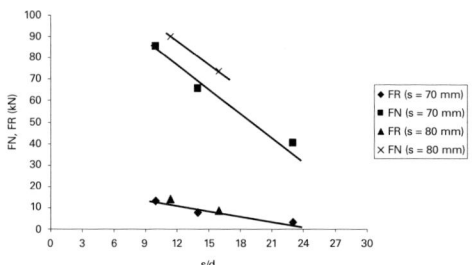

Figure 2.48 Disc cutting forces in Arcose in relieved mode [14].

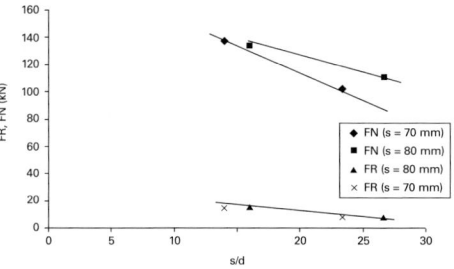

Figure 2.49 Disc cutting forces in Dolayoba limestone in relieved mode [14].

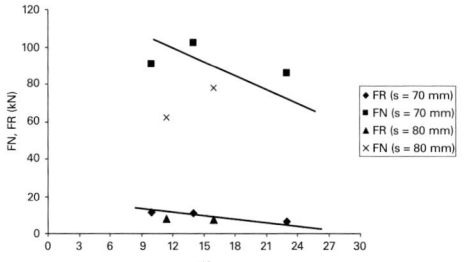

Figure 2.50 Disc cutting forces in Kartal limestone in relived mode [14].

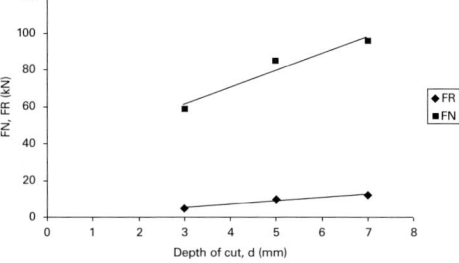

Figure 2.51 Disc cutting forces in Trakya siltstone in relived mode [14].

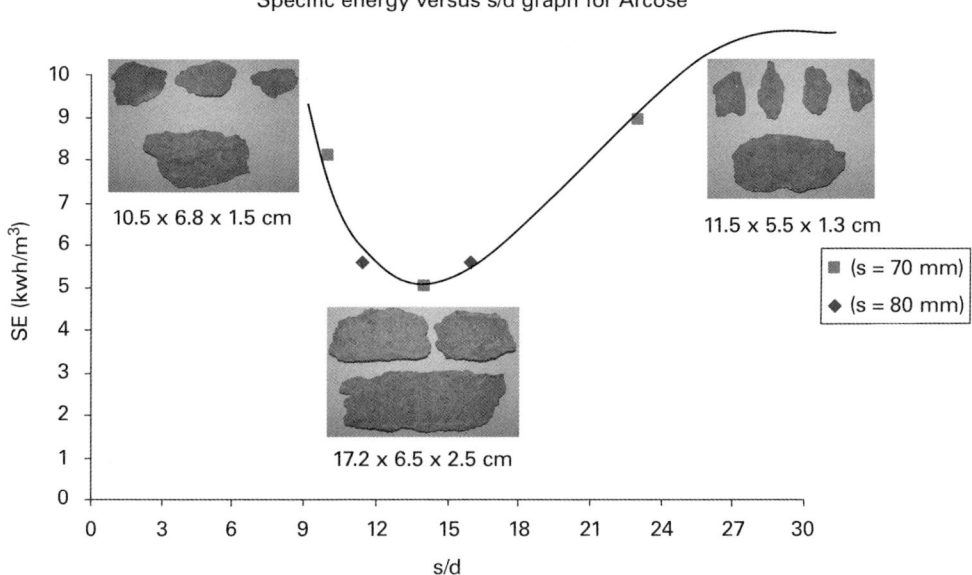

Figure 2.52 Variation of SE with s/d in arcose [14].

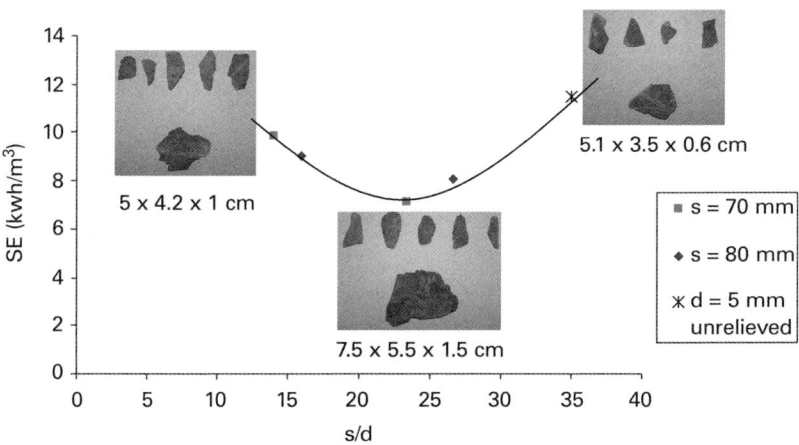

Figure 2.53 Variation of SE with s/d in Doloyaba limestone [14].

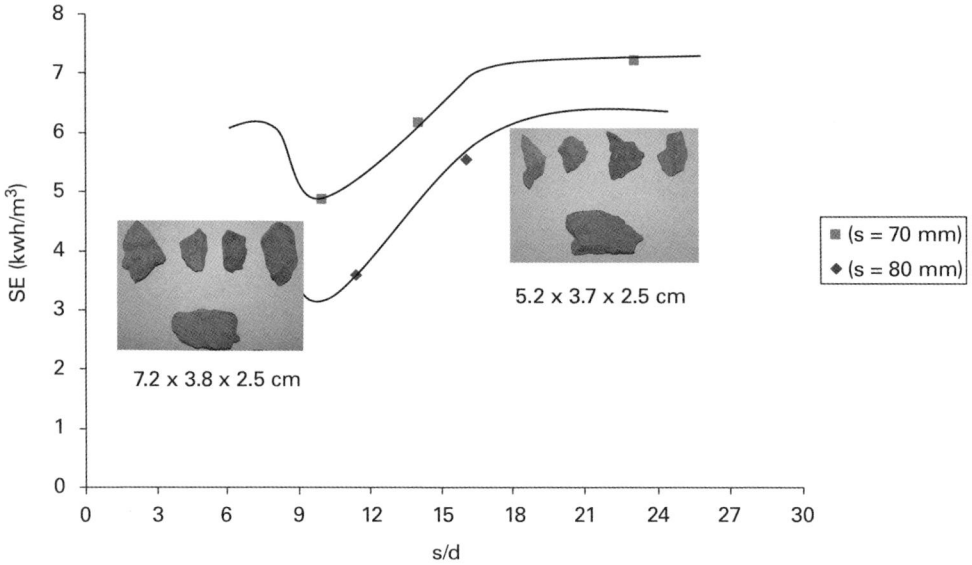

Figure 2.54 Variation of SE with s/d in Kartal limestone [14].

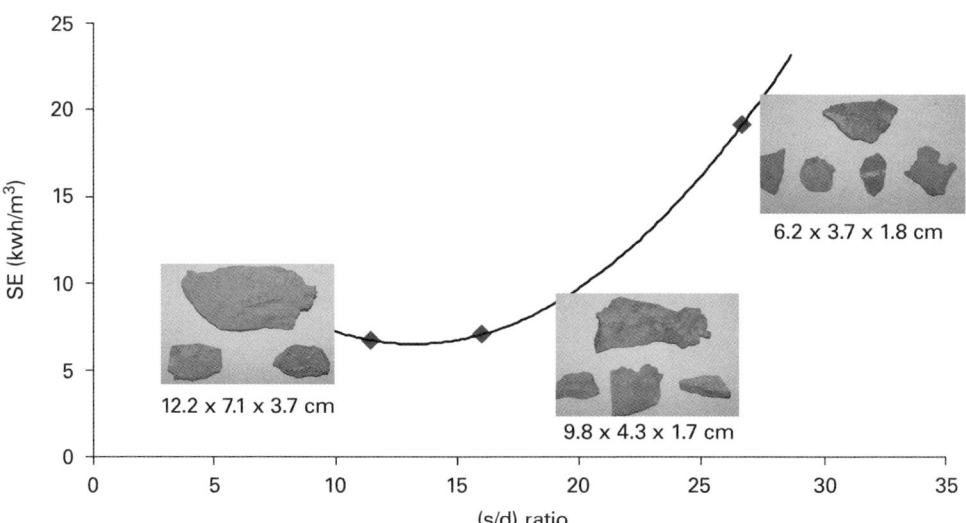

Figure 2.55 Variation of SE with (s/d) in Trakya siltstone [14].

The optimum specific energy values are obtained as 5.6 kWh/m^3 at s/d = 13.1 for Kurtkoy formation (Arcose), 4.0 kWh/m^3 at s/d=10 in Kartal formation, 7,1 kWh/m^3 at s/d=23 in Doloyoba formation and 5.4 kWh/m^3 at s/d of 14 in silt.

The mechanical and physical properties of the rock samples subjected to rock cutting test are given in Table 2.25.

The parameters given in Tables 2.17–2.24 are described as given below.

s = cutter spacing

d = depth of cut

FR = mean rolling force

FN = mean thrust force

FR' = maximum rolling force

FN' = maximum rolling force

SE = specific energy

These parameters are separately shown in Figure 2.56.

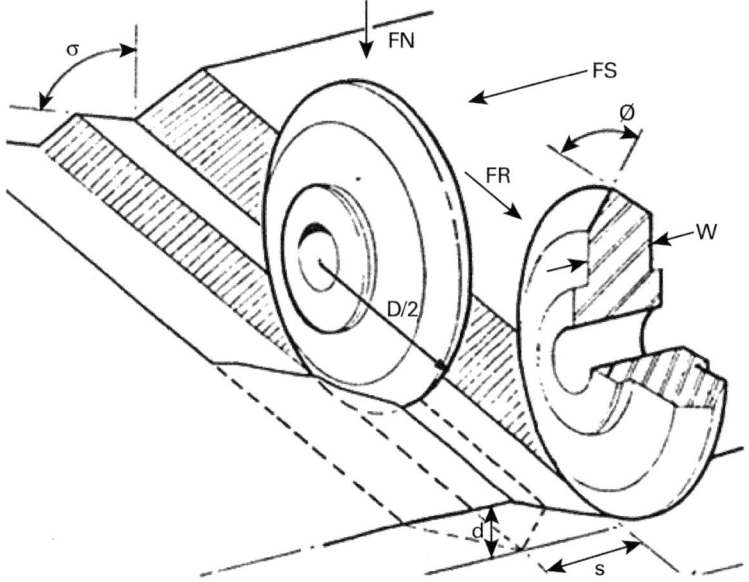

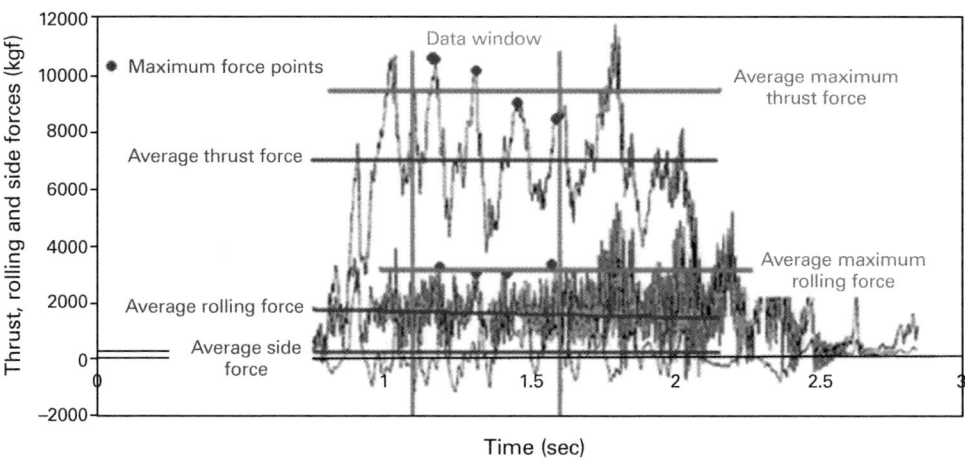

Figure 2.56 Typical variations of disc cutting forces.

Table 2.17 Summary of unrelieved cutting tests in Arcose [14].

d	FR	FR′	FR′/ FR	FN	FN′	FN′/ FN	FR/ FN	FR′/ FN′	SE
(mm)	(kN)	(kN)		(kN)	(kN)				(kWh/ m^3)
3	4.37	9.17	2.14	50.68	84.42	1.73	0.09	0.11	15.3
5	10.14	19.22	1.90	74.71	128.26	1.73	0.14	0.15	6.1
7	16.60	28.20	1.69	104.28	165.64	1.59	0.16	0.17	5.8

Table 2.18 Summary of unrelieved cutting tests in Doloyaba limestone [14].

d	FR	FR′	FR′/ FR	FN	FN′	FN′/ FN	FR/ FN	FR′/ FN′	SE
(mm)	(kN)	(kN)		(kN)	(kN)				(kWh/ m^3)
3	10.15	19.64	1.94	128.15	191.65	1.49	0.08	0.10	11.5
4	11.14	19.99	1.81	135.52	203.35	1.50	0.08	0.10	10.9
5	13.87	22.07	1.61	145.05	210.10	1.47	0.10	0.10	9.7

Table 2.19 Summary of unrelieved cutting tests in Kartal limestone [14].

d	FR	FR′	FR′/ FR	FN	FN′	FN′/ FN	FR/ FN	FR′/ FN′	SE
(mm)	(kN)	(kN)		(kN)	(kN)				(kWh/ m^3)
3	4.17	10.49	2.64	57.56	117.66	2.08	0.07	0.09	13.9
5	7.69	19.80	2.54	86.93	164.85	1.91	0.09	0.12	12.0
7	12.62	25.97	2.05	109.09	191.45	1.75	0.12	0.14	9.8

Table 2.20 Disc cutting test results in Trakya formation (siltstone) in unrelieved mode [14].

d	FR	FR′	FR′/FR	FN	FN′	FN′/FN	FR/FN	FR′/FN′	SE
(mm)	(kN)	(kN)		(kN)	(kN)				(kWh/m³)
3	2.57	3.60	1.97	27.40	46.77	1.71	0.09	0.08	11.42
	3.85	10.61	2.76	42.96	85.26	1.98	0.09	0.12	8.86
	5.77	13.23	2.29	81.20	131.93	1.62	0.07	0.10	26.59
	2.78	6.79	2.44	30.66	65.44	2.13	0.09	0.10	16.36
	4.02	13.06	3.25	58.89	119.46	2.03	0.07	0.11	24.13
	8.56	17.68	2.07	111.50	173.10	1.55	0.08	0.10	27.47
Average	4.59	10.83	2.46	58.77	103.66	1.84	0.08	0.10	19.14
5	7.96	19.57	2.46	79.34	189.47	2.39	0.10	0.10	5.25
	10.80	22.73	2.10	112.56	203.14	1.80	0.10	0.11	9.10
	7.55	23.59	3.12	70.17	150.76	2.15	0.11	0.16	4.75
	8.73	22.99	2.63	88.93	228.38	2.57	0.10	0.10	6.59
	8.71	1959	2.25	101.59	184.38	1.81	0.09	0.11	7.44
Average	8.75	2169	2.51	90.52	191.23	2.14	0.10	0.12	6.63
7	15.50	3516	2.27	110.24	191.86	1.74	0.14	0.18	6.62
	14.55	2708	1.86	126.83	202.37	1.60	0.11	0.13	6.22
	13.97	3059	2.19	112.32	229.41	2.04	0.12	0.13	5.97
Average	15.03	3112	2.06	118.54	197.12	1.67	0.13	0.16	6.27

Table 2.21 Summary of relieved cutting tests in arcose [14].

s	d	s/d	FR	FR′	FR′/FR	FN	FN′	FN′/FN	FR/FN	FR′/FN′	SE
(mm)	(mm)		(kN)	(kN)		(kN)	(kN)				(kWh/m³)
70	3	23.3	3.32	6.35	1.97	40.22	67.23	1.72	0.08	0.09	8.9
	5	14.0	7.88	14.49	1.84	65.43	114.83	1.76	0.12	0.13	5.0
	7	10.0	13.30	23.66	1.78	85.53	136.57	1.59	0.15	0.17	8.1
80	5	16.0	8.81	15.01	1.72	73.91	122.08	1.66	0.12	0.12	5.6
	7	11.4	14.23	29.22	2.08	90.07	182.47	2.05	0.16	0.16	5.6

Table 2.22 Summary of relieved cutting tests in Dolayoba limestone [14].

s	d	s/d	FR	FR′	FR′/FR	FN	FN′	FN′/FN	FR/FN	FR′/FN′	SE
(mm)	(mm)		(kN)	(kN)		(kN)	(kN)				(kWh/m³)
70	3	23.3	7.97	16.58	2.12	102.06	160.10	1.62	0.08	0.10	7.1
	5	14.0	14.48	25.12	1.75	137.50	202.49	1.48	0.11	0.12	9.9
80	3	26.7	8.01	15.63	1.98	110.36	172.43	1.57	0.07	0.09	8.1
	5	16.0	14.19	24.94	1.76	133.55	200.27	1.51	0.11	0.13	9.0

Table 2.23 Summary of relieved cutting tests in Kartal limestone [14].

s	d	s/d	FR	FR′	FR′/FR	FN	FN′	FN′/FN	FR/FN	FR′/FN′	SE
(mm)	(mm)		(kN)	(kN)		(kN)	(kN)				(kWh/m³)
70	3	23.3	6.60	15.09	1.98	86.15	138.81	1.62	0.08	0.11	7.2
70	5	14.0	10.71	21.45	2.01	102.42	185.16	1.81	0.11	0.12	6.2
70	7	10.0	11.14	24.34	2.21	90.66	166.37	1.84	0.12	0.15	4.9
80	5	16.0	7.47	17.21	2.35	78.15	149.97	1.97	0.10	0.11	5.5
80	7	11.4	7.74	17.67	2.27	62.26	122.29	1.96	0.12	0.14	3.6

Table 2.24 Disc cutting test results in Trakya formation (siltstone) in relived mode [14].

d	s/d	FR	FR′	FR′/FR	FN	FN′	FN′/FN	FR/FN	FR′/FN′	SE
(mm)		(kN)	(kN)		(kN)	(kN)				(kWh/m³)
5	16	8.51	24.39	2.87	93.48	180.35	1.93	0.09	0.14	3.76
		5.36	13.66	2.55	43.87	83.01	1.89	0.12	0.16	3.31
		6.76	16.84	2.49	66.79	130.87	1.96	0.10	0.13	6.55
		14.15	25.78	1.82	109.79	230.12	2.10	0.13	0.11	11.07
		14.39	25.78	1.79	110.66	230.12	2.08	0.13	0.11	10.59
Average		9.84	21.29	2.30	84.92	170.90	1.99	0.11	0.13	7.06

d (mm)	s/d	FR (kN)	FR' (kN)	FR'/ FR	FN (kN)	FN' (kN)	FN'/ FN	FR/ FN	FR'/ FN'	SE (kWh/ m^3)
7	11.4	15.11	30.34	2.007	99.01	219.81	2.22	0.152	0.138	5.51
		16.45	25.78	1.567	123.01	198.24	1.611	0.133	0.13	7.87
		13.23	30.46	2.302	87.29	146.06	1.673	0.151	0.208	4.85
Average		11.53	23.81	2.19	90.98	176.61	1.94	0.12	0.14	6.73

Table 2.25 Mechanical and physical properties of the rock samples subjected to rock cutting tests [14].

Engineering properties	Kurtkoy formation (arcose)	Kartal formation (limestone)	Dolayoba formation (limestone)
Uniaxial compressive strength (MPa $\pm$ sd)	34.1 $\pm$ 10.3	65.6 $\pm$ 6.7	119. $\pm$ 19
Brazilian tensile strength (MPa $\pm$ sd)	4.2 $\pm$ 0.8	7.4 $\pm$ 2	7.5 $\pm$ 1.7
Poisson ratio	0.26	0.35	0.36
Static elastic modulus (GPa)	6.4 $\pm$ 1.5	12.6	15.5 $\pm$ 0.9
Cerchar abrasivity index	2	1.5	1.5
Schmidt hammer (N-24 tip)	39 $\pm$ 5	40 $\pm$ 4	57 $\pm$ 3
Schmidt hammer (L tip)	30 $\pm$ 4	35 $\pm$ 3	45 $\pm$ 2
Density (kN/m^3)	26.8	26.2	27.2
RQD	—	—	—

2.6.3   Comparison of laboratory full-scale linear rock cutting test results with in-situ cutter performance – the effect of rock discontinuities

Rock samples collected from different tunnel projects around Istanbul and Turkey before the projects were subjected laboratory full-scale linear rock cutting experiments using different disc cutters at different penetration (p) and s/p ratios (where s is cutter spacing and p is depth of cut or penetration). The relationship between cutter force and specific energy value and s/p ratio were found for every tunnel projects for different rock types. Force and specific energy values are used in estimating the performance and design parameters of TBM used in different tunnel projects in different formations. Different zones were chosen for in situ observation of the TBM performance in different tunnels and compared with the laboratory results. Some comparisons between laboratory and field performance values of TBM projects are given here as case studies.

2.6.3.1 Tuzla–Dragos sewerage tunnel in Istanbul

The Tuzla–Dragos sewerage tunnel, situated on the Asian side of Istanbul, is part of a complex project covering the area of Kartal–Pendik and Tuzla. Tunnels completed by STFA Construction Company are found in X–X3 shaft and Kemiklidere pump station and have a total length of 6,490 m and final diameter of 4.5 m. The plan of the project area is shown in Figure 2.57.

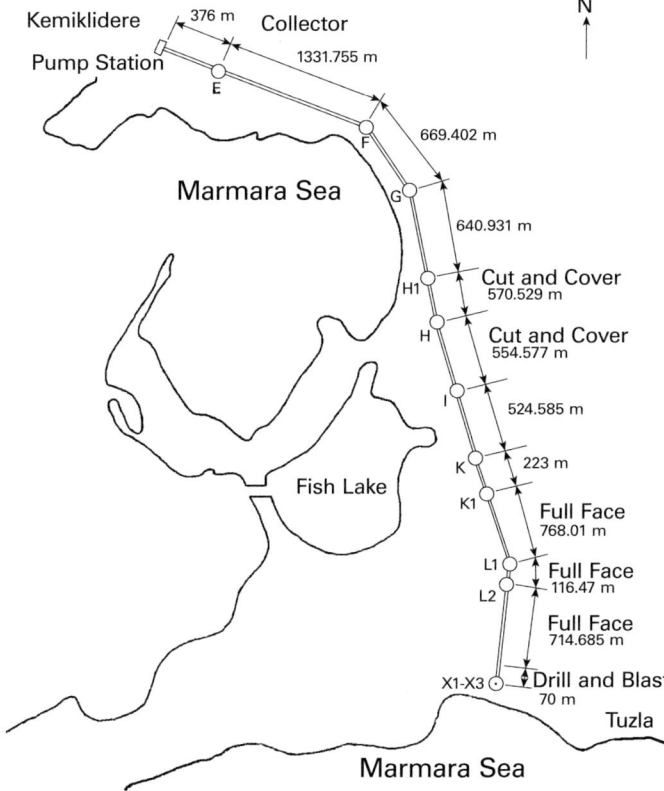

Figure 2.57 The Plan of the tunnel project [15].

The tunnels between X1–X3 and K1 shafts were opened with Robbins TBM Model 165-162, having a diameter of 5 m. The specifications of the Robbins 165-162/E1080 TBM is given Table 2.26 [15].

Table 2.26 Specifications of the Robbins 165-162/E1080 TBM [15].

Property	Values
Machine diameter	5.0 m
Number of cutters	36
Rotational speed	6 rpm
Normal thrust force	471 tons

Property	Values
Maximum thrust force	785 tons
Cutterhead power	600 HP
Power of the miscellaneous pumps etc.	285 HP
Conveyor belt capacity	476 m³/h
Electrical transformer	1,000/380V-50Hz

The main rock formation in the area is Lower Devonian aged Kartal formation composed of shale and limestone. In the upper levels there is Pliocene aged Belgrad formation of sand, gravel and silt, and Quaternary aged sediments. Diabase dykes of Cretaceous ages cut the Kartal formation in several places. The tunnels in X1–X3 and K shafts are mainly driven within limestone and shale. The limestone, gray-blue in color, is in medium thickness, sandy and fractured in several places, the shale is of dark gray-black and is carbonated, mainly jointed and laminated.

Laboratory full-scale rock cutting tests
Two different types of disc cutters were used throughout the experiments for comparative studies. The compressive strength of the rock sample tested was 57.9 ±5.6 MPa, and tensile strength was 3.6 ±0.3 MPa. The relationships between rolling force, cutting force, thrust force and depth of cut, for unrelieved and relieved cutting, for different cutters and specific energy values and s/d (cutter spacing/cutting depth) ratios are explained in detail in Chapter 5.

Field observations and comparison of performance values
Eleven different zones were chosen for in situ observation of the TBM performance in the Tuzla–Dragos tunnel. Special attention was paid to the fact that the rock formation in selected zones should have similar mechanical properties with those tested in the full scale cutting rig. The predicted excavation rate is for competent rock, and it is obvious that the geological discontinuities will increase the net excavation rate to a certain level, and the high amount of RQD or water inflow in the rock formation with a large amount of clay will decrease the daily advance rate due to regional collapses, face instability and chocking of the cutters etc. These factors affected tremendously the advance rate of the TBM during the Tuzla–Dragos tunnel driving. Measured and predicted values are compared in Table 2.27.

Table 2.27 Comparison of predicted and measured TBM performance values [15].

Selected	Zone	Disc cutting	Measured	Predicted	Measured	Predicted
Date	Tunnel (km)	Depth (mm)	FN kN/disc	FN kN/disc	Net cutting rate m³/h	Net cutting rate m³/h
14.10.1997	172.7	11	73.9	91.3	71	80
15.10.1997	175.3	7	69.8	58.1	49	53

Selected	Zone	Disc cutting	Measured	Predicted	Measured	Predicted
Date	Tunnel (km)	Depth (mm)	FN kN/disc	FN kN/disc	Net cutting rate m^3/h	Net cutting rate m^3/h
16.10.1997	190.2	8	69.9	66.4	57	61
17.10.1997	194.0	9	72.3	75.1	64	60
28.10.1997	225.6	7	59.3	58.1	49	53
30.10.1997	227.9	10	117.2	83.0	71	70
04.11.1997	251.3	8	111.2	66.4	57	61
11.10.1997	272.1	11	123.7	91.3	71	80
13.11.1997	275.4	9	89.0	75.1	64	60
14.11.1997	275.4	13	126.2	107.8	92	74
18.11.1997	284.5	8	96.0	66.4	57	61

2.6.3.2 Kadikoy–Kartal Metro tunnel in Istanbul

The Kadikoy–Kartal project, located on Asian Side of Istanbul, Turkey, is a twin metro line tunnel excavated by earth pressure balance (EPB) TBMs, which may be operated in closed or open mode. A line from Kozyatagi to Yenisahra station, which is approximately 1 km in length, is analyzed using field data obtained from one TBM operated in open mode (like hard rock TBM) and compared with laboratory rock cutting test data. The main geological formation in the tunnel route is Kartal formation, consisting of sedimentary rock formations of Triassic and Tertiary age, shale and limestone as seen in Figure 2.58 [12, 16].

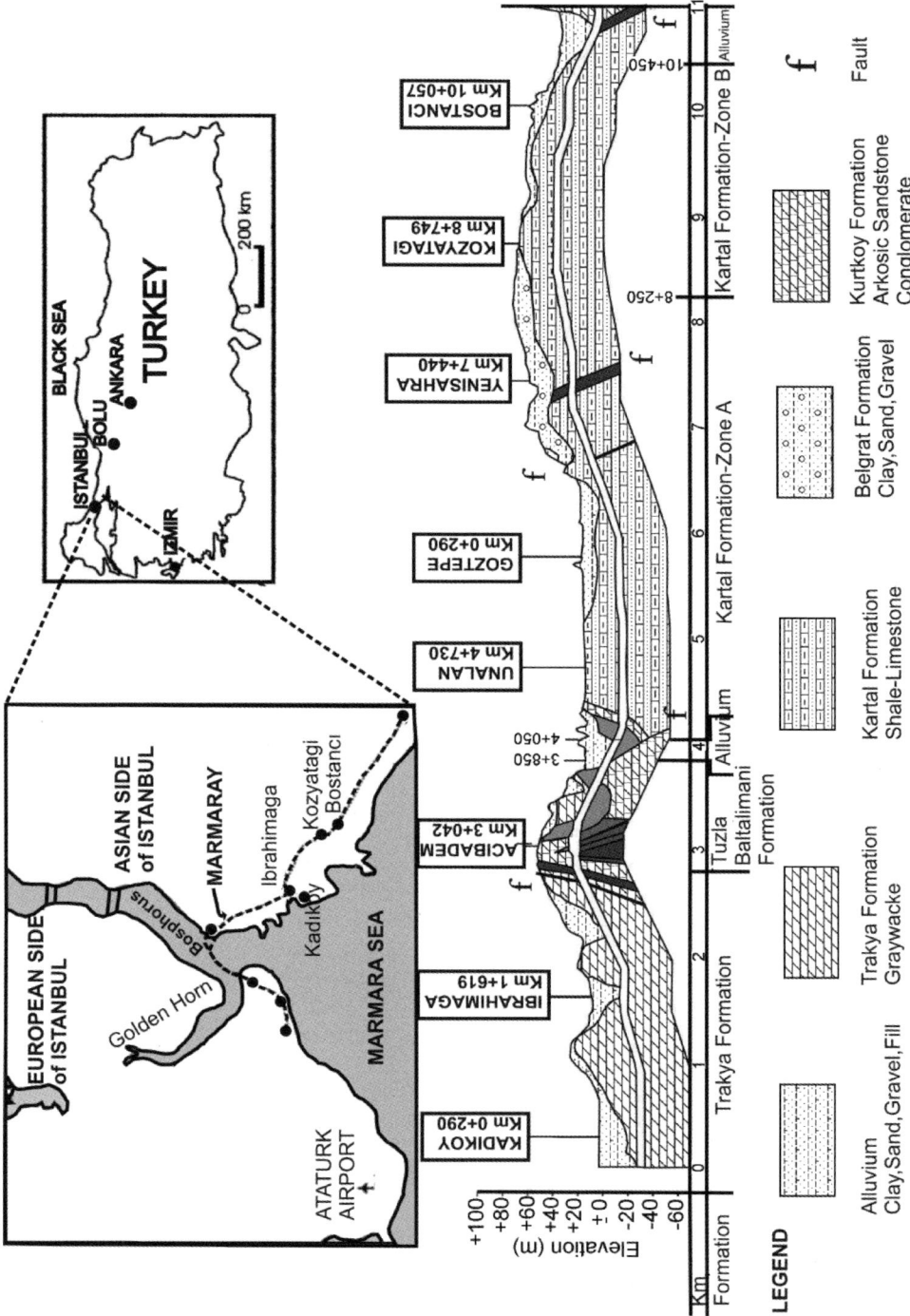

Figure 2.58 Location map of the project area and the geological cross-section of Kozyatagi–Kadikoy section [12].

The rock formation in Kozyatagi–Kadikoy metro line is highly fractured with mean RQD values of 21%. Magmatic rock inclusions such as andesite and dacite dykes frequently intrude through sedimentary rock formations. The thickness of the dykes ranges from 10 or 20 cm to 10–11 m [17]. These intrusions disturb the country rocks and generate many fractures, especially in the contact zones. The frequency of these types of intrusions was about 50–70 m along the tunnel alignment. The performance prediction of TBMs using full-scale rock cutting tests in the laboratory showed reliable results in competent rocks in the past [12], but comparative studies in fractured rocks are limited. It is known that laboratory full-scale rock cutting testing is a very useful tool in determining design parameters and performance prediction of a TBM for a specific job.

Laboratory full-scale rock cutting tests
The intact rock samples having minimum sizes of $1.0 \times 0.7 \times 0.7$m are obtained from shale and limestone (Kartal formation) along the tunnel line, and geomechanical parameters of the two rock samples used in laboratory tests are given in Table 2.28. The rock samples are subjected to full-scale laboratory rock cutting tests with different depth of cut and cutter spacing values, using a constant cross-section (CCS) disc cutter of 330 mm in diameter. Cutter forces (thrust force and rolling force) and specific energy values are recorded for each cut. The results of the tests are first used to calculate the TBM design and performance parameters such as torque and thrust requirements and cutting rates.

A summary of laboratory rock cutting tests with CCS type disc in unrelieved and relieved modes in two different formations (limestone and shale) are presented in Table 2.29.

Table 2.28 Geomechanical parameters of the two rock types used in laboratory tests [12].

Geomechanical parameters of the rocks	Kartal formation	
	Shale	Limestone
Uniaxial compressive strength	31.3 ± 3.0 MPa	65.6 ± 6.7 MPa
Brazilian tensile strength	5.5 ± 0.6 MPa	7.4 ± 2 MPa
Cerchar abrasivity index	1.4 ± 0.5	1.5 ± 0.3
Static Young's modulus	4.5 ± 0.6 GPa	12.6 ± 1.2 GPa
Static Poisson's ratio	0.25 ± 0.01	0.35
p-wave velocity	$3,623 \pm 174.6$ m/s	$7,385 \pm 123.9$ m/s
s-wave velocity	$2,051 \pm 8.3$ m/s	$3,263 \pm 46.6$ m/s
Dynamic Young's modulus	28.6 ± 0.6 GPa	77.3 ± 2.3 GPa
Dynamic Poisson's ratio	0.26 ± 0.03	0.38 ± 0.01
Density	27.0 ± 0.1 kN/m^3	26.3 ± 0.1 kN/m^3
Schmidt hammer hardness (N-24)	47.6 ± 3.4	40.0 ± 5.3

Table 2.29 Summary of rock cutting tests in unrelieved and relieved mode in two different rock samples [12].

Rock formation	Line spacing (mm)	Depth of cut (mm)	FR (kN)	FN (kN)	F'R (kN)	F'N (kN)	SE (kWh/m^3)
Limestone	unrelieved	3	4.1	56.5	10.3	115.4	13.9
		5	7.5	85.2	19.4	161.7	12.0
		7	12.4	107.0	25.5	187.7	9.8
	80	3	6.5	84.5	14.8	136.1	7.2
		5	8.9	88.5	19.0	164.3	5.8
		7	9.3	89.2	20.6	192.1	4.2
		8	10.3	91.0	22.5	206.1	4.1
Shale	unrelieved	5	4.6	43.8	9.5	67.4	6.6
		7	7.7	51.4	11.8	72.2	5.1
		9	11.2	66.1	16.2	92.9	3.3
	80	4	4.2	38.2	9.0	71.0	3.9
		6	5.7	45.0	11.3	87.9	3.3
		8	6.9	45.5	13.8	83.5	3.0

Field Observations and Comparison of Performance Values

The line between the Kozyatagi and Yenisahra stations is being constructed using a 6.57 m diameter Herrenknecht high-power EPB-TBM (operated in open mode), utilizing 432 mm disc cutters capable of operating at 267 kN cutter load. Table 2.30 lists the basic technical specifications of the TBM used in this project.

Table 2.30 Basic technical specifications of the TBM [12].

Parameter	Value
Machine diameter	6.57 m
Number of cutters	26 single + 6 double (12) = 38
Maximum thrust force per disk	267 kN
Thrust capacity of the main bearing	20,000 kN static
Cutterhead drive	Electric motors
Cutterhead power	1,260 kW (4 × 315)
Cutterhead rotational speed	1.6–5.5 rpm

Parameter	Value
Cutterhead torque	5,200 kNm at 1.6 rpm
	1,515 kNm at 5.5 rpm
Capacity of the thrust cylinders	42,575 kN at 350 bar
Thrust cylinder stroke	1.5 m
TBM weight (approx.)	350 tons
Average cutter spacing	90 mm

The machine parameters calculated after the full-scale rock cutting tests and manufacturer design values are compared in Table 2.31. A Herrenknecht TBM no. 360 started
excavating the tunnel line 2 on 20 August 2007 in open mode. The predicted TBM's
cutterhead design parameters from full-scale rock cutting tests such as cutterhead
torque and power are very close to the manufacturer's values in Kozyatagi–Kadikoy
metro tunnel project as seen in Table 2.32.

Table 2.31 Comparison of the machine parameters calculated after the full-scale cutting tests and
manufacturer design values [12].

		Number of disk	Thrust force (kN)	Cutterhead torque (kNm)	Cutterhead power (kW)
Manufacturer design values		38	42,575 kN at 350 bar	1,515 kNm at 5.5 rpm 5,200 kNm at 1.6 rpm	1,260
Predicted after rock cutting tests at 6 rpm and at 8 mm/rev	Limestone	44	13,261	1965	1,309
	Shale	44	6,643	1,318	878

The TBM performance data is recorded continuously by the contractor, Anadoluray
Joint Venture, using a data acquisition system equipped within the TBM. The entire
field data analyzed in this study was collected while the TBM was operating in open
mode. The operational parameters of the TBM such as penetration per revolution, rotational speed, torque, advance rate and thrust were recorded by the machine data
logger and analyzed ring by ring. Analyzed data includes a total of 574 rings (861 m)
in between Kozyatagi and Yenisahra stations. All the data is sorted by penetration per
revolution and reduced as seen in Table 2.32. The operational parameters in Table 2.32
are calculated based on average values for each penetration value. Field normal (thrust)
force (FN_f) and field rolling force (FR_f) per disc cutter is calculated. BI_f, known as the
boreability index, is calculated by dividing field thrust per cutter in kN/disc by field
penetration in mm/rev. Field net cutting rate (ICR_f) is calculated by using advance
rate and the tunnel face area. The field specific energy (SE_f) is determined by dividing
power consumed for excavation by the field net cutting rate (ICR_f), assuming energy
transfer ratio of k is to be 0.85.

Table 2.32 TBM (open mode) field data parameters [12].

Data #	Penetration (mm/rev)	Rotation (rev/min) Ave.	StDev.	Torque (MNxm) Ave.	StDev.	Advance rate (mm/min) Ave.	StDev.	FT$_f$ (kN/disc) Ave.	StDev.	BI$_f$ (kN/disc)/(mm/rev) Ave.	StDev.	FR$_f$ (kN/disc) Ave.	StDev.	Power consumed (kW) Ave.	StDev.	Field net cutting rate (m³/h) Ave.	StDev.	Field specific energy (kWh/m³) Ave.	StDev.	Field s/d ratio Ave.
1	4*	2.5	x	1.5	x	10.0	x	200.7	x	50.2	x	22.4	x	395.1	x	20.3	x	19.4	x	22.5
1	6	2.9	x	1.8	x	17.0	x	110.1	x	18.3	x	27.1	x	555.5	x	34.6	x	16.1	x	15.0
12	7	2.6	0.2	1.8	0.4	18.2	1.5	174.1	71.3	24.9	10.2	26.6	5.8	489.8	117.2	36.9	3.0	13.2	2.9	12.9
24	8	2.7	0.2	1.9	0.3	22.4	1.6	158.5	43.3	19.8	5.4	28.2	4.3	546.7	99.1	45.6	3.3	12.0	1.9	11.3
42	9	2.7	0.2	1.9	0.3	24.4	2.1	143.1	32.8	15.9	3.6	28.5	4.6	544.1	101.4	49.5	4.2	11.0	1.9	10.0
53	10	2.7	0.2	2.1	0.4	27.6	2.7	139.4	39.4	13.9	3.9	30.5	6.3	593.0	138.4	56.2	5.5	10.5	2.2	9.0
75	11	2.7	0.3	2.0	0.5	29.5	2.9	141.5	38.5	12.9	3.5	29.8	7.7	574.1	182.5	60.1	5.9	9.4	2.4	8.2
59	12	2.7	0.3	2.1	0.4	32.1	3.1	131.8	33.4	11.0	2.8	31.4	6.6	600.8	161.1	65.2	6.2	9.1	1.9	7.5
81	13	2.7	0.3	2.2	0.5	35.3	3.3	142.5	35.0	11.0	2.7	32.0	7.1	620.3	168.8	71.7	6.8	8.6	1.9	6.9
61	14	2.7	0.3	2.0	0.5	37.7	4.5	122.3	32.7	8.7	2.3	29.9	8.1	579.9	197.7	76.6	9.2	7.4	2.0	6.4
57	15	2.6	0.4	2.0	0.5	38.8	5.6	135.4	48.5	9.0	3.2	29.4	7.5	550.5	189.6	79.0	11.3	6.8	1.8	6.0
33	16	2.5	0.4	2.0	0.6	40.1	6.3	135.0	37.0	8.4	2.3	29.5	8.4	536.3	199.8	81.5	12.8	6.4	1.8	5.6
17	17	2.3	0.4	2.0	0.5	39.5	7.2	121.9	31.1	7.2	1.8	29.0	6.8	491.5	183.5	80.4	14.6	6.0	1.4	5.3
12	18	1.9	0.2	1.2	0.5	34.6	4.3	163.3	73.3	9.1	4.1	18.5	6.8	256.9	123.6	70.3	8.8	3.6	1.3	5.0
10	19	1.9	0.3	1.5	0.4	36.3	5.9	126.6	23.2	6.7	1.2	22.7	6.2	319.9	130.5	73.8	12.0	4.2	1.2	4.7
8	20	1.9	0.3	1.6	0.6	36.8	5.8	167.3	52.9	8.4	2.6	23.9	8.8	325.1	167.3	74.7	11.8	4.2	1.5	4.5
10	21	1.7	0.2	1.4	0.6	36.1	4.9	161.4	50.2	7.7	2.4	21.2	9.6	268.2	163.1	73.4	9.9	3.5	1.6	4.3
3	22	1.6	0.1	1.1	0.3	36.0	1.0	105.8	9.0	4.8	0.4	15.9	4.2	182.4	45.7	73.2	2.0	2.5	0.6	4.1
9	23	1.6	0.2	1.2	0.3	35.7	3.4	136.9	27.1	6.0	1.2	17.7	3.7	199.0	40.8	72.5	6.9	2.7	0.5	3.9
4	24	1.6	0.1	1.4	0.3	37.3	2.6	156.2	29.2	6.5	1.2	21.5	4.2	236.5	32.9	75.7	5.3	3.2	0.7	3.8
2	25	1.5	0.1	1.3	0.2	36.5	3.5	157.1	84.6	6.3	3.4	19.4	2.7	198.1	18.2	74.2	7.2	2.7	0.5	3.6
Average	15.5	2.3		1.7		31.5		144.3		12.7		25.5		431.6		64.1		7.7		7.6
Min	4.0	1.5		1.1		10.0		105.8		4.8		15.9		182.4		20.3		2.5		3.6
Max	25.0	2.9		2.2		40.1		200.7		50.2		32.0		620.3		81.5		19.4		22.5
Stdev	6.3	0.5		0.3		8.4		22.5		10.0		5.0		157.6		17.2		4.7		4.6

FT$_f$: Field Thrust Force, BI$_f$: Field Penetration Index, FR$_f$: Field Rolling Force

s:Spacing, d: Depth of cut, Min: Minimum, Max: Maximum, Stdev: Standart deviation, Ave: Average

* High thrust force for low penetration may be due to the cutting of high strength dykes

Figure 2.59 shows the variation of the field and the laboratory rock boreability index (ratio of the average net thrust per cutter to the penetration per cutterhead revolution) with different penetrations for the Kozyatagi–Kadikoy metro tunnel.

As explained by Gong et al. [17], the boreability index for a rock mass is not a constant. The rock mass boreability index decreases with increasing penetration rate, as also seen in Figure 2.59. The correlation between rock mass boreability index (at the field and from linear rock cutting tests at the laboratory) and penetration is fitted well to the power function, as shown in Figure 2.59. Gong et al. [17] defined the specific rock mass boreability index (SRMBI) at penetration 1 mm/rev, which is independent of the machine operational conditions (thrust force, RPM and torque) and eliminates the influence of the operation uncertainties on the rock mass boreability. They concluded that this index could be used to evaluate the rock mass boreability. There is a strong statistical relationship found between boreability index and penetration for the Kozyatagi–Kadikoy metro tunnel in Istanbul. The SRMBI values in a highly fractured rock mass in the field and a competent rock in the laboratory rock cutting tests for limestone and then shale are calculated as 172.9, 78.4 and 27.1 kN/disc/mm/rev, respectively [12].

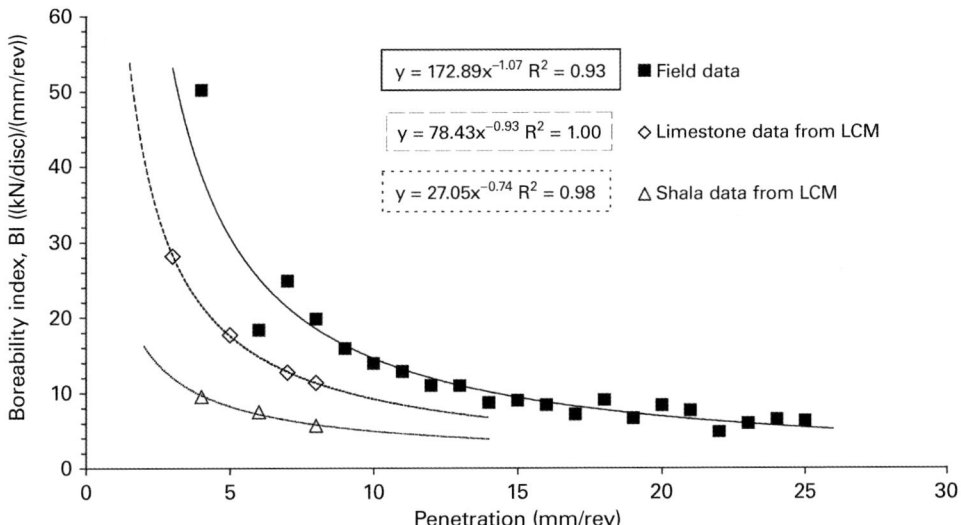

Figure 2.59 Variation of the field and the laboratory rock boreability indices with different penetrations for Kozyatagi–Kadikoy metro tunnel [12].

The results show that the specific rock mass boreability index is higher for field data in a highly fractured rock mass than in the laboratory rock cutting data in competent rock. This means that SRMBI is a good indicator of rock mass boreability, since it results in different values in different rock mass conditions. It proves that using normalized parameters such as SRMBI makes it easier to compare the effects of geological parameters. In order to use SRMBI as a performance predictor parameter, a large database needs to be accumulated and different rock masses need to be classified in terms of boreability [12].

The predicted performance, based on laboratory rock cutting experiments and the field performance of TBM, is compared by considering the optimum cutting conditions. The optimum cutting condition obtained from full-scale laboratory rock cutting tests indicates 8 mm/rev penetration at optimum s/d ratio of 10. The field data indicates that upper limit penetration is around 16 mm/rev which gives s/d ratio of around 5. This s/d ratio value of 5 is also an optimum value for field data. Therefore, predicted performance based on laboratory experiments (at 8 mm/rev of optimum penetration value) and field performances (at 16 mm/rev of optimum penetration value) are compared in Table 2.33.

Table 2.33 Predicted and the field TBM performance values [12].

		Mean thrust force FN (kN)	Consumed power (kW)	SE (kWh/ m^3)	Net cutting rate (m^3/h)
LCM cutting test results at optimum d = 8 mm/rev (Competent rock samples) (at 2.7 rpm)	Limestone	109.2	218	4.1	53
	Shale	54.7	148	3.0	49
Average of LCM results	Mixed (limestone + shale)	81.9	182	3.6	51
Average field results at d = 8 mm/rev at 2.7 rpm (using Table 2.32)	Mixed (limestone + shale)	158.5	547	12.0	46
Average field values at d = 16 mm/rev at 2.5 rpm (using Table 2.32)	Mixed (limestone + shale)	135.0	536	6.4	82

2.6.3.3 Tarabya sewerage tunnel in Istanbul

The Tarabya tunnel forms a major part of a sewerage project in Istanbul, Turkey, designed to clean up the unacceptable pollution present in the Istinye–Tarabya and Buyukdere bay areas of the city's Bosphorus River. The tunnel is 13,270 m long with an inside diameter of 2 m and is situated between Sariyer in the north and Baltalimani in south. The construction contract was awarded to a consortium made up of contractors Tinsa/Oztas/Hazinedaroglu/Simelko, which used a 2.9 m diameter, 560 kW Herrenknecht TBM to bore the tunnel. The performance of the TBM, which began boring in July 2000 and finished in November 2004, was recorded for detailed shift analysis [18].

The tunnel passes through limestone and shale rock formations of the Silurian–Devonian age, and sandstone, siltstone rock formations from the Carboniferous period.

Some magmatic intrusions and sediment fillings are also found along the alignment. A summary of the rock properties obtained from the tender documents is given in Table 2.34. These figures correspond to the initial 8,847 m of the tunnel. The rest was driven mainly through sediment fillings.

Table 2.34 Geomechanical parameters of the rock types used in laboratory tests [18].

Rock formation of the total (%)	Compressive strength (MPa)	Tensile strength(MPa)	Elastic modulus (GPa)
Limestone 65%	44–81	4–5	9–15
Shale 17%	55–59	2.4	9–10
Sandstone-siltstone12%	59	—	—
Dykes 1%	32–40	3	6–7
Sediment fillings 5%	—	—	—

Laboratory Full-Scale Rock Cutting Tests
The disc cutter from the TBM used in the rock cutting tests on a limestone having a compressive strength of 100 ± 8 MPa had a diameter of 305 mm and an edge width of 10 mm.

The optimum specific energy for the limestone rock excavated along the Tarabya tunnel was calculated to be 5.7 kWh/m^3 for an s/d ratio of 14. Cutter spacing of the face cutters in the TBM was 86 mm, given an optimum depth of cut of 6 mm. Rolling force FR for one disc for 6 mm depth of cut was calculated to be 9.82 kN from Figure 2.60. The torque and the cutting power for optimum depth of cut are calculated as 142.4 kNm and 239 kW respectively.

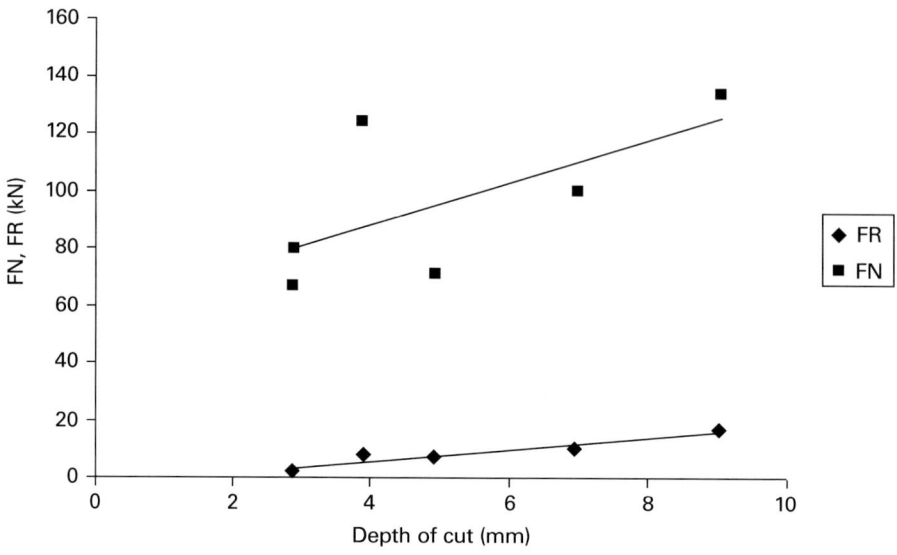

Figure 2.60 Relationships between depth of cut and cutter forces obtained in the laboratory [16].

Field observations and comparison of performance values

The actual cutter forces and measured values of thrust forces are given in Table 2.35. As seen, the field values were very close to the lab results, meaning that the thrust force values obtained in the laboratory could predict the actual field depth of cut values for 1 rev/mm, and subsequently the TBM net cutting rate.

Table 2.35 The actual and measured values of thrust force values [18].

Depth of cut (mm) % of the total	Measured in situ FN (kN)	In situ FN with %20 less due to friction factors (kN)	Measured laboratory values FN (kN)
2	100	81	80
3	110	87	88
4	120	94	96
5	135	101	108

General characteristics of Herrenknecht TBM used in tunnel drive are given in Table 2.36.

Table 2.36 General characteristics of TBM used in Tarabya tunnel [18].

Parameter	Value
TBM excavation diameter	2,915 mm
TBM shield diameter	2,870 mm
Disc number	20

Parameter	Value
Disc diameter	305 mm (edge width 10 mm)
Maximum rotational speed	16 rpm
Maximum torque applied	725 kNm
Number of thrust cylinders	6
Maximum thrust force	3,750 kN (250 bar)
Total power	620 kW

Excavation through sedimentary rocks, limestone, sandstone–siltstone and andesite dykes, started in July 2000 and finished in November 2004. A mean monthly advance rate of 377 m was obtained with a machine utilization of 35%. It has been shown that full-scale lab cutting tests can be as a useful guide for TBM efficiency. Optimum cutting conditions were obtained using a total machine thrust force of 259 t given an optimum disc penetration of 6 mm/revolution.

2.6.3.4 Esenler–Otogar metro tunnel in Istanbul

The Otogar Bagcilar metro tunnel is a part of the transport project of the European part of Istanbul. The 6.52 m diameter mixed ground EPB tunnel boring machine was used to excavate at Guney Sanayi station in Basaksehir–Bagcilar metro tunnel during this study. The geological conditions of the general area consisted of soft ground, primarily clays, silts and sands. Between Basaksehir and Bagcilar stations the geology is called Kirklareli formation. The Kirklareli formation is represented by hard and dense limestone and includes some fossils. The tunnel was excavated using a shielded Lovat TBM the technical specifications of which are given Table 2.37 [19].

Table 2.37 Basic technical specifications of the TBM [19].

Parameter	Value
Machine diameter	6.52 m
Number of cutters	8 center + 30 single + 8 corner = 46
Total thrust capacity	780–1,355 tons
Cutterhead drive	Electric motors
Cutterhead power	1160 kW
Cutterhead rotational speed	0–3.2 rpm
Cutterhead torque	2,610–3,597 kNm
Thrust cylinder stroke	1.5 m
TBM weight (approx.)	567 tons
Average cutter spacing	75 mm

Geomechanical parameters of the sample are given in Table 2.38.

Table 2.38 Geomechanical parameters of Kirklareli formation with fossil [19].

Uniaxial compressive strength (MPa)	31.9 ± 14.8
Brazilian tensile strength (MPa)	3.94 ± 1.5
Density (kN/m^3)	24.1 ± 0.4
Cerchar abrasivity index	0.75 ± 0.4
Uniaxial compressice strength estimated from point load strength (MPa)	51.7 ± 2.9

Laboratory full-scale rock cutting tests

Summary of laboratory rock cutting tests with V-type disc in unrelieved and relieved mode in Kirklareli formation are given in Table 2.39. The correlations of normal force, rolling force and specific energy with different depth of cut in relieved mode are given in Figure 2.61 and Figure 2.62 gives the relationship between optimum specific energy and ratio of cutter spacing/depth of cut (s/d). As seen from Figure 2.61, normal and rolling forces increase with increasing depth of cut and as is seen from Figure 2.61, specific energy decreases with increasing depth of cut. As seen from Figure 2.62 optimum (s/d) ratio is found to be 6.3. This determines that the optimum penetration for 75 mm of line spacing is 12 mm.

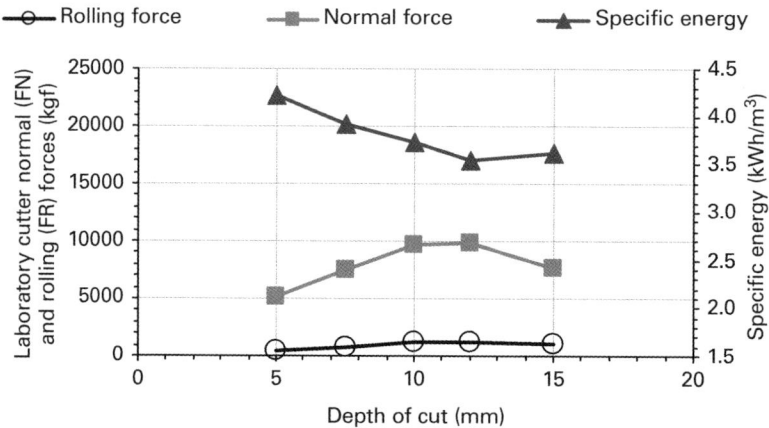

Figure 2.61 Relationships between depth of cut and laboratory normal force, rolling force and specific energy for cutting limestone in relieved mode (spacing=75 mm) [17].

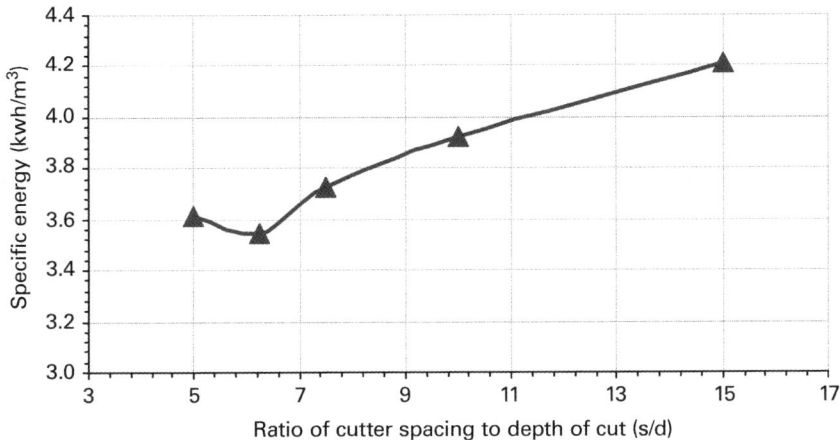

Figure 2.62 Relationship between s/d (cutter spacing / depth of cut) ratio and specific energy for cutting limestone (spacing = 75 mm) [19]

Table 2.39 Summary of laboratory rock cutting tests with V-type disc in unrelieved and relieved mode [19].

Rock formation	Line spacing (mm)	Depth of cut (mm)	FR (kN)	FN (kN)	F'R (kN)	F'N (kN)	SE (kWh/m³)
Limestone	unrelieved	5	5.8	72.6	11.0	112.1	5.1
	unrelieved	7.5	9.7	102.8	19.3	160.4	4.9
	unrelieved	10	17.3	126.5	31.3	198.6	5.8
	unrelieved	12	12.6	105.6	28.8	172.1	4.3
	75	5	4.6	52.1	8.8	98.2	4.2
	75	7.5	7.7	74.3	16.4	136.1	3.9
	75	10	12.4	97.4	22.6	175.6	3.7
	75	12	12.2	98.4	26.1	171.1	3.5
	75	15	11.6	76.2	24.2	147.5	3.6

Field observations and comparison of predicted performance values

The TBM performance data is recorded continuously by the contractor, Gulermak-Dogus Joint Venture, using a data acquisition system equipped within the TBM. The field data analyzed in this study was collected during excavation of the tunnel in the Kirklareli formation.

The operational parameters of the TBM such as rotational speed, torque, advance rate and thrust are recorded by the machine data logger and analyzed ring by ring. Analyzed data includes a total of 262 rings in 16 days (366.8 m) in Guney Sanayi station. All the analyzed data is summarized in Table 2.40.

Table 2.40 TBM field data parameters [19].

Date	Total ring	Total excavation time (minute)	RPM (rev/min)	Torque (kNm)	Thrust (kN)	ICR (m³/h)
2008-10-25	18	821	2.7	2,514	14,638	62
2008-10-26	12	537	2.6	2,639	15,904	63
2008-10-27	7	153	2.8	2,030	10,769	130
2008-10-31	17	718	2.9	1,933	11,576	67
2008-11-01	22	904	3.0	2,002	12,634	69
2008-11-02	7	238	3.0	2,074	14,391	83
2008-11-03	22	702	2.9	1,789	10,431	89
2008-11-04	25	718	2.9	1,918	10,646	99
2008-11-05	19	655	3.0	1,968	13,631	82
2008-11-06	7	307	3.0	2,164	15,044	65
2008-11-08	12	417	3.0	1,818	13,008	81
2008-11-09	22	758	3.0	1,979	12,495	82
2008-11-10	19	759	3.0	2,075	13,307	71
2008-11-11	17	808	3.0	2,184	14,764	60
2008-11-12	18	849	3.0	1,890	15,562	60
2008-11-13	18	877	3.0	1,872	8,015	58
Average			**2.9**	**2,053**	**1,2926**	**76**

The average instantaneous cutting rate is found to be 76 m³/h in the field at 2.9 rev/min of the cutterhead. The net cutting rate is very close to the predicted TBM performance values of the laboratory linear rock cutting tests. The average torque and thrust force are found to be 2,053 kNm and 12,926 kN, respectively, in the field. Peak torque and thrust values of the LCM are close to the field values of the TBM. The laboratory cutting test results in competent rock formations are in good agreement with actual values.

2.7 Conclusions

Tunneling activities have been carried out in Turkey from antiquity up to now, and more recently in Istanbul in great quantity. Turkey's varied landscapes are the product of a wide variety of tectonic processes that have shaped Anatolia over millions of years. Palaeozoic, Mesozoic and Cenozoic formations are all seen in Istanbul. The expected problems in tunneling are directly related to the complexity of the geology of Turkey, frequent changing of strata, dykes, high water ingress, resulting frequent face collapses, squeezing of the TBM and excessive surface deformations as encountered in Istanbul in the Otogar–Esenler metro tunnels which caused an extra cost of 35.6 million USD of the project. For an efficient mechanized tunneling operation in Istanbul it is essential to understand the behavior of TBMs in different geological formations. That is why, within the scope of this book, a brief summary of the geology of Turkey and Istanbul are given first, describing the main geological formations in Istanbul with physical, mechanical and cuttability characteristics of the rocks obtained in the laboratory using a full-scale linear rock cutting machine. The rock cutting experiments are the most reliable and economic solution for selection, designing and predicting/optimizing performance of TBMs. Full-scale testing minimizes the uncertainties of scaling and any unusual rock cutting behavior not reflected in its physical properties. Results of this test can be used as input for selection, designing and predicting/optimizing the costs and performance (excavation/production/cutting rate) of TBMs for feasibility purposes. This test along with deterministic computer simulation is accepted as the most reliable and economical method for these purposes. The predicted TBM performance using full-scale linear rock cutting tests in (a) Tuzla–Dragos sewerage tunnel, (b) Kadikoy metro tunnels (c) Tarabya sewerage tunnel and (d) Esenler–Otogar tunnels are compared with TBM performance values obtained in the field. The predicted values are for competent rock, but it is obvious that geological discontinuities will increase the net excavation rate to a certain level, and a large amount of RQD or water ingress in the rock formation with a large amount of clay will decrease the daily advance rate due to regional collapses, face instability and chocking the cutters etc. These factors affected tremendously the advance rate of the TBM in complex geology. The results show that the specific rock mass boreability index is higher for field data in a highly fractured rock mass than in the laboratory rock cutting data in competent rock. This means that SRMBI is a good indicator of rock mass boreability, since it results in different values on different rock mass conditions. This proves that using normalized parameters such as SRMBI can make it easier to compare the effects of geological parameters. It is hoped that the information given in this chapter will help job owners and contractors to make rational decisions in designing and executing tunneling projects.

References

[1] Kusch, H. (2009) *Tore zur Unterwelt (German) (Secrets of the Underground Door to an Ancient World),* V.F. Samler, Graz.

[2] Okay, A.I. (2008) Geology of Turkey: A synopsis. *Anschnitt,* **21**, 19–42.

[3] Bozkurt, E., Mittwede, S.K. (2001) Introduction to the geology of Turkey—a synthesis, *International Geology Review,* **43** (7), 578–594.

[4] Bilgin, N., Balci, C., Copur, H., Akyuz, S., Namli, N., Tuysuz, L. (2013) *Future tunnelling projects in Istanbul and some considerations for using mechanized tunnelling based on previous experiences.* World Tunnel Congress, Underground – the way to the future, May 31-June 07, Geneva, Switzerland.

[5] Undul, O., Tugrul, A. http://iaeg2006.geolsoc.org.uk/cd/PAPERS/IAEG_392.PDF (14 April 2016).

[6] Akyuz, H.S. (2007) *Stratigraphy of Istanbul.* Proceeding of 3th Symposium of Istanbul Geology, Istanbul Branch of the Chambers of Engineers of Geology, Symposium date 7–9 December 2007, January 2010, Istanbul ISBN 978-9944-89-887-4.

[7] Yuksel, A. (2014) The effect of geological discontinuities on the performance of TBMs in a Metro Project, Istanbul Technical University, PhD Dissertation.

[8] Yuksel, A., Yesilcimen, O., Arioglu, E. (2006) Geological and geotechnical investigations realized for in Kadikoy-Kartal Metro Project, *Rocmec 2006, VIII Regional Rock Mechanics Symposium,* 2–3 November, Istanbul.

[9] Tuysuz, L. (2012) Performance prediction of a TBM in a complex geology. Istanbul Technical University. MSc Dissertation.

[10] Sonmez, H. and Ulusay, R. (1999) *Modifications to the geological strength index (GSI) and their applicability to the stability of slopes.* Int J Rock Mech Min Sci, **36** (6), 743–760.

[11] Bilgin, N., Copur, H., Balci, C. (2014) Mechanical *Excavation in Mining and Civil Industries,* CRC Press, Taylor and Francis Group, London.

[12] Balci, C. (2009) Correlation of *rock cutting tests with field performance of a TBM in a highly fractured rock formation: A case study in Kozyatagi-Kadikoy Metro Tunnel,* Turkey. Tunn. Undergr. Sp. Technol. **24**, 423–435.

[13] Bilgin, N., Copur, H., Balci, C. (2013) TBM Performance prediction for Uskudar–Umraniye–Cekmekoy–Sancaktepe metro tunnels and risk analysis. Report of Investigations, Istanbul Technical University.

[14] Bilgin, N., Copur, H., Balci, C., Feridunoglu, C., Tumac, D. (2006) The cuttability characteristics of the rock formations in Kadikoy-Kartal metro line for TBM Performance prediction. Research report prepared for Anadoluray Joint Venture, Istanbul Technical University.

[15] Bilgin, N., Balci, C., Tuncdemir, H., Eskikaya, S., Akgul, M., Algan, M. (1999) The performance prediction of a TBM in Tuzla–Dragos sewerage tunnel, Proceedings of the World Tunnel Congress on Challenges for the 21st Century. Oslo, pp. 817–827.

[16] Yuksel, A., Sozak, N.N., Gulle, G. (2005) Engineering Geology report prepared for Kadikoy-Kartal Metro Project: *KK-GE-TR-GN-004, Anadoluray Joint Venture*, Istanbul.

[17] Gong, Q.M., Zhao, J., Jiang, Y.S. (2007) *In situ TBM penetration tests and rock mass boreability analysis in hard rock tunnels*. Tunn. Undergr. Sp. Technol. **22** (3), 303–316.

[18] Bilgin, N., Feridunoglu, C., Tumac, D., Cinar, M., and Ozyol, L. (2006) *TBM cutting performance in Istanbul*, Tunnels & Tunnelling International, pp. 17–19.

[19] Balci, C., Tumac, D., Copur, H., Bilgin, N., Yazgan, S., Demir, E., Aslantas, G. (2009) Performance Prediction and Comparison with In-Situ Values of A TBM: A Case Study of Otogar-Bagcilar Metro Tunnel in Istanbul, *World Tunnel Congress*, 23–28 May, Budapest, Hungary.

3 Difficult ground conditions dictating selection of TBM type in Istanbul

3.1 Introduction

Istanbul has a very complex geology, and in the near future almost the majority of TBM tunneling projects of Turkey are planned to be carried out in this fast growing city. Bearing in mind this reality, the main objective of this chapter is oriented to show how the optimum selection of TBM type in Istanbul, gradually changed from open type TBM (Baltalimani Tunnel), to double shield TBM (Moda-Tuzla Tunnel), to slurry type TBM (Marmaray Tunnels) and finally to EPB-TBMs over the past 25 years. This gradually progressing selection based on the complex geology of Istanbul is a typical example to the concept of 'learning costs'.

3.2 Case study of open TBM in complex geology (1989), in Baltalimani tunnel: Why open type TBM failed

Collector tunnels designed to renew Istanbul's inadequate sewerage network and to clean the polluted Golden Horn (Halic) were constructed within an ambitious pollution abatement program involving several hundred kilometers of circular sewerage tunnels of diameters 2.2–5 m. Baltalimani tunnel, having a diameter of 4.5 m and a length of 2,318 m was primarily intended to be excavated by a Robbins 145–168 gripper-open type full face tunnel boring machine. It was one of the first experiences of excavating a tunnel with a TBM in Istanbul [1]. The machine, as seen in Figure 3.1, was equipped with 35 disc cutters (four discs of 12″ diameter and 31 at 15.5″); it had a cutting power of $7 \times 125 = 875$ HP. The geological difficulties forced the withdrawal of the machine after a drivage of 1230 m and it was replaced by a shielded Herrenknecht roadheader, as seen in Figure 3.2. A typical tunnel support, as used in Baltalimani, is shown in Figure 3.3.

Figure 3.1 Robbins 145-168 open type TBM [1].

TBM Excavation in Difficult Ground Conditions. Case Studies from Turkey. First Edition. Nuh Bilgin, Hanifi Copur, Cemal Balci.
© 2016 Ernst & Sohn GmbH & Co. KG. Published 2016 by Ernst & Sohn GmbH & Co. KG.

Figure 3.2 Herrenknecht shielded roadheader.

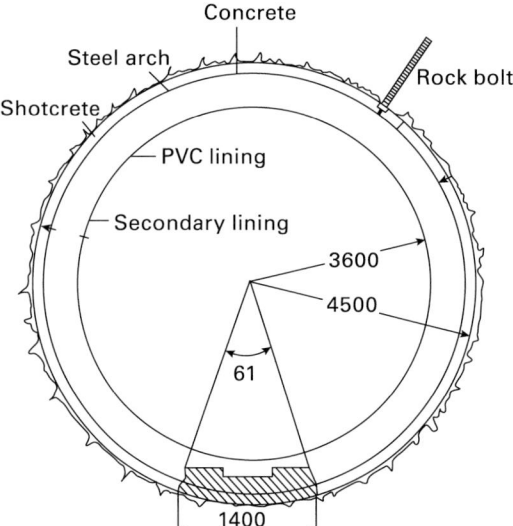

Figure 3.3 The support used in the Baltalimani tunnel [1].

Rock bolts, steel arches and shotcrete were immediately installed after the excavation. Later, the secondary lining was given a PVC liner to protect the concrete from corrosion by hydrogen sulfide gas. Design of the machine was governed by geological requirements and specifications of the project such as tunnel diameter, and rock properties. The tunnel was opened in Buyukada and Trakya formations frequently cut by andesite and diabase dykes of 1–30 m thickness. Buyukada formation, of Upper Devonian Age, consists of micritic and nodular limestone and carbonate rich shale. It is strongly folded, slightly jointed and has a massive appearance. The joints are generally perpendicular to the bedding and perpendicular to the sub-vertical dip. Trakya formation consists of mudstone, shale-graywackee and conglomerate units. It is closely jointed and strongly folded. RQD in Trakya formation is comparatively very low.

Machine performance in Buyukada and Trakya formations is summarized in Table 3.1 and Figure 3.4. Some data on geological conditions such as RQD, RMR, Q, overbreak and rock mass quality values is also given in Figure 3.4. As one can easily see from this table, the machine utilization in Trakya formation is four times less than Buyukada formation, main because of the very bad ground conditions. The characteristics of these two complex geological formations were outlined in previous chapter.

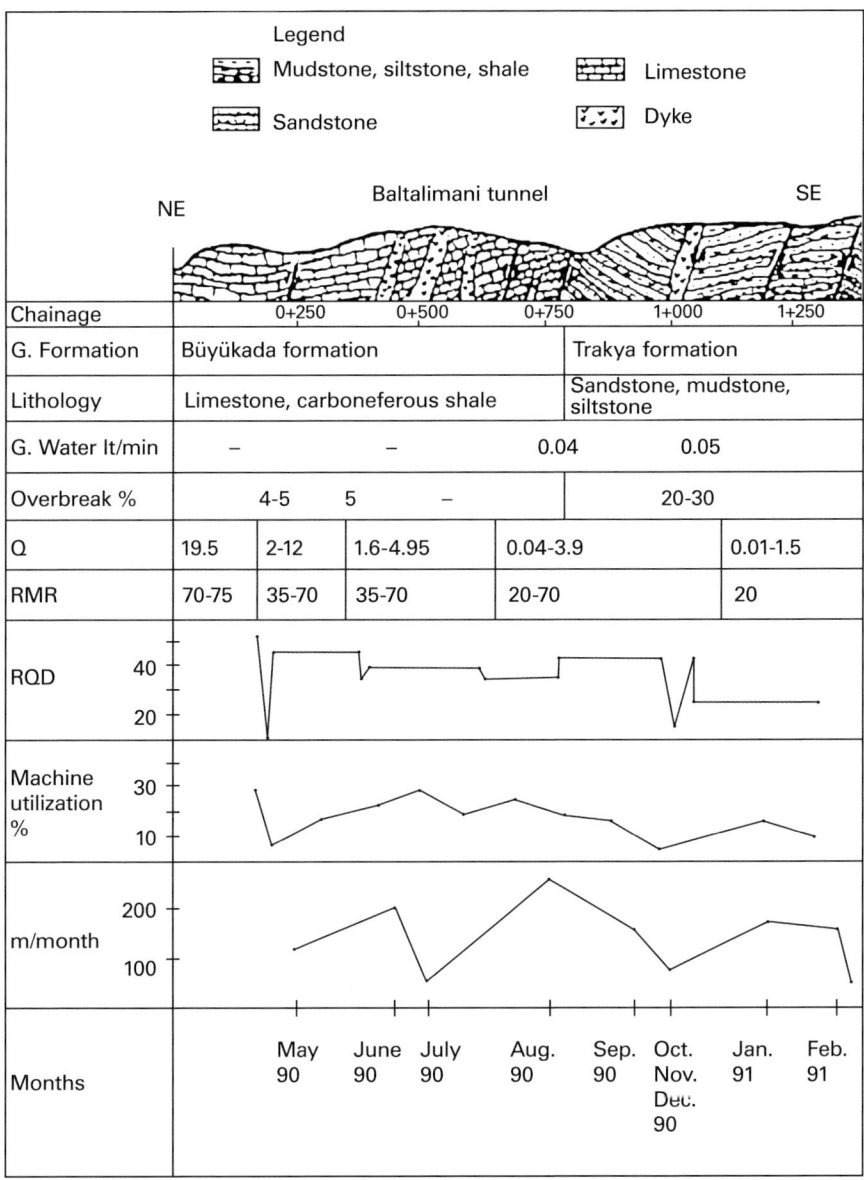

Figure 3.4 Geology of the excavated area and general performance of TBM [1].

The support of the caverns at crown and sidewalls took most of the machine down-time during the excavation through Trakya formation. High compressive loads of the gripper pads of 690–1,100 tons also caused the deformation of the support system and cracks on the shotcrete cover. The pads went into the sidewalls when they encountered highly crushed zones or clay bands, so wooden blocks were placed between the gripper pads and the support system to distribute the loads on the sidewalls.

Table 3.1 Machine performance in Buyukada and Trakya formations [1].

Parameter	Buyukada formation	Trakya formation
Machine utilization %	28.5	7.2
Machine downtime %	71.50	92.6
Net cutting rate m/h	1.22	1.7
Advance rate m/h	0.35	0.13
Average shift advance m/shift	3.15	1.24
Best shift advance m/shift	11.50	9.57
Lowest shift advance m/shift	0.60	0.2
Average daily advance m/day	7.18	3.12
Best daily advance m/day	20	16.5
Lowest daily advance cm/day	0.22	0.5
Average weekly advance m/week	43	21
Best weekly advance m/week	46	66
Lowest weekly advance m/week	9.95	1.9
Average monthly advance m/month	197	84
Best monthly advance m/month	261.4	177.65
Lowest monthly advance m/month	56.2	17.33

Table 3.2 Machine downtime in Buyukada and Trakya formation [1].

Parameter (%)	Buyukada formation	Trakya formation
Disc replacement	16.84	0.8
Disc control	3.05	0.17
Support	14.04	44.58
Mucking and waiting for wagons	14.47	4.34
Mechanical breakdown	5.11	3.0
Electrical breakdown	1.21	1.39
Ground conditions	3	29.39
Crane failure	4.20	0.25

Parameter (%)	Buyukada formation	Trakya formation
Conveyor failure	0.88	1.75
Ventilation failure	0.31	0.52
Maintenance	2	0.58
Other	6.39	5.83
Total	71.50	92.60

Collapses and overbreak were major problems faced during the tunnel drivage through Trakya formation. Collapses took place in different rock conditions such as faulted, crushed, fractured or highly altered zones. Special support systems were applied in each case according to the form of caverns. These are briefly summarized below.

3.2.1   Collapse between chainage 0+920 and 0+935 km

The collapse occurred in very crushed and fractured interbedded siltstone–mudstone zone. It started from the right shoulder and enlarged towards the left shield of the machine. The support system used in this collapse area is shown in Figure 3.5.

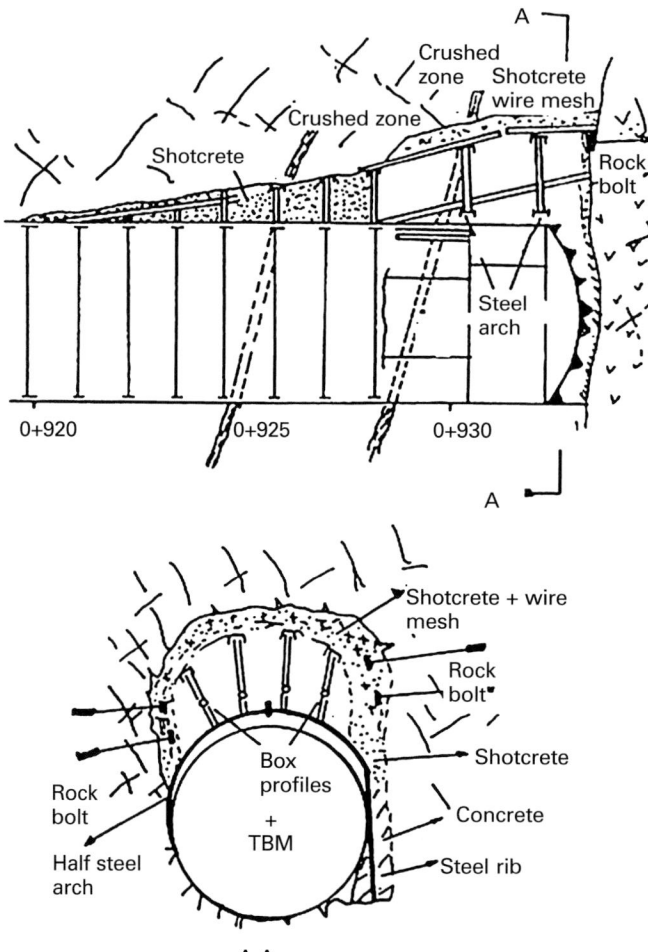

Figure 3.5 Collapse between chainage 0+920 and 0+935 km [1].

3.2.2 Collapse between chainage 0+965 and 0+982 km

Geological and structural conditions produced severe stability problems within 10 m of the collapsed area. Extensive overbreak from left wall and roof caused some caverns of 1–1.5 m in dimension. Well-developed medium closely spaced joints (20–50 cm max.) and other nonsystematic discontinuities separated the rock mass into rock elements of 5–20 cm. The support system used was similar to the previous one (see Figure 3.6)

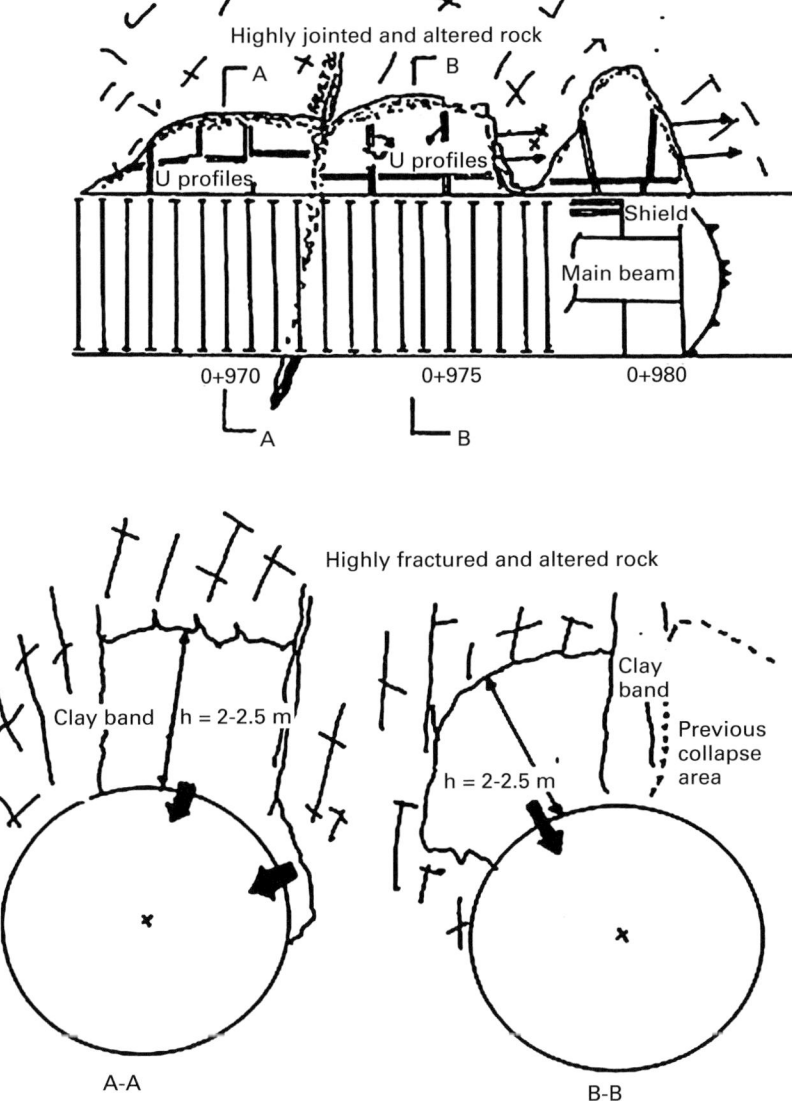

Figure 3.6 Collapse between chainage 0+965 and 0+982 km [1].

3.2.3   Collapse between chainage 1+148 and 1+155 km

Highly jointed and fractured rock with clay-filled siltstone, and mudstone of Trakya formation were encountered in this area. Rock blocks collapsed onto the shield due to the presence of joint sets.

3.2.4   Collapse between chainage 1+220 and 1+235 km

The main rocks at the collapsed area are siltstone and mudstone, and although they were not obvious, some sandstone bands were recorded. Trakya formation in this area is intensely jointed and fractured. Joints are generally filled with clay or calcite. At the chainage of almost the 1+210 km an overbreak occurred due to the fractured rock on the left-hand wall, and later a major collapse occurred when a fault zone was encountered [1].

3.3   Double shield TBM in the Istanbul–Moda collector tunnel, 1989/90

A double shield TBM was used in the Moda collector tunnel having a length of 1226 m within the years 1989/90. The specifications of the machines are given in Table 3.3. The rock formation was interbedded sandstone, mudstone and claystone of Trakya formation [2]. Segment lining was used in the tunnel. The TBM was launched into the tunnel within a circular shaft. After only 80 m of advance, the TBM unexpectedly entered a fault zone and the TBM jammed due to a tunnel face collapse. A sink hole formed on the road above 5 m of the tunnel. The TBM was rescued by opening a shaft above the tunnel [3]. The mean machine utilization time was as low as 20%. This TBM was later used in the Tuzla tunnel.

Table 3.3 Specifications of the Robbins 165-162 / E1080 TBM [4].

Property	Values
Machine diameter	5.0 m
Number of cutters	36
Rotational speed	6 rpm
Normal thrust force	471 tons
Maximum thrust force	785 tons
Maximum thrust force	600 HP
Power of the miscellaneous pumps etc.	285 HP
Conveyor belt capacity	476 m³/h
Electrical transformer	1,000/380V, 50Hz

3.4   Double shield TBM working without precast segment, difficulties in difficult ground: Tuzla-Dragos tunnel in Istanbul

The Tuzla–Dragos sewerage tunnel is situated in a highly populated area of the Asian side of Istanbul. Tunnels having a total length of 6,490 m and final diameter of 4.5 m were completed by STFA Construction Company. The tunnels in X1–X3 and K1 shafts having a length of 1,670 m and a depth changing between 6 m and 17 m were opened with a Robbins TBM, Model 165-162 having a diameter of 5 m. The tunnel was opened without segmental lining due to the restrictions of TBM and tunnel diameters [4]. A general view of the machine is illustrated in Figure 3.7 and the final cross-section of the tunnel is given in Figure 3.8.

Figure 3.7 TBM used in Tuzla-Dragos Tunnel [3].

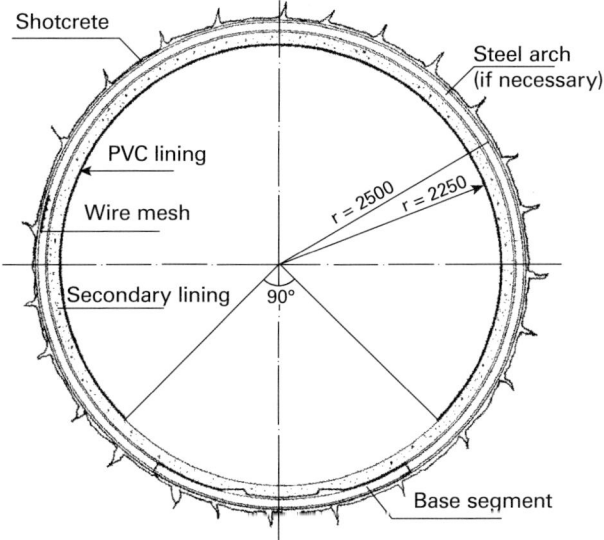

Figure 3.8 Cross-section of Tuzla-Dragos Tunnel [4].

The main rock formation in the area is Lower Devonian aged Kartal formation, composed of shale and limestone. In the upper levels there is Pliocene aged Belgrad formation of sand, gravel and silt, and Quaternary aged sediments. Diabase dykes of Cretaceous ages cut Kartal formation in several places. The tunnels in X1–X3 and K1 shaft are mainly driven through limestone and shale. Limestone of gray-blue color was of medium thickness, sandy and fractured in several places, and shale of dark gray-black color was carbonated, mainly jointed and laminated.

Tuzla-Dragos district is highly populated area of Istanbul. This fact limited the number of drill holes for side investigations. Geotechnical and other studies that were carried out before starting the excavation, indicated that highly fractured clayey zones with excessive water inflow should be expected during tunnel drivages, resulting in lower machine utilization time and lower daily advance rates than expected for competent rock. The performance of the TBM is given in Table 3.4.

Table 3.4 Performance of the TBM in the Tuzla-Dragos tunnel [3].

Starting date	6 October 1997
Finishing date	31 July 1998
Length of tunnel	1,600 m
TBM diameter	5 m
Final tunnel diameter	4.5 m
Average machine utilization in fractured rock	10%
Machine utilization in competent rock	35%
Average net cutting rate	50 m^3/h
Average progress rate	5.1 m/h
Best daily advance	15.2 m/day
Average daily advance	6.2 m/day
Best weekly advance	69 m/week
Average weekly advance	33 m/week
Best monthly advance	253 m/month
Average monthly advance	135 m/month
Cutter cost	4 $/m^3

A database, accumulated during Tuzla-Dragos tunnel excavation, gave the opportunity to formulate a rating system to predict the machine utilization; this is given in Table 3.5. Anyone using this table must remember that this is only valid for a double-shielded TBM advancing without segmental lining, using a support system with wire mesh, shotcrete, steel arch and a secondary lining [1]

Figures 3.9 and 3.10 illustrate two cases for the application of this rating system.

Case 1 represents a highly clayey zone with water inflow, where a collapse in tunnel chainage 0+102.8 occurred on 1 May 1998, RQD value of the zone was 0–5% and RMU was calculated as 10. Daily advance before the collapse occurred was 1.23 m.

Table 3.5 Rating for machine utilization (RMU) using double-shield TBM and wire mesh, shotcrete, steel arch and secondary lining system [3].

RQD	>70	70–50	50–30	30–20	<20
Ratings	10	8	4	2	0
Water inflow	Non	Wet	Drop	Leakage	High inflow
Ratings	4	3	2	1	0
Clay Content	Non	Very Little	Filling between fractures	High amount	Excessive
Ratings	4	3	2	0	0
Maintenance facilities	Extremely good	Very good	Good	Fair	Poor
Ratings	6	5	4	2	0
Contractor experience	Extremely good	Very good	Good	Fair	Poor
Ratings	6	5	4	2	0

RMU < 16   Machine utilization 3–5%
RMU < 16–20   Machine utilization 5–10%
RMU < 20–22   Machine utilization 10–20%
RMU < 22–24   Machine utilization20–30%
RMU < 24–26 Machine Utilization%30–45

Case 2 represents a highly fracture mudstone and siltstone zone with water inflow, collapse in tunnel chainage 0+1008 which occurred on 19 March 1998, when the RQD value of the zone was around 25% and RMU was calculated as 19. Daily advance before the collapse occurred was 7.8 m.

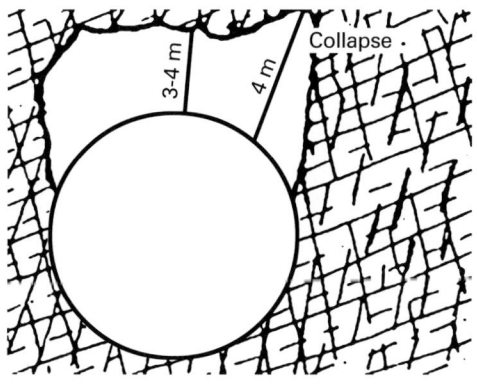

Figure 3.9 Case 1, collapse at 0+1028 km [4].

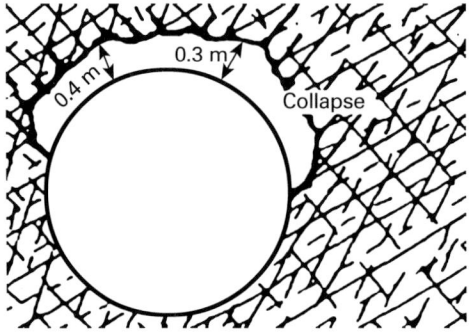

Figure 3.10 Case 2, Collapse at 0+1008 km [4].

3.5   Difficulties in using slurry TBMs in complicated geology, Marmaray tunnel project

The Bosphorus rail link beneath the Istanbul Strait, developed as part of Turkey's Marmaray Project, represents the first physical rail link between Europe and Asia. The 76.3 km rail link was inaugurated in October 2013 on the 90th anniversary of the Turkish Republic. Lines 3, 4, 5 and 6 were excavated with 7.5 m diameter Hitachi-Zozen slurry TBMs. The mean advance rates of slurry TBMs were found to be much lower compared to those of EPB-TBMs used in Istanbul. The main problems of using slurry TBMs were clogging the return lines of the slurry pipes due to the highly fracture characteristics of the rock formations and the high quartz content of rocks wearing the slurry pipes. Typical performance of slurry TBM No. 5 over a period of two years is given in Table 3.6. As can be seen, the TBM excavated 2,415 m in two years, with a mean daily advance rate of 3.5 m, and the TBM stopped for two periods owing to damage of the cutterhead. A fault zone, as seen in Figure 3.11, caused face collapse and damaged the cutterhead. This was a turning point in the design of the cutterheads of TBMs used in Istanbul. Grizzly bars were added to others TBMs also to stop face collapses in fractured rocks, one in Beykoz sewerage tunnel and the other one in Goztepe–Kadikoy metro tunnels.

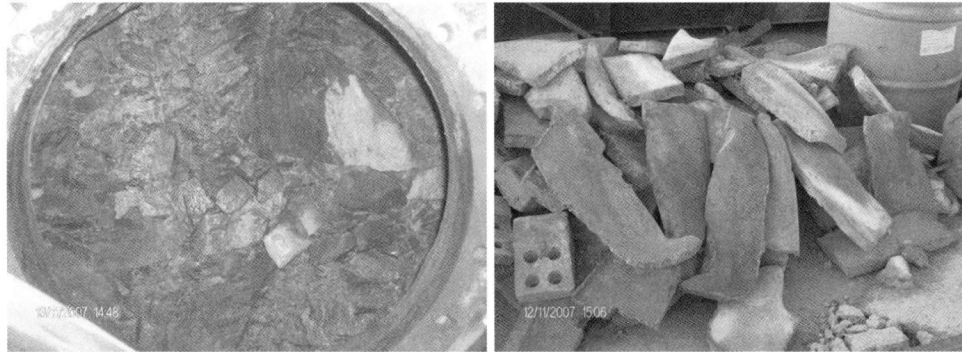

Figure 3.11 Typical fault breccia causing face collapse and damaging the cutterhead.

Large diameter slurry TBMs were never used thereafter in Istanbul due to bad experiences in the Marmaray tunnels. However, the two-deck Eurasia roadway tunnel, facing high face pressures up to 11 bars, was an exception [5].

Table 3.6 Performance of TBM No. 5 in the Marmaray project.

Date	Rings	m/month	Total m	Geological formation
Jan 07	10	15	15	Tuzla, Baltalimani, Trakya
Feb 07	66	99	114	
Mar 07	12	18	132	
Apr 07	110	165	297	
May 07	77	115.5	412.5	
June 07	39	58.5	471	
July 07	110	165	636	Alluvium, Trakya
August 07	81	121.5	757.5	Trakya
September 07	15	22.5	780	
October 07	0	0	780	
November 07	0	0	780	
December 07	20	30	810	
Jan 08	98	147	957	
Feb 08	79	118.5	1,075.5	
Mar 07	124	186	1,261.5	
Apr 08	154	231	1,492.5	
May 07	144	216	1,708.5	
June 08	120	180	1,888.5	
July 08	62	93	1,981.5	
August 08	103	154.5	2,136	
September 08	91	136.5	2,272.5	
October 08	56	84	2,356.5	
November 08	0	0	2,356.5	
December 08	39	58.5	2,415	
The best advance per day, m			13.5	
The average advance per day, m			3.45	
The best advance per week, m			57	
The average advance per week, m			24.15	
The best advance per month, m			231	
The average advance per month, m			101	

3.6   Difficulties in single-shield TBM working in open mode in complex geology: An example from Kadikoy–Kartal metro tunnel

This section concerns about the performance of two TBMs working between Kozyatagi and Kadikoy. The project is a part of metro line starting from Pendik, integrating with the Marmaray Project and going to Halkali in the south-east of the city. The project was commissioned to Anadoluray Joint Venture, formed by the contractors, Yapi Merkezi, Yuksel, Dogus, Yenigun and Belen.

Sedimentary rock formations of Triassic and Tertiary ages with a complex geology are found in the area. The main geologic formation to be excavated was Kartal formation with several diabase and andesite dykes approximately every 70 m. The contact zones between dykes and sedimentary rocks are usually highly fractured, causing several problems in front of the cutting head of the TBM. Dolayoba limestone was also found between chainage 6+050 and 5+850 m [6].

Two identical Herrenknecht TBMs were chosen for Lines 1 and 2, because these machines can work in open and closed modes. Basic specifications of the two TBMs are given in Table 3.7.

Table 3.7 Basic technical specifications of the TBM [6].

Parameter	Value
Machine diameter	6.57 m
Number of cutters	26 single + 6 double (12) = 38
Maximum contact pressure per disk	267 kN
Maximum thrust capacity of the main bearing	20,000 kN static
Cutterhead drive	Electric motors
Cutterhead power	1,260 kW (4×315)
Cutterhead rotational speed	1.6–5.5 rpm
Cutterhead torque	5,200 kNm at 1.6 rpm 1,515 kNm at 5.5 rpm
Maximum working capacity of thrust cylinders	42,575 kN at 350 bar
Thrust cylinder stroke	1.5 m

The excavation data was carefully collected during tunnel drives. At the beginning, it was observed that the contact zones between dykes and the main rock formation were highly fractured, and in some areas big rock blocks having sizes up to $30 \times 40 \times 50$ cm were ripped off by the disc cutters from the fractured zones, passing through the openings of the cutterhead and causing severe problems such as collapses of the tunnel face. Typical big blocks can be seen in Figure 3.12 [6].

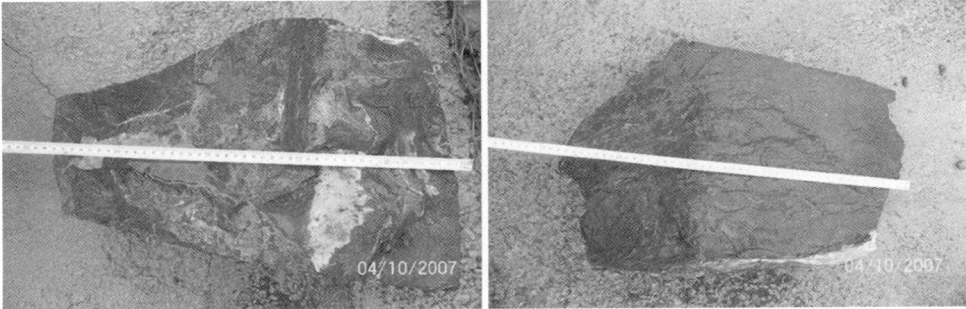

Figure 3.12 Big rock blocks coming from the tunnel face [6].

The face collapses dramatically decreased daily advance rates to around 2.5 m/day. After several technical discussions between project management staff and TBM manufacturers it was decided to install some grizzly bars to limit the big rock blocks passing through the openings of the cutterhead. The general view of the TBMs before and after the modification are seen in Figure 3.13. Daily advance rate increased to up to 6 m/day after the modification of the cutterhead as seen in Figure 3.14. This figure also shows the changes in the TBM after transition to EPB mode, in soft and fractured ground [6].

Figure 3.13 TBM-360, cutterhead with grizzly bars on the right and cutterhead before modification on the left [6].

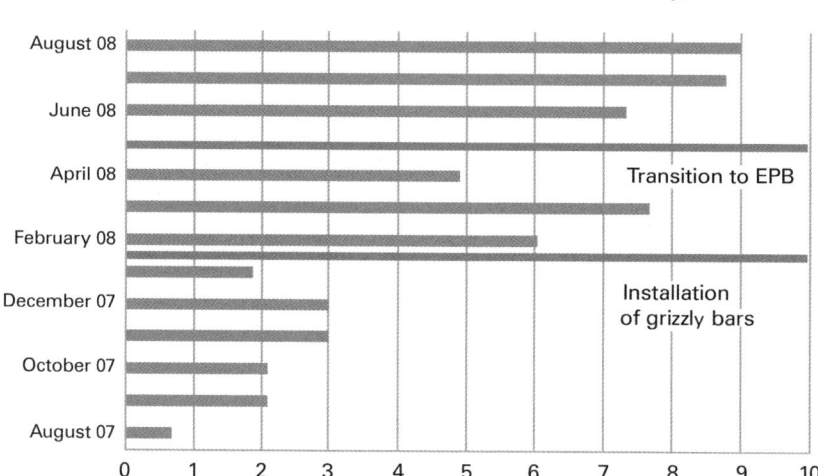

Figure 3.14 Daily advance rate of TBM, before and after replacing grizzly bars and screw conveyor [6].

3.7   Eurasia tunnel excavated by a large diameter slurry TBM

The Eurasia Tunnel Project (Istanbul Strait Road Tunnel Crossing Project) connects the Asian and European sides via a highway tunnel going underneath the Bosphorus. The project was constructed by a Joint Venture (YMSK-JV) formed under the leadership of a Turkish firm (Yapi Merkezi) and a South Korean firm, SK E&C. A two-deck tunnel was constructed underneath the seabed using special technology to successfully deal with the complex challenges faced. The 13.7 m Herrenknecht slurry TBM was designed exclusively for this joint venture and included a number of specially developed features such as 21″ disc cutters with internal pressure compensation device or a hyperbaric transfer shuttle. Specifically innovated, designed and developed seismic joints that can resist 12 bar pressure were installed to accommodate differential displacements at particular locations at the transition zones between rock and soft soils. The project was given an award by International Tunneling Association (ITA) as the "Major project of the year 2015" [5].

3.8   Conclusions

The experience gained in Istanbul with TBMs is a typical example to saying that "Learning costs". The first open-type TBM used in Istanbul in heavily fractured Trakya formation was a nightmare for the contractor, and the machine was withdrawn and replaced with a shielded roadheader. However, in competent rocks such as the Ermenek tunnel a 6.6 m diameter Wirth open-type TBM did a very good job with a mean daily advance rate of 13.4 m, including all stoppages and 20.6 m excluding stoppages [7].

Double-shield TBMs were unsuccessful in the geological conditions of Istanbul. However, in other tunnels in Turkey they proved to be a notable success in competent rock. In the Yamanli tunnel, using a Robbins double-shield TBM with a diameter of 4.3 m, an average daily advance rate of 16 m was obtained with a machine utilization time of 50% [8].

In the Mavi tunnel a 4.88 m Seli double-shield TBM reached an average daily advance rate of 13.3 m, mainly in dolomitic limestone and sandstone. However, in a transition zone, the tunnel collapsed and almost a month was spent on rescue operations [9]. In the Suruc water tunnel a mean daily advance rate of 15.7 m, including all stoppages, was obtained in Marl with 7.88 m diameter double-shield TBM, and the average machine utilization time was around 42% [10]. Big diameter slurry TBMs also were unsuccessful in Istanbul, due to high abrasivity and fracturing characteristics of the geological formations. Time spent on changing the worn slurry pipes and opening the blocked slurry return lines, considerably decreased machine utilization time. The other problem of using slurry TBMs is the need for a large area for slurry plant, and in the majority of tunneling project in Istanbul, there is only a limited space for this type of facility. The Eurasia tunnel was an exception to this because, for high face pressures of up to 12 bar, using a slurry TBM was a necessity.

As explained above, the use of TBMs in Istanbul gradually changed from open gripper-type TBMs to double-shield TBMs, slurry TBMs, open-mode single-shield TBMs and finally to EPB-TBMs equipped with grizzly bars, which is now a standard type of excavating machine.

References

[1] Bilgin, N., Nasuf, E., Cigla, M. (1993) *Stability problems effecting the performance of a full face tunnel boring machine in Istanbul Baltalimani Tunnel, Assessment and prevention of failure phenomena in rock engineering,* Balkema ISBN 90 5410 3094, pp. 501–506.

[2] Dalgic, S. (1997) *The comparison of predicted and encountered geologic conditions in tunnels.* Engineering Geology, November, number 51, pp 36–40 (in Turkish).

[3] Anon 2015, *Catalogue of notable tunnel failures (up to 2015), The Government of Hong Kong Special Administrative Region*, Civil Engineering and Development Department, p. 380.

[4] Bilgin, N., Balcı, C., Tuncdemir, H, Eskikaya, S., Akgul, M., Algan, M. (1999) Performance prediction of a TBM in difficult ground conditions. AFTES – Journées d'études Intenationals de Paris, 25–28 October, Paris, France.

[5] https://awards.ita-aites.org/winners.html. (2016)

[6] Bilgin, N., Ozbayır, T, Sozak, T., Eyigun, Y. (2009) Factors affecting the economy and efficiency of metro drivages with two TBMs in Istanbul in very fractured rock. World Tunnel Congress, 23–28 May, Budapest, Hungary.

[7] Kocbay, A. (2010) *A typical example to performance analysis of TBMs, Ermenek Tunnel (in Turkish),* Technical Bulletin of DSI, No 108, pp. 37–50.

[8] Akgul, M., Akgul, E., Bostanci, E., Copur, H. (2015) *Analysis of disc cutter consumption of a Double Shield TBM.* World Tunnel Congress, May 22–28, Dubrovnik, Croatia, pp 1–10

[9] Seven, S. (2010) Pre and post injections in TBM excavation in difficult ground conditions, two examples from Turkey. Istanbul Technical University. MSc Dissertation.

[10] Ilci, N., Temel, M., Sezgin, S., Akpinar, T., Guarasio, S., Polat, C., Bilgin, N. (2013) *Clogging and squeezing effect of marl-clayey limestone on the performance of a hard rock TBM in Suruc Tunnel, Turkey*, World Tunnel Congress, Geneva, Underground – the way to the future! G. Anagnostou & H. Ehrbar (eds)

4 Difficult ground conditions affecting performance of EPB-TBMs

4.1 Introduction

It is expected that in a near future more than 500 km of tunnels will be excavated in Turkey with EPB-TBMs in areas of complex geology. Surface collapses can sometimes occur in EPB-TBM excavations in soft ground and in complex geology. Excessive ground deformations can cause damage to the surrounding buildings as was experienced in the Otogar–Esenler metro tunnels which caused an extra cost of 35.6 million USD for the project [1].

Ancient water wells are a significant risk for surface collapses in areas like Istanbul. Surface collapse, which occurred in the Yenikapi and Mahmutbey–Mecidiyekoy metro lines is a typical example to this. It is reported that a severe surface collapse occurred during the Yenikapi–Istanbul metro tunnel construction resulting the death of five people living in a hostel [2]. Figure 4.1 is a typical example, showing the effect of passing an undetected water well when excavating under buildings.

Figure 4.1 Surface collapse in Mahmutbey Mecidiyekoy Metro tunnel on the 13.06.2015.

Given the considerations above, the main objective of this chapter is to discuss the factors affecting the performance of EPB-TBMs in complex geology and to present a performance prediction model, taking into account the experience gained in Istanbul in

TBM Excavation in Difficult Ground Conditions. Case Studies from Turkey. First Edition. Nuh Bilgin, Hanifi Copur, Cemal Balci.
© 2016 Ernst & Sohn GmbH & Co. KG. Published 2016 by Ernst & Sohn GmbH & Co. KG.

EPM TBMs excavations. This chapter summarizes the results published in this respect in different articles [3, 4, 5].

4.2　Factors affecting performance of EPB-TBMs

Earth pressure balance (EPB) shield tunneling is considered to be an effective tunneling method when surface settlements must be avoided by controlling face stability and underground water inflow while preventing the dispersion of dangerous pockets of explosive gas within the tunnel.

EPB tunneling requires the application of soil conditioning to obtain correct excavation control, particularly when working in complex cohesionless soils [6]. The laboratory research presented by Peila et al. [7] verified that suitable conditioning is possible when using EPB technology in rock masses – see Figure 4.2. Furthermore, their study showed clearly that both slump testing and extraction testing with a screw conveyor must be used to define the appropriate amount of conditioning agents. Slump testing was found necessary to define the minimum conditioning parameters, but only extraction testing allows the simulation of tunneling operations at an adequate scale to define the optimal conditioning parameters. Testing of samples from three rock mass formations showed that good EPB operation management requires the use of a more fluid mixture than defined by slump testing, especially in a schistose rock mass.

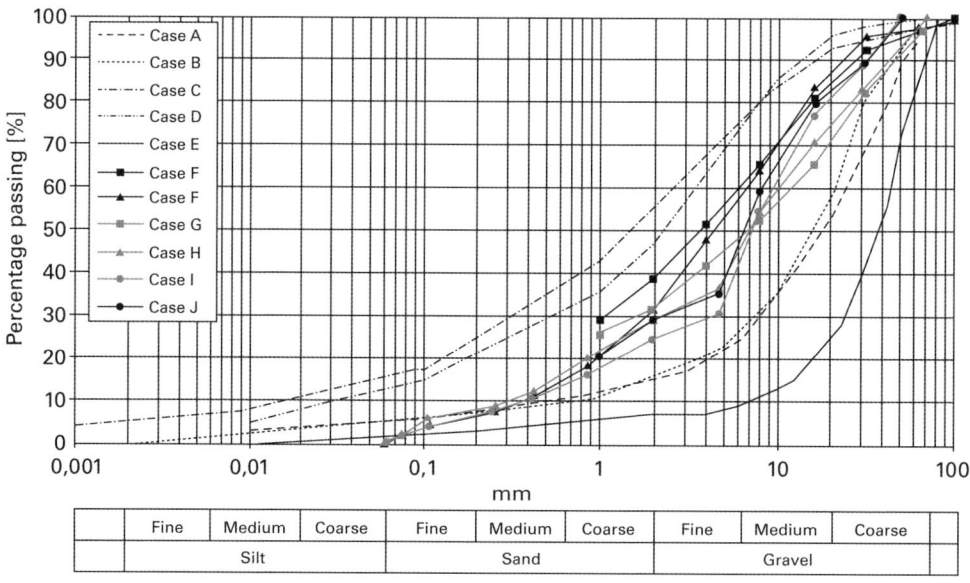

Figure 4.2 Grain size distribution in muck produced by rock TBMs reported by Peila et al. [7].

For successful TBM excavation in soil, using the correct chemicals for conditioning is of a prime concern [8]. However, the grain size of the soft ground dictates also the type of excavation as can be seen in Figure 4.3.

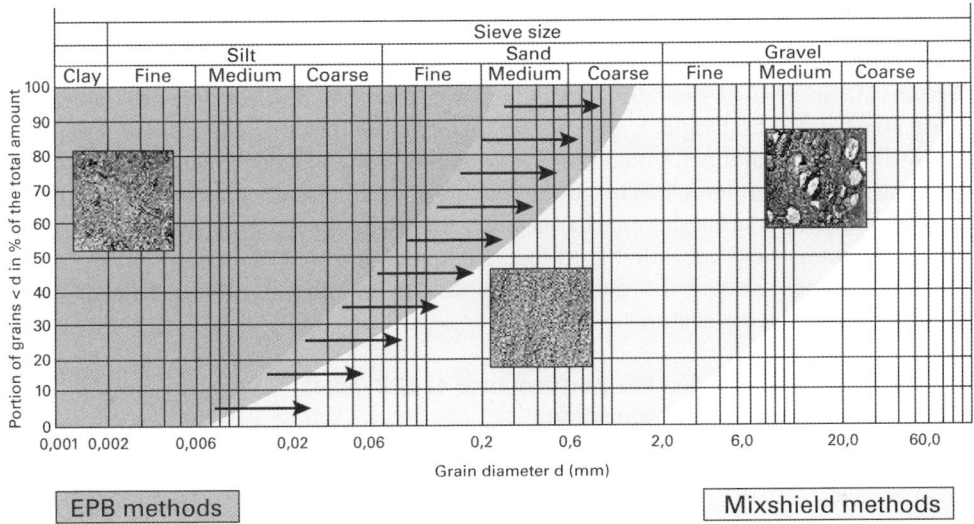

Figure 4.3 Effect of grain size on the selection of excavation type [9].

4.3   Performance prediction of EPM TBMs in difficult ground conditions

Selecting the appropriate tunneling equipment and organizing an effective operational management with the right tunneling crew is a major concern for efficient scheduling of future projects, and this necessitates the creation of a model to estimate daily advance rates. Namli et al. [3,4] and Bilgin et al. [5] suggested an empirical model for predicting daily advance rates of EPB-TBMs working in closed and semi-closed mode for excavation of fractured hard rocks in complex geology. Ates et al. [10] used data from 235 different tunneling projects to estimate thrust and torque requirements for different ground conditions, while Song et al. [11] and Shi et al. [12] suggested a theoretical model for estimation of cutterhead torque and thrust requirements of TBMs working in soft ground. Maidl and Wingmann [13], Frough et al. [14] and Farrokh et al. [15] suggested an empirical model for estimation of the machine utilization time for EPB-TBMs. Evaluation of EPB-TBM performance in mixed ground conditions was made by Toth and Zhao [16]. Recently, Copur et al. [17] used a stochastic model implemented into a deterministic model to predict the performance of EPB-TBMs.

However, as will be noticed above, there are only a limited number of research studies on predicting the performance of EPB-TBMs. Especially, predicting daily advance rates in complex geology, in mixed face, in fractured rocks, in transition zones etc. is a big concern. The initial experiences in Istanbul with a gripper type TBM were a nightmare, with roof collapses causing major delays in tunnel excavation, Bilgin [18]. A typical example of EPB-TBM was the Kadikoy-Kozyatagi tunnel, where changing TBM from open mode to closed mode with a few modifications of the cutterhead resulted in a tremendous increase in daily advance rates, Bilgin et al. [19]. The data collected from several tunneling projects served as the basis for creating a model to predict daily advance rates of EPB-TBMs. Within the scope of this research program, the validity of the model was tested in Uskudar–Umraniye–Cekmekoy–Sancaktepe metro project and more recently in the Mecidiyekoy–Mahmutbey project.

Uskudar–Umraniye–Cekmekoy–Sancaktepe metro line of 20 km, having two tubes, is being constructed by Dogus Construction Company with four EPB-TBMs. The geology of Palaeozoic age is highly complex with limestone, siltstone, mudstone, arcosic sandstone and conglomerate. The geological units are frequently cut by andesite dykes with significantly fractured contact zones. In more than 20 locations, the contact zones are the potential weak areas liable to face collapse. Overburden changes of 10–80 m, and old buildings situated above the tunnel lines, are the most significant risks encountered during tunnel excavation.

4.3.1   A model to predict the performance of EPB-TBMs in difficult ground conditions

In this section a model based on past experience obtained in Istanbul is used. The TBM data collected from different TBM projects was used to develop the model described in this chapter. The methodology is briefly given below.

Specific energy is the energy needed to excavate a unit volume of rock; it is one of the most important factors in determining the efficiency of rock excavation and it can be used to estimate net or instantaneous production rates of a mechanical excavator as given in Equation (4.1), Rostami et al. [20] and Copur et al. [21].

Field specific energy is calculated using Equation (4.2) with the aid of data obtained from the TBM's data acquisition system.

$$NPR = k \cdot P/SE \tag{4.1}$$

Where NPR is net production rate in (m^3/h), k is energy transfer ratio from the cutting head to the tunnel face (usually 0.8 for TBMs), P is power needed to excavate the rock for related SE, and SE is specific energy in (kWh/m^3). Power is directly related to rotational speed of the cutterhead, to torque (and hence the rolling force of the cutters), to geotechnical properties of the rock, to type or cutters and design of the cutterhead.

Specific energy of a TBM may be calculated as:

$$SE = 2 \cdot \pi \cdot N \cdot T/NPR \tag{4.2}$$

Where N is rotational speed in (rpm) of the cutterhead and T is TBM torque in (kNm) which is directly obtained from data acquisition system of TBM. The part $(2 \cdot \pi \cdot N \cdot T)$ of Equation (4.2) is the power spent during excavation for a given torque and rotational speed.

Equations (4.1) and (4.2) clearly indicate that if field specific energy and the power spent during excavation are predicted, net production rate of a TBM can be calculated. The following sections of the chapter will be oriented to this respect of the research work.

4.3.2   Estimation of optimum specific energy from full-scale laboratory cutting experiments

Specific energy is the energy to spend to excavate a unit volume of rock and may be obtained experimentally using a full-scale laboratory cutting rig. It levels off after a certain depth of cut or penetration, given an optimum value and is directly related to rock compressive strength, depending on rock grain size and fracture degree of the rock mass, as given in Figure 4.4, Bilgin et al. [22]. The effect of rock grain size and Rock Quality Designation (RQD) on SE will be explained in detail below.

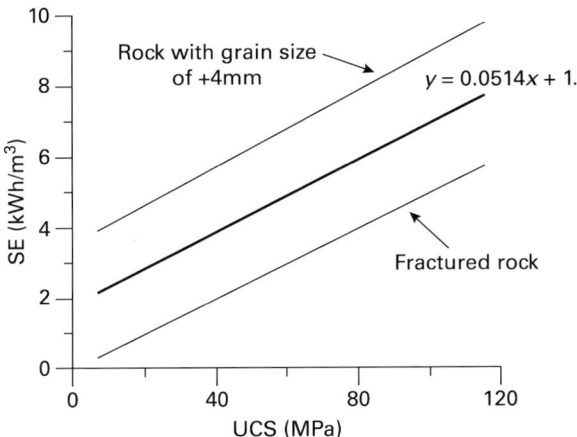

Figure 4.4 Relationship between uniaxial compressive strength and optimum specific energy obtained in the laboratory, Bilgin et al. [22].

4.3.3   Estimation of optimum field specific energy

A typical relationship obtained between field specific energy values and advance per revolution for Kadikoy–Kartal Metro tunnel TBMs is given in Figure 4.5. It can be clearly seen from this figure that field specific energy levels off and stays almost constant after a certain value of advance per revolution.

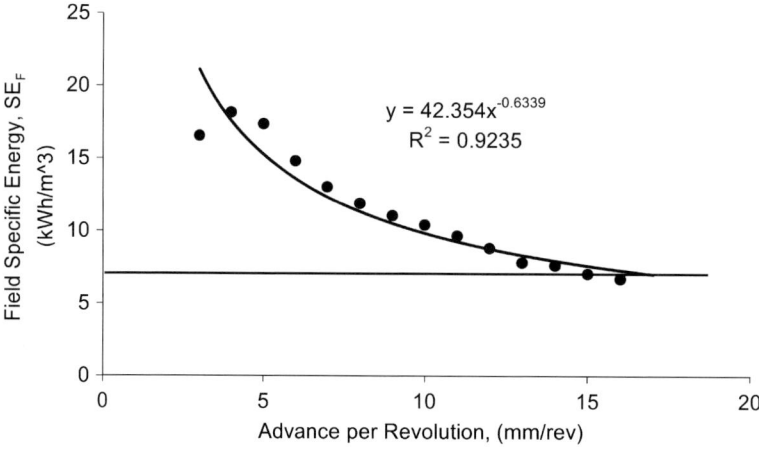

Figure 4.5 Relationship between advance per revolution and field specific energy in Kadikoy–Kartal metro tunnel for Kartal Dolayoba limestone, siltstone, carbonated shale (mean UCS = 45.8 MPa).

Table 1 tabulates the project description of each tunnel considered in the model, the mean compressive strength of the geological formations obtained from borehole samples, the field specific energy and the predicted specific energy values using the equation given in Figure 4.4. However, one important point is that sedimentary formations found in the Istanbul region are cut by dykes, making the geological formation highly fractured in some areas. For this reason, EPB-TBMs in Istanbul metro tunnels are used with a chamber with mean face pressure in the range 0.7–1.5 bar. This amount of face pressure increases specific energy values to 1.8 times higher than in open mode. However, in transition zones and around the contact zones between dykes and the main rock, especially if the ground water level is high, face pressure is usually increases up to 3–3.5 bar. As clearly seen from Table 4.1, the predicted specific energy values using Figure 4.4 are in good agreement with the field specific energy values.

Table 4.1 Description of tunnel projects with field and predicted specific energy values [3].

Project	Geology	D [m]	UCS [MPa]	SE_F [kWh/ m^3]	SE_P [kWh/ m^3]	$SE_F/$ SE_P	NPR_F [m^3/h]
Hard Rock TBMs							
Beykoz Utility T.	Limestone, sandstone, carbonated shale	3.2	96.3	5	6.75	0.75	14.2
Cayirbasi Water T.	Interbedded, sandstone, limestone, mudstone	3.1	119.3	9.5	7.93	1.20	16.5
Uluabat Power T.	Akcakoyun limestone	5.1	52.0	5	4.5	1.10	96

Project	Geology	D [m]	UCS [MPa]	SE$_F$ [kWh/m^3]	SE$_P$ [kWh/m^3]	SE$_F$/SE$_P$	NPR$_F$ [m^3/h]
Uluabat Power T.	Karakaya meta-sandstone, mudstone; graphitic schist	5.1	25.0	3	3.1	0.97	135
Hard Rock TBMs working in closed mode vary with face pressure general in the range 0.7–1.5 bar							
Kartal-Kadikoy Metro T.	Kartal-Dolayoba limestone, siltstone, carbonated shale	6.6	45.8	7	6.75	1.03	100
Pendik-Kaynarca Metro T.	Kartal form. limestone, shale, mudstone	6.5	42.0	6	6.2	0.97	105
Pendik-Kaynarca Metro T.	Dolayoba form. limestone	6.5	32.0	7	6.8	1.03	100

D = tunnel diameter, UCS = uniaxial compressive strength, SE$_F$ = field specific energy, SE$_P$ = predicted field specific energy, NPR$_F$ = field net production rate.

Uniaxial compressive strengths (UCS) are mean values, and for Beykoz and Cayirbasi tunnels they are obtained from the rock samples obtained from the tunnel face, while for the other projects UCS is the mean value obtained from core samples. SE$_P$ values are obtained using the relation given in Figure 4.3, for EPB mode the values are multiplied by 1.8 as explained in the text.

One important factor in calculating specific energy values using the equation given in Figure 4.4 is that if the rock is coarse grained around 0.4 cm (such as arcose, sandstone, conglomerate etc.), calculated specific energy should be increased by around 35%.

If the TBM is used in open mode, based on RQD range, a reduction factor as described below should be taken into consideration for specific energy calculations:

100% ≥ RQD ≥ 70%, no change in specific energy (rock mass behaves as massive rock)

70% > RQD ≥ 50%, decrease specific energy by around 5%

50% > RQD ≥ 30%, decrease specific energy by around 15%

20% > RQD, decrease specific energy by around 20% (risk of face–roof collapse)

If EPB-TBM is used, the estimated specific energy from laboratory rock cutting experiments should be multiplied by 1.8 for mean face pressures of 0.7–1.5 bar.

Cutting power of an EPB-TBM working in optimum specific energy conditions in hard fractured rocks up to 100 MPa of compressive strength may be calculated empirically as given below.

For TBMs working in closed mode in hard fractured rock:

$$P_{cutting\text{-}EPB} = K \cdot D \text{ (for EPB pressure 0.7–1.5 bar)} \tag{4.3}$$

K = 118.8 for UCS up to 100 MPa and (for N = 2–2.5 rpm, EPB pressure 0.7–1.5 bar)

K = 95 for UCS up to 70 MPa and (for N = 2–2.5 rpm, EPB pressure 0.7–1.0 bar)

For TBM working in open mode in hard rock:

$$P_{cutting\text{-}HardRock} = 70 \cdot D \tag{4.4}$$

Where, $P_{cutting}$ is the cutting power in kW and D is cutterhead diameter in m.

4.3.4 Estimation of machine utilization time

Machine utilization time, as defined in Table 4.2 and the learning curve as given in Figure 4.6, can be used to calculate daily advance rates.

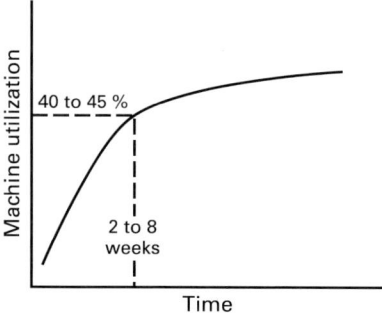

Figure 4.6 Learning curve [6].

Table 4.2 Guidelines for estimating TBM utilization time [3, 4].

Stoppage type		Stoppage Duration, % of one day shift time
Adverse ground	Contact zones between geological formations	A few days to 1 week
	Dykes	A few days to 1 week
	Faults	A few days to 2 weeks
	Water	A few days to 1 week
TBM breakdown	New TBM (experienced crew)	2–4%
	New TBM (inexperienced crew)	4–6%
	Refurbished TBM (experienced crew)	4–6%
	Refurbished TBM (inexperienced crew)	6–8%

Stoppage type		Stoppage Duration, % of one day shift time
Cutter replacement	Quartz content of 0–20%	5%
	Quartz content greater than 20%, weak and blocky ground	5–10%
	Quartz content greater than 20%, hard rock	10%
Muck transportation	By train: Transportation distance of 0–3 km	7%
	By train: Transportation distance greater than 3 km	10%
	By belt conveyor: Transportation distance of 0–3 km	5%
	By belt conveyor: Transportation distance greater than 3 km	7%
Maintenance	Experienced contractor and crew	10%
	Moderately experienced contractor and crew	15%
Setting the segments		20–25%
TBM mobilization at stations		2–3 weeks
Other stoppages		10–15%
Machine utilization		22–45%

A numerical example will be given below to clarify TBM performance prediction methodology.

4.3.5 Numerical example to estimate daily advance rate of an EPB-TBM

A numerical example to the calculation of daily advance rate of EPB-TBM having a diameter of 6.57 m in complex geology when passing conglomerate having a uniaxial compressive strength of 70 MPa and RQD of 55%.

Specific energy is found using the equation given in Figure 4.4 as 5.4 kWh/m^3. Conglomerate is a coarse-grained rock, so specific energy should be increased by 35% resulting in 7.3 kWh/m^3. This value should be corrected for RQD by decreasing 15% resulting in 6.5 kWh/m^3. Since an EPB-TBM is being used, specific energy is again corrected by multiplying by 1.8, resulting in 11.8 kWh/m^3.

The cutting power of the EPB-TBM is estimated by using Equation (4.3) resulting in 748 kW. The net production rate is estimated by using Equation (4.2) resulting in 53.2 m^3/h.

The working pattern is 20 h/day. Referring to Table 4.1, stoppage due to TBM breakdowns can be estimated as 7%, stoppage due to muck transportation by belt conveyor as 5%, stoppage due to maintenance as 10%, stoppage due to cutter replacement as 10%, stoppage due to the replacement of the segments as 20%, and stoppage due to other reasons as 8%. This sums up total 60% of stoppage. Therefore, machine utilization is estimated as 40%. Knowing the tunnel cross-section area of 34.2 m², the daily advance rate is estimated as:

$$\text{Daily advance rate} = \frac{53.2 \cdot 20 \cdot 0.4}{34.2} = 12.4 \text{ m/day} \tag{4.5}$$

4.3.6   Comparison of predicted and realized EPB-TBM performance values

The performance prediction model given above was prepared for Dogus Construction Company in January 2013 before the construction of tunnels was started. The TBM performance given below concerns Lines 1 and 2 opened between Umraniye and Carsi stations (shafts 8 and 7). TBM excavation between two stations along Line 2 was started on 4 March 2013 and terminated on 11 June 2013. TBM excavation on Line 1 was started on 21 June 2013 and finished on 21 August 2013. The geological profile of the area is given in Figure 4.7 where mudstone, sometimes sandstone of Gozdag geologic formation are dominant with a predetermined significant diabase dyke.

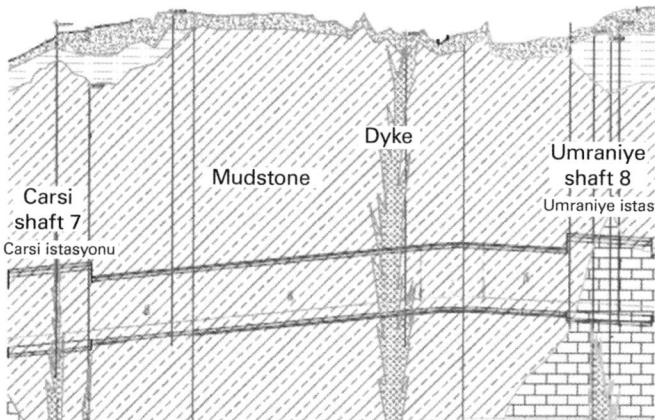

Figure 4.7 The geological profile between Carsi and Umraniye stations.

Table 4.3 gives the mean values of some rock properties between the two shafts.

Table 4.3 Mean rock properties between two stations [5].

Rock property	Value
UCS, MPa	36.2 ±12
E, MPa	3625–4120
Cohesion, kPa	89–284
RQD %	0–100, mean 36
GSI	25–50

Work distributions in Lines 2 and Line 1 are given in Figures 4.8 and 4.9, respectively. As seen from these figures, TBM utilization in Line is 2 is 16% and in Line 1 is 24%.

Bearing in mind that excavation in Line 2 started first, this is a typical consequence of the learning period.

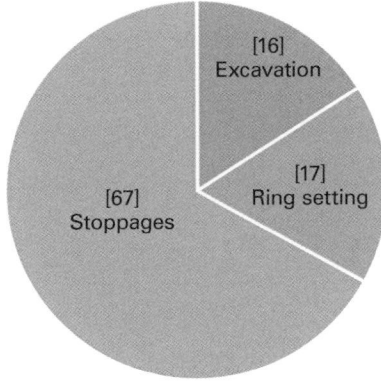

Figure 4.8 Work distribution in Line 2.

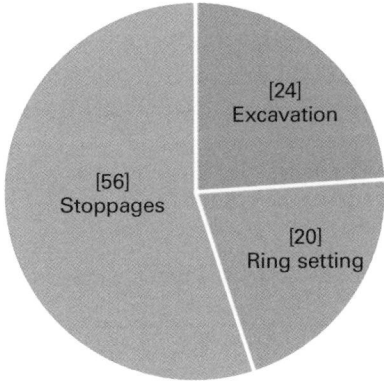

Figure 4.9 Work distribution in Line 1.

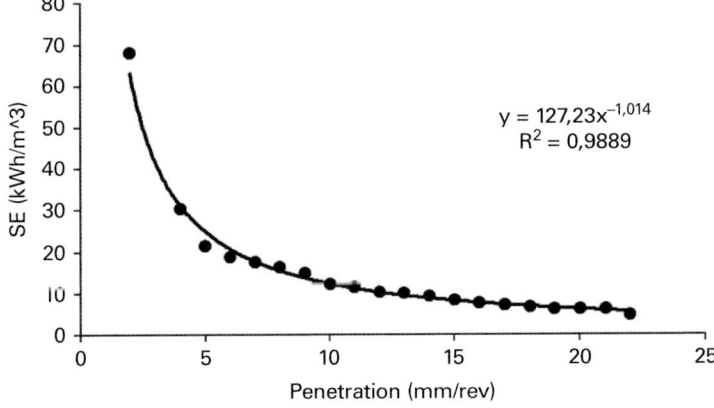

Figure 4.10 The variation of specific energy against penetration for Line 2.

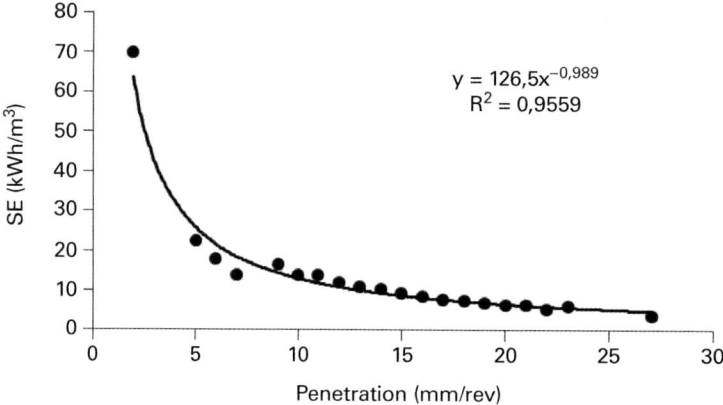

$$y = 126{,}5x^{-0{,}989}$$
$$R^2 = 0{,}9559$$

Figure 4.11 Variation of specific energy against penetration for Line 1.

The variation of field specific energy with penetration for Line 2 is given in Figure 4.10, and for Line 1 in Figure 4.11. The comparison of specific energy values for different values of penetration in mm/rev is made in Table 4.4, which shows that the specific energy obtained for the line 1 is around 10% higher than for Line 2 for the same rock formation and for identical TBM. Bearing in mind that Line 1 is excavated 80 m behind Line 2, the increase in SE in Line 1 may be explained by the fact that TBM in Line 1 worked in stressed and destructed geological zone.

Table 4.4 Variation of specific energy values in Lines 1 and 2 (between shafts 8 and 7) with TBM penetration [5].

Penetration mm/rev	SE kWh/m^3 in Line 1, which is 80 m behind the Line 2	SE kWh/m^3 in Line 2, which is 80 m ahead Line 1
3	42.7	41.8
6	21.5	20.7
9	14.4	13.7
12	13.6	10.2
15	8.7	8.2
18	7.3	6.8
21	6.3	5.8

Predicted and realized TBM performance values in Lines 2 and 1 are given in Table 4.5. Specific energy values in this table are taken from Figures 4.10 and 4.11 for penetration of 18 mm/rev, the point where the SE values level off almost to a constant value, which is recommended in the model developed, Namli et al. [3, 4].

As seen from this table, actual performance of the TBM in Line 1 is better than performance of the TBM in Line 2 and closer to predicted performance values.

Table 4.5 Predicted and realized TBM Performance in Lines 2 and 1 between shafts 8 and 7 [5].

Excavation parameter	Predicted	Realized Line 2	Realized Line 1
Machine utilization, %	29–38	16	24
Segment mounting, %	23–25	17	21
Specific energy, kWh/m^3	5.7–7.7	6.8	7.3
Waiting days	34–38	20	30
Mean m/day with stoppages	8.0–12.3	8.4	11.0
Mean m/day without stop.	15.3–28.0	10.6	21.2
Job termination days	62–99	97	62

Checking the validity of the model for the same project covering the period 12 February 2013 to 7 March 2015 for a length of 6.250 km is given below.

All shift data covering the period between 12 February 2013 and 7 March 2015 were carefully analyzed for Lines 1 and 2. The TBMs used in the two lines were identical, which allowed the results to be compared, as given in Table 4.6, which shows that predicted daily advance rates are very close to actual values.

Table 4.6 Predicted and actual daily advance rates covering the period 12 February 2013 to 7 March 2015 [5].

Shaft, stations and date	Predicted m/day	Realized m/day Line 2 TBM (Z363)	Realized m/day Line 1 TBM (Z360)
8–7; Umraniye-Carsi	8–12.5	6.6	8.1
7–6; Carsi-Libadiye	5.9–7.9	7.5	7.8
6–5; Libadiye-Kisikli	13.4–22	22	19.2
5–4;Kisikli-Altunizade	5.5–7	7.1	8.7
4–3; Altunizade-Baglarbasi	9.1–12.5	12.9	12.9
3–2; Baglarbasi-Fistikagaci	7–8.6	11.6	—
Mean values	8.2–11.8	11.3	11.3

Mobilization in the shafts was estimated to be 14–21 days, and the mean actual value was 34 days.

4.3.7 Verification and modification of the model for silty-clay and sand in the Mahmutbey–Mecidiyekoy metro tunnels

The Mahmutbey–Mecidiyekoy metro tunnels are currently being excavated by Guler-mak-Kolin-Kalyon Joint Venture with three EPB-TBMs. As it will be noticed above, the model proposed for estimating daily advance rate is developed for rocks on the basis of calculating SE from laboratory compressive strength of the rocks and modifying SE by taking into account the RQD and grain size of the rock formations. However, it is apparent that the proposed model needs a modification for soft ground as in silty clay, sand etc. The Mahmutbey–Mecidiyekoy tunnels gave the opportunity to develop the model for soft ground, since in some part of Lines 1 and 2, the TBMs are planned to pass through hard rock, silty clay and sand.

Some geotechnical properties of the geological formations in the studied area are given below. The rock formation has a mean compressive strength of 40 MPa and RQD of 45%. Silty clay has a natural water content of 27%, liquid limit of 58, plasticity Limit of 16 and plasticity index of 16. The size distribution of the sand is given in Figure 4.12.

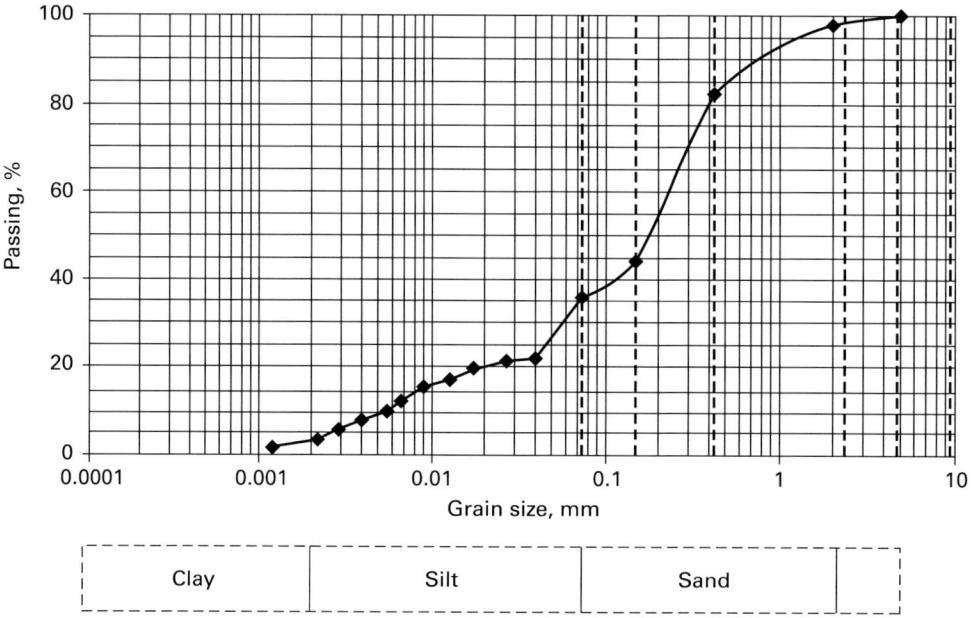

Figure 4.12 Size distribution of the sample taken in geological formation referred to as sand in the project geotechnical report.

Table 4.7 is a summary of the mean face pressure, optimum field specific energy and cutting power in different geological formations. As can be seen in this table, optimum field specific energy changes not only depending on different geological formations but also on the position of the TBM, being in front of or behind the other. Optimum field specific energy values are taken from Figures 4.13–4.18, where they level of after certain values of penetration. It is important to note that, for hard rock, optimum SE is measured as 5.5 kWh/m^3 and SE predicted using the criteria given in Section 2.2 (Chapter 2) is 5.9 kWh/m^3.

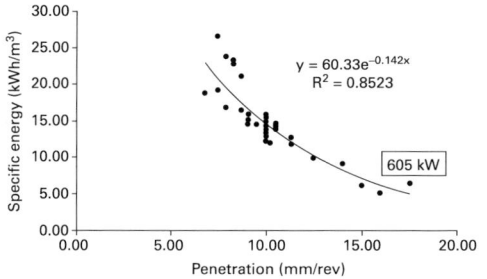

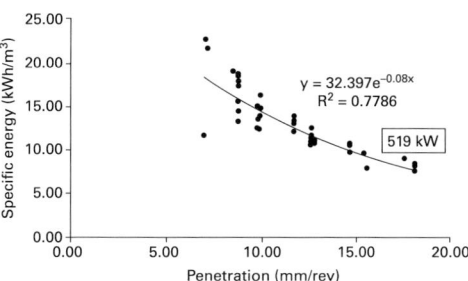

Figure 4.13 The variation of SE with penetration in rock, with the TBM 80 m in front of the second TBM in Line 1.

Figure 4.14 The variation of SE with penetration in rock, with the TBM behind the second TBM in Line 2.

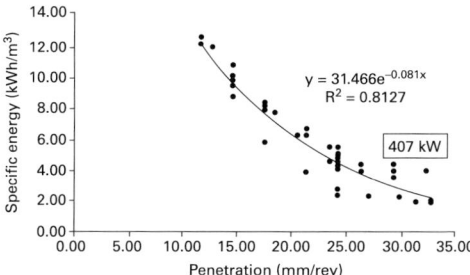

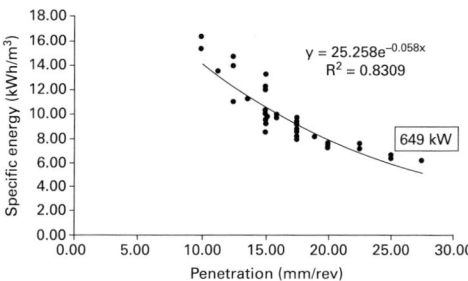

Figure 4.15 The variation of SE with penetration in silty clay with the TBM 80 m in front of the second TBM in Line 1.

Figure 4.16 The variation of SE with penetration in silty clay, with the TBM 80 m in front of the second TBM in Line 2.

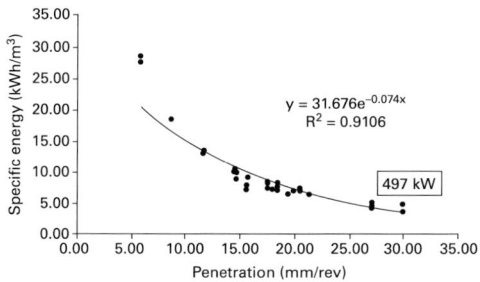

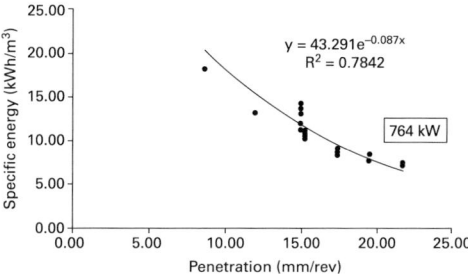

Figure 4.17 The variation of SE with penetration in sand, with the TBM 80 m in front of the second TBM in Line 1.

Figure 4.18 The variation of SE with penetration in sand, with the TBM 80 m in front of the second TBM in line 2.

Table 4.7 Variation of mean face pressure, optimum field specific energy and cutting power in different geological formations for the Mahmutbey–Umraniye metro tunnels [5].

Geologic formation and position of TBMs	Mean face pressure, bar	Optimum SE kWh/m^3	Cutting power kW at optimum SE
Hard rock, TBM in front, in Line 1	0.5–07	5.5	605
Hard rock, TBM behind in Line 2	0.5–07	7	519
Silty clay, TBM in front in Line 1	1.0–1.5	3	649
Silty clay, TBM in behind in Line 2	1.0–1.5	6	407
Sand, TBM in front in Line 1	1.0–1.5	4	497
Sand, TBM in behind in Line 2	1.0–1.5	7	764

Optimum field SE value for rock formations having UCS of 40 MPa and RQD of 45% is found to be 5.5 kWh/m^3 and the predicted value of SE using Figure 4 and the criteria in Section 2.2 (Chapter 2) is 5.9 kWh/m^3. For mean SE, the cutting power is found to be 605 kW and the predicted value using Equation (4.3) is 618 kW. However, the field measurements made in Uskudar–Umraniye–Cekmekoy–Sancaktepe and Mahmutbey–Mecidiyekoy metro tunnels revealed that the TBM which is behind the first one (within 80 m) generates around 15% higher SE values. This may be explained by the fact that the TBM that is behind is affected by the disturbance and the stress generated by the first TBM. However, the rock formation that is disturbed and fractured requires 15–25% less cutting power, which is also true for stiff silty clay. Silty clay has a plastic behavior, and the SE of the second machine is almost double that of the first one. The behavior of TBMs in sand is completely different from other geological formations. The cutting power of the trailing TBM is almost 55% higher than the leading one and its SE is almost 80% higher.

4.4   Conclusions

TBM performance estimation is essential for time scheduling in mechanized tunneling. There are several methods published on the subject. However, in complex geology, especially where EPB-TBMs are used, a methodology based on past experience is necessary for estimating tunneling performance and daily advance rates in highly fractured rock and sandy-silt, silty-sand formations. This chapter has given a summary of such methodology and a comparison of the predicted and actual results in Uskudar–Umraniye–Cekmekoy–Sancaktepe metro tunnels for complex geology with dyke inclusions. The methodology is based on predicting field specific energies and comparing cutting power of EPB-TBMs. The predicted daily advance rates are compared with field values. The model developed is also verified using some field results from Mahmutbey–Mecidiyekoy metro tunnels.

References

[1] Ocak, I. (2009) Environmental effects of tunnel excavation in soft and shallow ground with EPBM: the case of Istanbul. *Environ Earth Sci* **59**, 347–352.

[2] Chan, J. (2015) *Catalogue of notable tunnel failures*, The Government of Hong Kong. Special Administrative Region, Civil Engineering and Development Department.

[3] Namli, M., Cakmak, O., Pakis, I.H., Tuysuz, L., Talu, T., Dumlu, M., Balci, C., Copur, H., Bilgin, N. (2013) *A methodology of using past experiences in the performance prediction of a TBM in a complex geology and risk analysis.* World Tunnel Congress, Geneva, Switzerland.

[4] Namli, M., Cakmak, O., Pakis, I.H., Tuysuz, L., Talu, T., Dumlu, M., Şavk, S., Bilgin, N., Copur, H., Balci, C. (2014) *The performance prediction of a TBM in a complex geology in Istanbul and the comparisons with actual values.* Proceedings of the World Tunnel Congress – Tunnels for a better Life. Foz do Iguaçu, Brazil.

[5] Bilgin., N, Namli, M., Copur, H., Balci., C., Shaterpour Mamaghani., A. (2016) *A model to predict the performance of EPB-TBMs in a complex geology in Istanbul*, World Tunnel Congress, 22–28 April, San Francisco, USA.

[6] Vinai., R, Oggeri, C., Peila, D. (2008) Soil conditioning of sand for EPB applications: A laboratory research. *Tunn Undergr Sp Technol,* **23**, 308–317.

[7] Peila, D., Picchio, P., Chieregato, A. (2013) Earth pressure balance tunnelling in rock masses: Laboratory feasibility study of the conditioning process. *Tunn Undergr Sp Technol,* **35** (2013) 55–66.

[8] Langmaack., L. and Lee, K.F. (2016) Difficult ground conditions? Use the right chemicals! Chances–limits–requirements. *Tunn Undergr Sp Technol,* On line in January 2016.

[9] Bappler, K. (2006) *Mechanical tunneling, full face tunnel boring machines, Herrenknecht A.G. ITA/AITES training Course.* World Tunnel Congress, Seoul, South Korea.

[10] Ates, U., Bilgin, N., Copur, H. (2014) Estimating thrust, torque and other design parameters of different type TBMs with some criticism of TBMs used in Turkish Tunnelling Projects. *Tunn. Undergr. Sp. Technol.* **20**: 46–63.

[11] Song, X., Liu, J., Guo, W. (2010) *A cutter-head torque forecast model based on multivariate nonlinear regression for EPB shield tunneling.* Proc. Int. Conf. Artif. Intell. Comput. Intell, 104–108.

[12] Shi, H., Yang, H., Gong, G., Wang, L. (2011) Determination of the cutterhead torque for EPB shield tunneling machine. *Automat. Constr.* **20**, 1087–1095.

[13] Maidl, U., Wingmann, J. (2009) Predicting the performance of earth pressure shields in loose rock. *Geomech. Tunn.* **2**, 189–197.

[14] Frough, O., Torabi, S. R., Tajik, M. (2012) Evaluation of TBM utilization using rock mass rating system: a case study of Karaj-Tehran water conveyance tunnel. *Journal of mining and Environment,* **1** (3), 89–98.

[15] Farrokh, E., Rostami, J., Laughton, C. (2012) Study of various models for estimation of penetration rate of hard rock TBMs. *Tunn. Undergr. Sp. Technol.* **30**, 110–123.

[16] Toth, A., Zhao, J. (2013) *Evaluation of EPB-TBM performance in mixed ground conditions.* In: Anagnostou, G., Ehrbar, H. (Eds.), World Tunnel Congress, May 30–June 6, Geneva, Switzerland, pp. 1149–1156.

[17] Copur, H., Aydin, H., Bilgin, N., Balci, C., Tumac, D., Dayanc, C. (2014) Predicting performance of EPB-TBMs by using a stochastic model implemented into a deterministic model. *Tunn. Undergr. Sp. Technol.*, **42**, 1–14.

[18] Bilgin, N., Nasuf, E., Cigla, M. (1993) *Stability problems effecting the performance of a full face tunnel boring machine in Istanbul-Baltalimanı Tunnel*, In: Proceedings of Pasamehmetoglu et all. Assessment and prevention of Failure Phenomena in rock engineering, International Symposium, Istanbul, Turkey. Balkema.

[19] Bilgin, N., Ozbayir, T., Sozak, N., Eyigun, Y. (2009) *Factors affecting the economy and the efficiency of metro tunnel drivage with two TBMs in Istanbul in very fractured rock.* ITA AITES World Tunnel Congress, May 23–28, Budapest, Hungary.

[20] Rostami, J., Ozdemir, L. & Neil, D.M. (1994) Performance prediction: A key issue in mechanical hard rock mining. *Mining Engineering* **11**, 1263–67.

[21] Copur, H., Tuncdemir, H., Bilgin, N. & Dincer, T. (2001) Specific energy as a criterion for used of rapid excavation systems in Turkish mines. *Trans. Inst. Min. Metall. Section A*, **110**, A149–157.

[22] Bilgin, N., Copur, H., Balci, C. (2014) *Mechanical excavation in mining and civil industries*, ISBN-13:978-1-4665-8474-7, CRC Press, Taylor and Francis Group, London.

5 Selection of cutter type for difficult ground conditions

5.1 Introduction

V-type disc cutters were the only type of disc cutters used in TBMs up to 40 years ago, then it changed gradually to CCS disc cutters since they were more resistant to abrasive characteristics of the rocks and they lasted longer than V-type discs. Tungsten carbide studded disc cutters are also used in abrasive rocks. In mixed grounds it is essential to use disc cutters and chisel-rippers together. Bilgin et al. [1] give more detailed information on cutter types used in TBMs. This chapter is aimed at explaining the range of cutter types used in TBMs in terms of the cutting efficiency in difficult grounds. First, we will summarize the comparative study of V- and CCS type disc cutters and conical cutters used for a double-shield TBM in the Dragos sewerage tunnel in Istanbul [2]. Then we will look at the eficiency of chisel cutters against disc cutters, the inefficient use of tungsten carbide studded disc cutters in the Marmaray–Istanbul project and the methodology of changing disc cutters to chisel cutters in the Beykoz–Istanbul tunnel [3, 4, 5, 6].

5.2 Comparative studies of different type of cutters for Tuzla–Dragos tunnel in Istanbul – test procedure and results

A Sandvik S35/80H type conical bit and two different types of disc cutters were used throughout the experiments for these comparative studies [2]. The profiles of V-type and CCS discs tested on a full scale linear cutting rig are shown in Figure 5.1.

The compressive strength of the rock sample tested was 58 MPa, and tensile strength was 3.6 MPa. The relationships of rolling force, cutting force and thrust force with depth of cut, for unrelieved cutting for different cutters, are shown in Figures 5.2 and 5.3 and the mean values are summarized in Table 5.1.

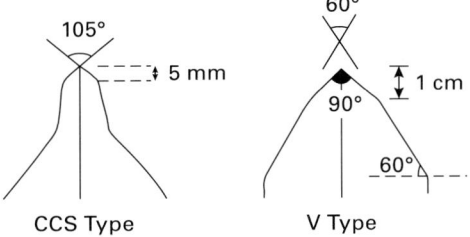

Figure 5.1 The profiles of V-type and CCS discs tested [2].

TBM Excavation in Difficult Ground Conditions. Case Studies from Turkey. First Edition. Nuh Bilgin, Hanifi Copur, Cemal Balci.
© 2016 Ernst & Sohn GmbH & Co. KG. Published 2016 by Ernst & Sohn GmbH & Co. KG.

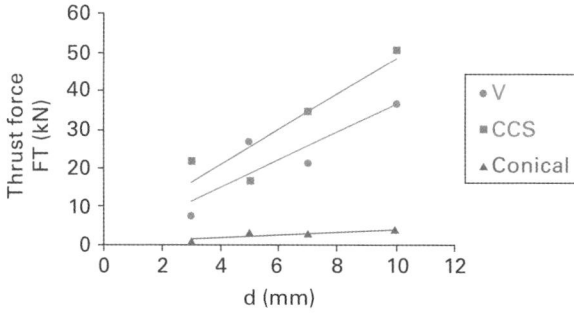

Figure 5.2 The relationship between thrust force and depth of cut for different cutters [2].

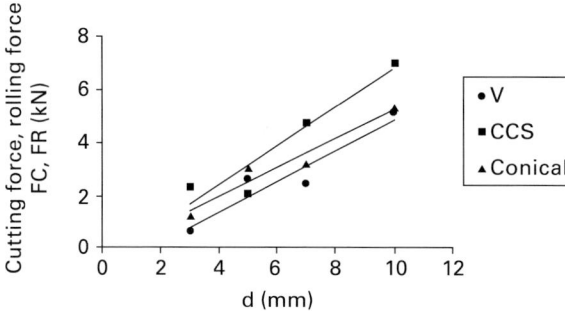

Figure 5.3 The relationship between rolling force and depth of cut for different cutters [2].

Table 5.1 The summary of the cutting test results [2].

Type of cutters:	Conical bit	V-type	CCS type
F_T (kN/mm)	0.44	3.65	5.21
F'_T (kN/mm)	1.78	5.84	8.34
F_R, (kN/mm)	0.49	0.40	0.64
F'_R, (kN/mm)	1.57	0.72	1.09

In Table 5.1, F_T is mean thrust force, F'_T is the peak thrust force, F_R is the mean rolling force, F'_R is the peak rolling force for disc cutters, F_C and F'_C are the mean and peak cutting forces for the conical bit tested. Thrust force may be defined as vertical cutter force; rolling force and cutting force may be defined as horizontal cutter forces. As seen in Table 5.1, thrust force in the V-type disc cutter is less than the CCS disc cutter for a given depth of cut. This shows clearly the reason why CCS disc cutters are being changed to V-type disc cutters when the capacity of thrust cylinders is beyond the capacity to supply the necessary penetration for an efficient cutting process in a specific TBM excavation.

The relationships between specific energy values and the s/d (cutter spacing/cutting depth) ratios are given in Figure 5.4.

As seen in Figure 5.4, the optimum specific energy values for conical bits are obtained for s/d ratio of 4, and for V-profile and CCS profile discs the optimum values are obtained for s/d ratios of 8–10, with relevant specific energy values of 5.8 kWh/m^3 and 2.1 kWh/m^3. After a long discussion between the responsible persons of the construction company and the research team, CCS disc cutters were chosen for the tunnel excavation since they have longer cutting life and had lower specific energy values than the other cutters.

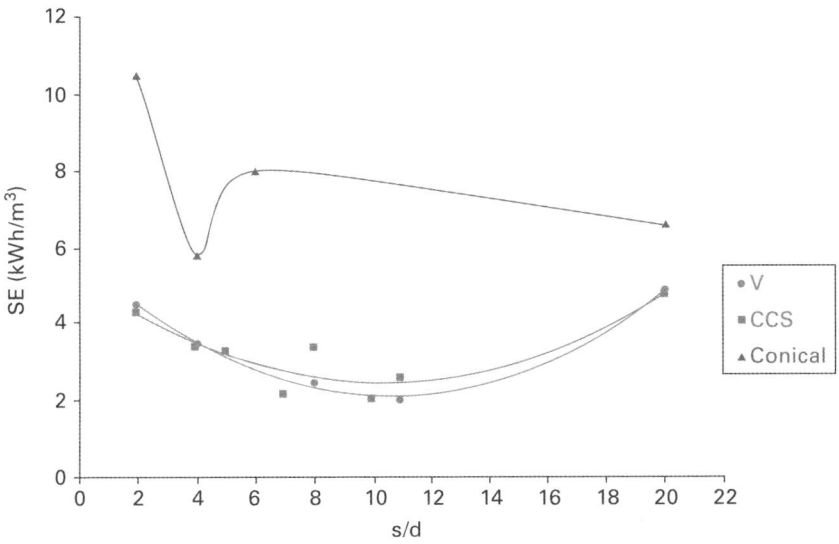

Figure 5.4 The relationship between specific energy and s/d ratios [2].

5.2.1 Efficiency of chisel cutters as against disc cutters

Disc cutters are undoubtedly necessary cutting tools for TBMs. However, Bilgin [3] Roxborough [4] and Evans [5] all reported that the pick cutters were 3–4 times more efficient than the disc cutters in medium-strength and non-abrasive rocks, verifying the theoretical findings of Evan. This is true for short cutting lengths, but after a certain length of cut, the pick cutters will deteriorate quickly in hard and abrasive rocks, making disc cutters the most popular cutting tools for hard and abrasive rocks. In soft clayey formations, the discs would not turn freely, resulting in local wearing, making the tunnel excavation quite inefficient. That is why pick cutters are preferred in special cases with soft grounds, and this necessitates an understanding of their cutting mechanics in detail. Nowadays, in complex geologies that may cover a few kilometers of hard or medium-strength rocks, sometimes with frequent boulders, and a few kilometers of soft ground, contractors use a combination of discs (CCS or V-type) and chisel cutters together. A good example for this type of geology and tunnel project is explained by Bilgin et al. [6] for Beykoz–Istanbul sewerage tunnel. V-type disc cutters in a specific zone have proven to be the most effective tools for rock excavation using a TBM for this project, Guclucan et al. [7, 8, 9]. In this project, the thrust of the TBM was found to be inefficient (either not enough or beyond the thrust capacity of the machine) for excavating very hard and abrasive quartzite over a few hundred meters of the tunnel line.

Although high cutter wear was expected in quartzite, CCS disc cutters were changed to the V-type disc cutters in order to get sufficient depth of cut per cutterhead revolution. Another important point for this project is that, for one particular length of tunnel, the geological formation changed from very soft to soft mudstone, siltstone. Disc cutters could not turn properly due to the lack of friction between the rock and the disc cutter and this resulted in discs being badly flattened during tunnel excavation. All the disc cutters were changed to chisel cutters (Figure 5.5) giving the opportunity to compare the efficiency of chisel cutters against disc cutters in certain specific conditions. Chapter 8 provides detailed information on the geology of the Beykoz tunnel.

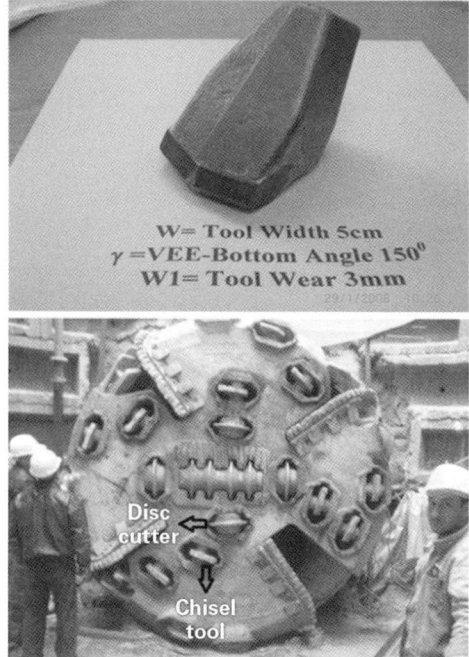

Figure 5.5 Process of replacing disc cutters with chisel cutters (top) and chisel cutters used (bottom) [6].

The efficiency of chisel tools in medium-strength rocks is clearly seen in Figure 5.6, which shows the relation between compressive strength of rock and TBM penetration index (thrust force per unit cutting depth per revolution) [6]. It is also apparent in Figure 5.6 that the chisel tools demand around 3.5 times less thrust than the disc cutters for a given penetration. However, this advantage is not the same if the torque is considered for both cutters. As seen in Figure 5.7, the torque required for a unit depth of cut per revolution (torque index) is only around 1.4 times less for chisel tools compared to disc cutters. One of the most important factors determining the efficiency of an excavation process is the energy spent on excavating a unit volume of rock (specific energy). To estimate the specific energy, the cutting power of the TBM is first calculated by using recorded torque values, and then the cutting power is divided by the instantaneous production rate. Figure 5.8 shows clearly how the specific energy decreases with penetration for both chisel tools and disc cutters [6]. Figure 5.8 reflects clearly the efficiency of chisel tools against disc cutters, since they have better penetrability characteristics for a given thrust. It should be concluded that the disc cutters are unavoidable tools for cutting especially hard and abrasive rocks, since they have high thrust and wearing life capacity. However, it may be better to use chisel/ripper tools in relatively softer rocks in order to reduce the torque and thrust requirements of the TBMs, resulting in immediately higher production rates. The most important disadvantage of chisel tools is their limited durability and wearing life in relatively harder rocks.

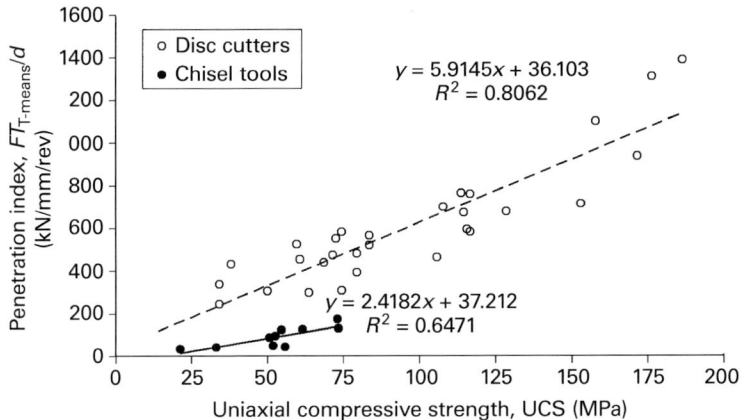

Figure 5.6 Relationship between uniaxial compressive strength of rocks and penetration index of TBM for chisel tools and disc cutters [6].

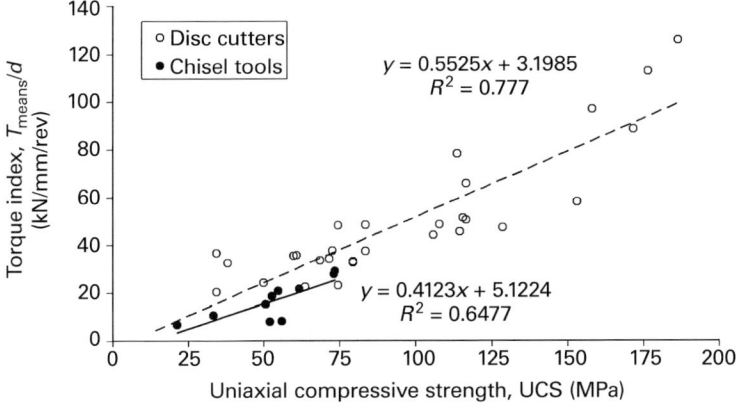

Figure 5.7 Relationship between uniaxial compressive strength of rocks and torque index (torque of TBM per unit depth of cut per revolution) of TBM for chisel tools and disc cutters [6].

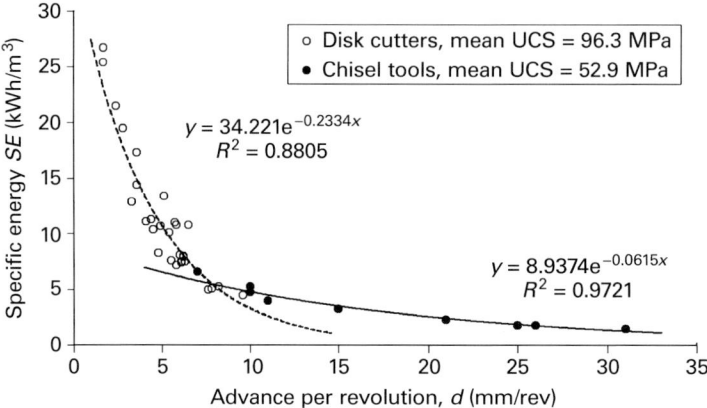

Figure 5.8 Relationship between advance per revolution of TBM cutterhead and consumed specific energy for chisel tools and disc cutters for mean values of compressive strength of rocks [6].

5.3 The inefficient use of tungsten carbide studded disc cutters in the Marmaray–Istanbul project

The Marmaray–Istanbul tunnels projects were mostly excavated in Trakya formation, which is very abrasive. The characteristics of this formation are broadly explained in Chapter 2. Worn tungsten carbide studded disc cutters in this project are shown in Figures 5.9 and 5.10. During the excavation it was noticed that the steel hubs of the disc were wearing quicker than tungsten carbide studs, causing early failure for disc cutters and later increase of TBM cutterhead vibration. This type of cutter was found to be inefficient in very abrasive rock formations in Istanbul and never recommended thereafter for the excavation of the Marmaray tunnels.

Figure 5.9 Worn tungsten carbide studded disc cutters from the Marmaray project.

Figure 5.10 Worn tungsten carbide studded disc cutter from the Marmaray Project.

5.4   Conclusions

First, the results were summarized for a comparative study of V-type, and CCS disc cutters and conical cutters using full-scale rock cutting experiments for a double-shield TBM used in the Dragos sewerage tunnel in Istanbul [2]. It is concluded that V-type disc cutters are more efficient in terms of cutter force than CCS disc cutters, but CCS cutters are preferred in practice since they are more resistant to abrasive characteristics of the rocks and lasted longer than the V-type discs. The methodology for changing disc cutters to chisel cutters in the Beykoz–Istanbul tunnel and the inefficiency of using the tungsten carbide studded disc cutters in Marmaray–Istanbul project were also explained.

References

[1] Bilgin, N., Copur, H., Balci. C. (2014) *Mechanical excavation in mining and civil industries*, CRC Press, Taylor and Francis Group, London.

[2] Bilgin, N., Balci, C., Tuncdemir, H., Eskikaya, S., Akgul, M., Algan, M. (1999) *The performance prediction of a TBM in Tuzla–Dragos sewerage tunnel*. Proceedings of the World Tunnel Congress on Challenges for the 21st Century. Oslo, Norway.

[3] Bilgin, N. (1977) Investigation into mechanical cutting characteristics of some medium and high strength rocks. University of Newcastle upon Tyne, Dissertation.

[4] Roxborough, F.F. (1985) Research in mechanical excavation, progress and prospects. Proceedings of Rapid Excavation and Tunnelling Conference, Las Vegas, USA.

[5] Evans, I. (1972) Relative efficiency of picks and discs for cutting rock. MRDE Report No. 41, National Coal Board, UK.

[6] Bilgin, N., Copur, H., Balci, C. (2012) Effect of replacing disc cutters with chisel tools on performance of a TBM in difficult ground conditions. *Tunn Undergr Sp Technol*, **27**, 41–51.

[7] Guclucan, Z., Meric, S., Gursoy, C., Algan, M., Bilgin, N., Balci, C., Tumac, D. (2007) *The use of a TBM in difficult ground conditions*. Proceedings of the 2nd Symposium on Underground Excavations for Transportation, Istanbul, Turkey (in Turkish).

[8] Guclucan, Z., Meric, S., Algan, M., Palakci, Y., Bilgin, N., Bilgin, A.R., Balci, C., Tumac, D. (2008) *The use of a TBM in difficult ground conditions in Beykoz – Kavacik Sewerage Tunnel*. Proceedings of the World Tunnel Congress, Underground Facilities for Better Environment and Safety, Agra, India.

[9] Guclucan, Z., Meric, S., Palakci, Y., Bilgin, N., Balci, C., Copur, H., Namli, M., Bilgin, A.R., Kandemir, E. (2009) *The use of theoretical rock cutting concepts in explaining the cutting performance of a TBM using different cutter types in different rock formations and some recommendations*. Proceedings of World Tunnel Congress, Safe Tunnelling for the City and for the Environment, Budapest, Hungary.

6 Effects of North and East Anatolian Faults on TBM performances

6.1 Introduction

Turkey is in a tectonically active region that experiences frequent destructive earthquakes. At a large scale, the tectonics of the region are controlled by the collision of the Arabian Plate and the Eurasian Plate. At a more detailed level, the tectonics become quite complicated. A large piece of continental crust almost the size of Turkey, called the Anatolian block, is being squeezed to the west. The block is bounded to the north by the North Anatolian Fault and to the south-east by the East Anatolian fault. The East Anatolian Fault (EAF) is a major strike-slip fault zone in eastern Turkey. It forms the transform type tectonic boundary between the Anatolian Plate and the northward-moving Arabian Plate. The North Anatolian Fault (NAF) is an active right-lateral strike-slip fault in northern Anatolia which runs along the transform boundary between the Eurasian Plate and the Anatolian Plate [1]. Five tunneling projects affected directly by the EAF and NAF faults are the Uluabat, Kargi, Dogancay, Gerede and Nurdagi tunnels, which are shown in Figure 6.1.

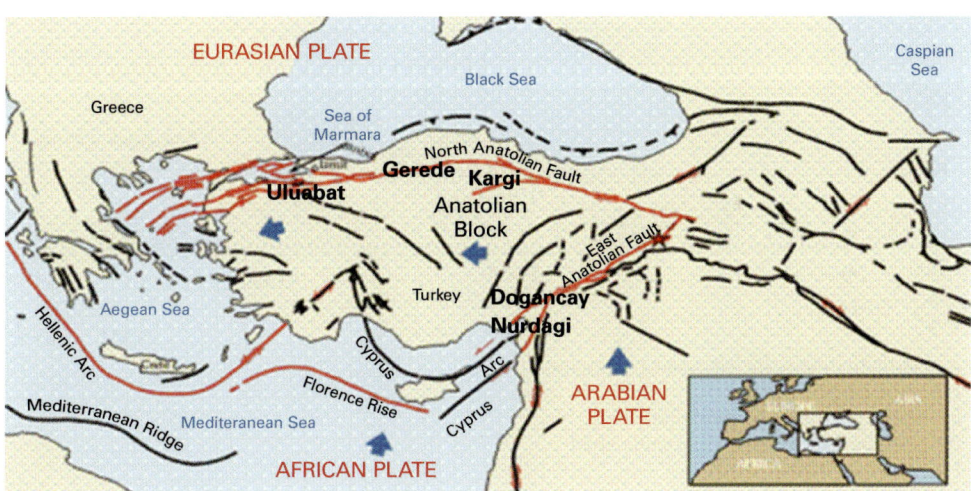

Figure 6.1 North and East Anatolian Faults in Turkey [1].

6.2 Kargi tunnel

The Kargi hydropower project was developed to utilize the energy potential of the Kizilirmak River between the towns of Osmancik and Boyabat. The excavation of an 11.8 km tunnel has been recently finished, of which 7.8 km were excavated with a double-shield Robbins TBM of 9.84 m diameter, and the 4 km from inlet part were opened using NATM [2a, 2b, 3]. The TBM was launched in April 201, but after only boring 80 m, it became stuck in a section of collapsed mixed ground face of hard rock and running ground. In the first 2 km of the tunnel, the TBM jammed seven times due to face collapses, and bypass tunnels were required to free the TBM [3]. After a few collapses the cutterhead drive torque capability was increased by more than 50% and the thrust capacity of the TBM was also increased by hydraulic power pack cylinder jacks including overbore capabilities, shield lubrication and belt scales. The average advance was 144.2 m/month before modification and 407.7 m/month after modification. The performance values of the TBM and the drill and blast excavations are given in Table 6.1 [2a, 2b].

Table 6.1 Performance values of TBM and drill and blast excavations [2a, 2b].

Parameter	TBM	Drill and blast
Boring length (km)	7.8	4.0
Months to complete	28.0	23.0
Average advance m/month	271.0	173.8
Best month (m)	723.2	281.5
Best day (m)	39.6	12.0
Monthly average (m)	271.4	173.8
Time to mobilize, months	16.0	4.0

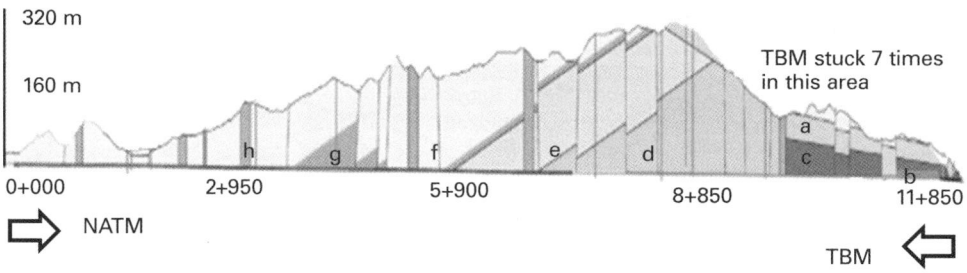

a = sandstone gravels, silty sand and clay
b = ultramaphic rocks
c = ophiolitic rocks, serpentinite
d = metapelites, micashist, grapitic shist
f = andesitic basalt
g = agglomerate
h = dykes

Figure 6.2 The geological map of Kargi Tunnel [2a, 2b].

The geological map of the Kargi tunnel is given in Figure 6.2. Continuous probe drilling was carried out for 5 km along the tunnel and umbrella arch was applied in nine different areas to stop face collapse.

6.3   Gerede tunnel

The purpose of the project is to supply drinking water to Ankara, via a tunnel having a length of 31.6 km. Under the project, water from the Gerede Creek, which is an offshoot of the Filyos River, will be transmitted to Camlidere Dam via the Isikli Regulator. The tunnel will transfer water from Gerede Creek valley to Camlıdere dam reservoir. The geology of the Gerede tunnel consists mainly of volcanic units (Ilicadere, Uludere, Deveoren), Markusa formation, and a local alluvial plain, which is composed predominantly of the Sogukcam formation. Deveoren volcanics have a dacitic character, and spans most of the south of the study area. Uludere volcanics consist of volcanic breccia, tuff and basalt units. Depending on the quality and quantity of the discontinuities, there can be some groundwater circulation. Ilicadere volcanics generally comprise tuff and volcano-clastics. In this section, basalt, andesite and volcanic breccia are also seen. The Markusa formation usually consists of greenish-brown, thick-bedded sandstone to clay stone. The Sogukcam formation mainly consists of white, rough, moderately weathered, and moderate to weak strength beige units. RQD values in this unit are between weak and very poor [4]. The schematized geological cross-section of the tunnel is seen in Figure 6.3.

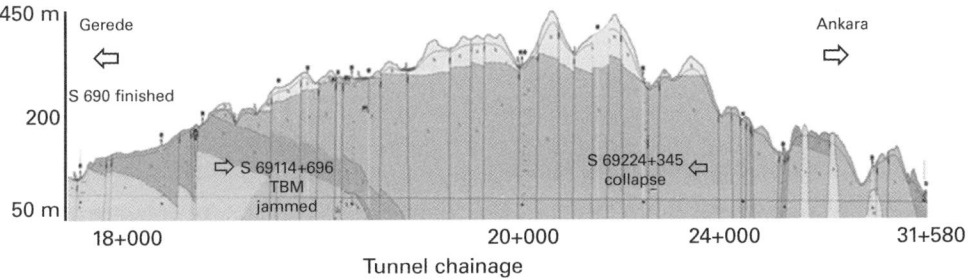

Figure 6.3 The geological cross-section of Gerede (Isiklar) tunnels.

The tunnel, which has a final diameter of 4.5 m and a length of 31,580 m will transfer the water with a capacity of 40 m^3 per second. This project is one of the longest water conveyance tunnels in the world with two shafts at depths of 424 m and 72 m having diameters of 1.9 m and 10 m respectively. The tunnel was intended to be completed by using three double-shielded TBMs of 5.5 m diameter. According to the planning program, tunnel excavation began in 2010 simultaneously in three points: entrance portal, shaft and output portal, with S-690, S-691 and S-692 TBMs. The TBM S 690 excavated a length of 9588 m and the first part was finished. However, this has been one of the most problematic TBM tunneling operations in Turkey. The first drive, downstream from Gerede to the intermediate shaft, was completed without any significant problems, but the second drive, which started from the intermediate shaft towards Ankara, is being excavated in downstream direction by the S-691 DS TBM. The TBM became stuck in smectite clay, which has a tremendous swelling characteristic. Swelling stresses started breaking the segments, and the broken segments were then supported by steel arches. It is thought that the smectite zone is behaving like a pillar zone, protecting the tunnel from the highly stressed pressurized water reservoir ahead of the tunnel and so the current plan is to remove the machine from this side, and to continue from the Ankaran side to take advantage of the dip of the tunnel for water removal. The third drive, which was being excavated by the S-692 DS, upstream of the

Ankaran side, has been constantly hindered by the complex existing geological conditions of heavily altered and weathered volcano-clastic rocks under very high water tables, which even caused one 12-month stoppage and required a bypass tunnel (Ch. 27+582.6). The S-692 shield was trapped at chainage 24+344.86 km after the tunnel suffered a collapse in July 2014. The pressure deformed the telescopic shield and about 20 m of segmental lining, as can be seen in Figure 6.4, causing a huge water and material inflow, estimated to be around 1,250 m^3 in 15 minutes [5]. Currently, it has been decided to continue the excavation with another machine, a specially designed crossover Robbins TBM.

Figure 6.4 Broken segments and the material inflow into the tunnel advancing from the Ankara side in Gerede tunnel.

6.4   Dogancay energy tunnel

The Dogancay tunnel is located in the north-east of Turkey, affected by North Anatolian Fault, and it is a part of a hydroelectric project licensed by Enerji Sa. The tunnel length is 6,655 m, and the excavation started in September 2012 and ended in July 2015. The tunnel was excavated by a double-shield Herrenknecht M-1665, having a diameter of 4.1 m, max. thrust of 19,700 kN, auxiliary thrust of 18,260 kN, nominal torque of 1,749 kNm, gripper force at 400 bar of 25,334 kN and rotational speed 0–12 rpm. The tunnel route contains Carboniferous age limestone and shale, Upper Permian age limestone and Lower Permian age siltstone, claystone, sandstone and quartzite. The frequency of encountered rock types are 54.7% limestone, 33.4% interbedded sandstone-quartzite-claystone, 7% interbedded sandy limestone and shale and 5% shale. RMR varies between 10 and 70, and is mostly between 10 and 20 in shale. The overburden within 3 km of the tunnel route is around 1,000 m and only two boreholes could be opened in the tunnel route prior to starting the excavation. Tectonic stresses squeezed the TBM several times, causing considerable delays in tunnel drivage. Figure 6.5 shows the general layout of the tunnel route with the main faults (F1, F2 etc.) [5].

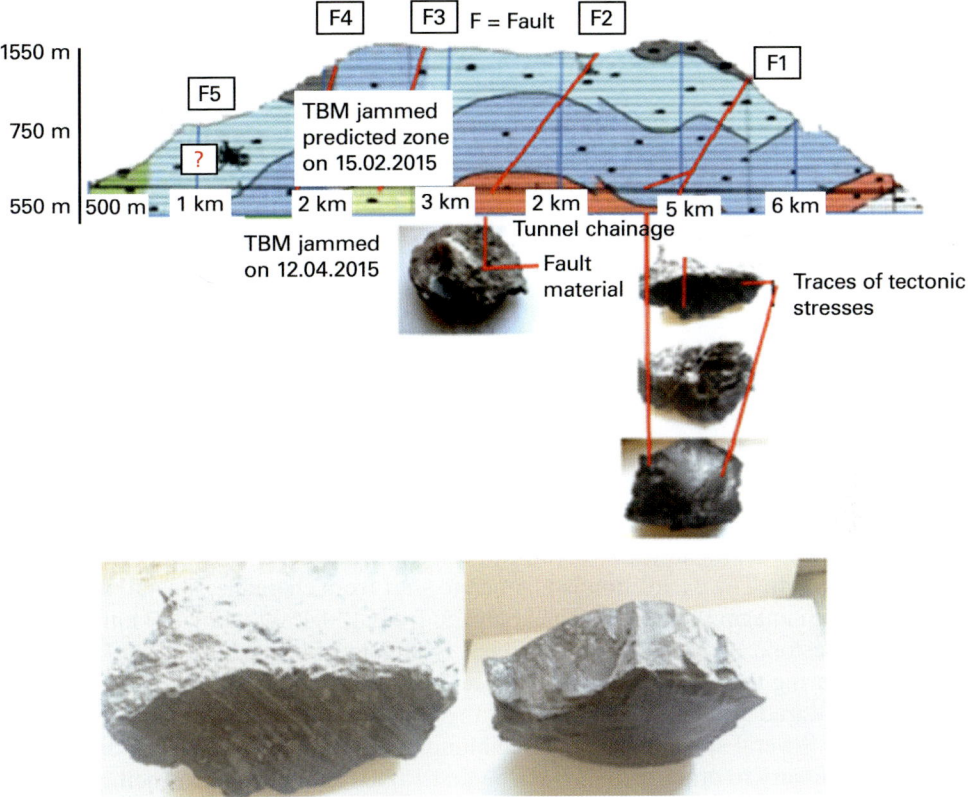

Figure 6.5 Geological profile of Dogancay project and the evidence of tectonic stresses on the rock samples [5].

The squeezing of the TBM occurred at the contact areas of two faults, F1 and F2. The rock samples taken from the squeezing zones are also seen in Figure 6.5, which clearly shows that rock samples bear the traces of tectonic stresses, suggesting that the main reason for the TBM jamming is tectonic stresses. Within a site investigation carried out by the author, it was also estimated that the same type of squeezing would also occur at the contact points of the faults with the tunnel route. The TBM passed fault F3 with some small problems, but it jammed around Fault F4 on 14 March 2015. However, the TBM was rescued in 21 days by opening two rescue galleries. It is important to note that the orientation and magnitude of the tectonic stresses have seldom been measured in Turkish tunneling projects. It is highly recommended to include such measurements in tunnels which are planned to be opened around North Anatolian and East Anatolian projects [5].

The ratio of torque to thrust force is a good indicator of the squeezing of a TBM. Figure 6.6 shows the variation of this ratio between 12.04.2015 and 14.04.2015. As can clearly be seen from this figure the ratio started at 2, and dropped to 0.5 by the 7th ring, then later increased up to 1.5 and dropped again to 0.5, and the TBM was jammed within the 25th ring.

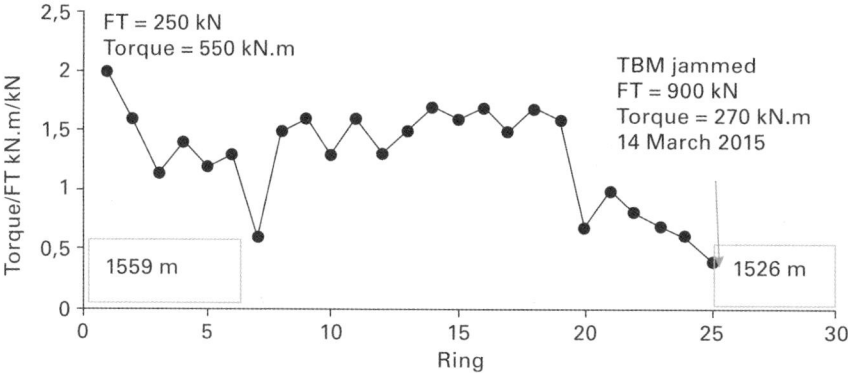

Figure 6.6 The variation of the TBM torque/thrust ratio with ring numbers within squeezing zone in Dogancay tunnel [5]

6.5   Nurdagi railway tunnel

The tunnel is for railway transportation, and the project involves two tubes, each with a length of 9,750 m. The excavation is planned to start from chainage 13+450 km and to terminate at chainage 3+700 km. The chainage from 13+450 to 12+400 km involves Karadag limestone of Mesozoic age, which is affected by the East Anatolian Fault (EAF), fracturing the rock formation to a great extent. High water ingress is expected in this area. Karadag limestone discharges the water at the toe of the mountain at the Nurdagi site. Several springs are available along the EAF. Due to technical difficulties and time necessary to procure the TBM, the first 1,050 m in limestone is being currently opened using NATM. The geological cross-section of this area, which is planned to be opened by drill and blast method, is seen in Figure 6.7 [5].

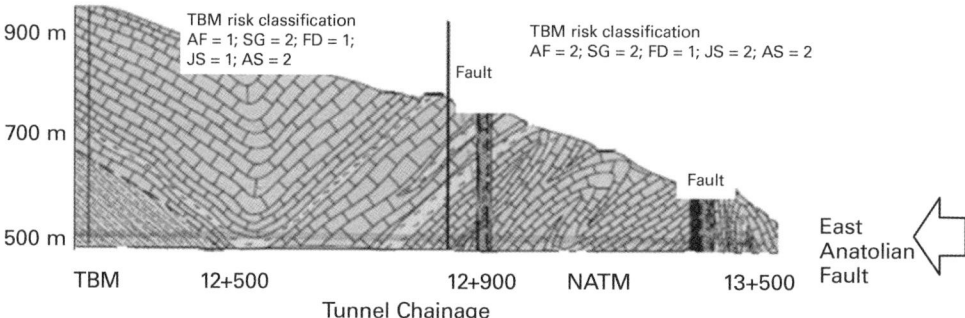

Figure 6.7 Geologic cross-section of the tunnel between chainage 13+450 and 12+400 km in Nurdagi tunnel [6]

The geological formation from 12+400 to 4+850 km consists of middle Ordovician aged Kizlac formation of very massive interbedded meta-sandstone, meta-quartzite and meta-mudstone, having very high strength and abrasivity characteristics. This section is planned to be excavated with a single-shield hard rock TBM. However, the massive characteristics of the rock formation may change at 4+850 to 3+700 km, being affected by local faults and shear zones; in this zone, RQD values are very low, and high water ingress is also expected, so this section is also planned to be excavated by NATM.

The tunnels excavated close to the NAF and EAF led to the development of a risk classification method defined in Table 6.2. According to this table, the use of a TBM in the Nurdagi tunnel at 13+500 to 12+800 km is very risky, 12+800 to 12+500 km is risky and it is favorable up to 4+850 km [5].

Table 6.2 A TBM risk classification for tunnels to be excavated close to NAF and EAF [5].

Factors effecting the risk of using TBMs	Classification
1. Distance of the tunnel to NAF and EAF, the possibility of tectonic stresses. AF	1. Within 0.5–2 km of NAF and EAF 2. Very close to NAF and EAF
2. The possibility of large amounts of water ingress into the tunnel. Detailed geological reports and careful observation of drilling logs are necessary. SG	1. Less than 100 lt/sec 2. More than 100 lt/sec
3. The possibility of seeing geological discontinuities in front of tunnel face. The criterion is that in NATM it is easy to see and control geological discontinuities in the tunnel face. FD	1. Easy 2. Difficult
4. Geological discontinuities, RMR, Q, JS	1. Q, RMR 2. Q, RMR
5. The presence of anticlinal and synclinal AS	1. One per 1 km 2. More than one per 1 km

If the total mark is 8–10, it is very risky to use TBM; if the total mark is 5–8, it is risky; if the total mark is 2–5, the risk of using TBM is in medium level; if the total mark is 0–2, using TBM is not risky.

6.6   Uluabat energy tunnel

The working area is situated on the southern part of Uluabat-Bursa (Apolyont) Lake, Turkey. The tunnel, with a length 11.4 km, started to be excavated with a 5.05 m diameter EPB-TBM from chainage 11+465 km in June 2006, and terminated in March 2010, at chainage 1+792 km. The tunnel route within chainages 11+465 to 7+750 km and 6+000 to 1+792 km consisted of Karakaya formation of Triassic age with meta-detritic rocks such as fine grained meta-claystone, meta-siltstone, meta-sandstone and graphitic schists. A geological map of the area is given in Figure 6.8.

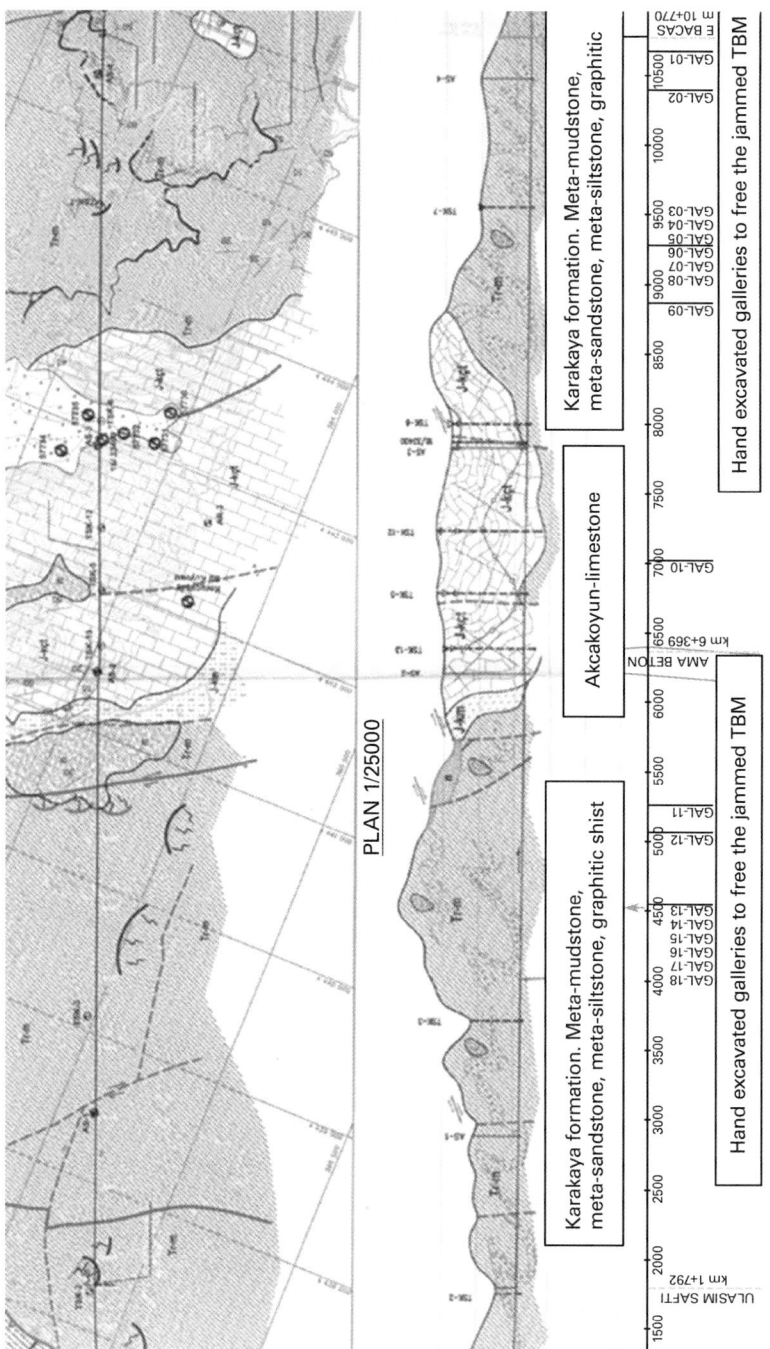

Figure 6.8 Geological map of the Uluabat energy tunnel and the areas where bypass tunnels were opened to rescue the TBM [6].

Physical and mechanical properties of meta-detritic rocks have a range of values: compressive strength 1–96 MPa, elasticity modulus 1.53–39.5 GPa. Rock formation may be classified according to RMR as III–V. The tunnel route between chainage 7+750 and 6+000 km consists of Akcakoyun formation of Jurassic aged limestone with crystallized calcite fillings. Compressive strength is in the range 24–127 MPa, tensile strength 3.5–5.4 MPa, elastic modulus 9.1–126.6 GPa, and the rock formation can be classified according to RMR as III [6].

During the tunnel excavation, the TBM was jammed at different places within the tunnel, mainly in meta-detritic rock formations due to the highly squeezing characteristics of the Karakaya formation. Figure 6.9 shows a general view of the squeezing ground around the TBM shield, and gallery No 3, which was opened to rescue the jammed shield [6].

Figure 6.9 General view of the squeezing ground around the TBM shield in Uluabat tunnel [6].

Eighteen rescue galleries were constructed to free the trapped TBM, and 192 days were spent in TBM rescue operations. Detailed study of the TBM performance data showed that overburden, RMR values, the increase of machine thrust for a given tunnel length and time, the variation in the torque/thrust ratio can all be used as a reliable basis to alert the practicing engineers to implement some mitigating measures, such as using bentonite injection around the TBM shield – see Bilgin and Algan, [6]. This was accomplished by pumping a bentonite mixture between the shield skin and the ground, with the aim of reducing frictional forces. The bentonite mixture was delivered to the shield in a mixing car, to keep it fluid. Injection through 2 inch diameter ports, six in total, was done using an electrically operated screw pump. Delivery pressure depended on the proximity of the ground at the injection points, and varied between 0.2 bar and 2.5 bar, when ground squeezing had reduced the clearance gap (overcut). Two to three cubic meters of the thin Bentonite solution was used for each advance. Steel plates of 20 mm in thickness were welded to the TBM skin surface around the injection ports to facilitate the injection. The aim was to produce a groove to permit bentonite injection because excessive ground squeezing sealed the injection ports so strongly that bentonite could not be pumped even at pressures in excess of 10 bar. The exception to this statement occurred when the TBM was stopped for a long time, and the ground squeezed down even around the raised plates, consequently sealing the outlet. The study also shows that waiting time has a large effect on TBM jamming when excavating in weak rock. The geology at various chainages presented squeezing or settlement at extreme rates up to 60 mm per hour. The TBM shield length of 12 m, in the prevailing geology, proved to be a critical factor. This was because the ground moved so quickly, squeezing the rear edge of the shield body and reducing advance rates, culminating in a chain of events that eventually stopped the TBM. For a first initiation, the fixed overcut was increased from 35 mm eventually to 95 mm on the radius. However, this was not found sufficient in some areas, and the bentonite injection was utilized at critical points. Bentonite delivery pressure depended on the proximity of the ground at the injection points, and varied between 0.2 bar and 2.5 bars when ground squeezing had reduced the clearance gap (overcut). Two to three cubic meters of the thin bentonite solution was used for each advance. Steel plates of 20 mm in thickness were welded to the TBM skin surface around injection ports to facilitate the injection. The aim was to produce a groove to permit bentonite injection; since excessive ground squeezing sealed the injection ports so strongly that bentonite could not be pumped even at pressures in excess of 10 bar [6].

6.7   Tunnels excavated by drill and blast methods

Although this book is devoted to tunnels driven by TBMs, it is of prime importance to cite also some tunnels excavated by conventional tunneling methods and affected by the North Anatolian Fault – Ayas and Bolu tunnels being the most critical examples to this.

6.7.1   Ayas Tunnel: the most difficult tunnel in Turkey affected by North Anatolian Fault

The construction of the railway tunnel between Beypazari and Istanbul (Arifiye–Sincan, started in 1976, length of 10.064 km). The lined inner diameter of the tunnel is 9.60 m. The tunnel was excavated by the NATM and about 230 million USD has been spent on the tunnel construction and it is still not finished. The main reason is the highly complicated geology and the difficult ground conditions. Ayas Tunnel was the longest railway tunnel planned in Turkey. The project involved 1,200,000 m^3 of tunneling excavation, 2,600,000 m^3 of open excavation, 450,000 tons of shotcrete, 6,700 tons of steel ribs, 6,100 tons of anchorage, 1,500 tons of wire mesh and 210,000 m^3 of B225 tunneling concrete works.

6.7.2 Bolu tunnel

This tunnel is part of the Gumusova–Gerede Highway O-4 within the Trans-European Motorway project, and tunnel exaction started on 16 April 1993. The total cost of the tunnel is about 300 million USD. It has twin 17 m wide bores, carrying three lanes of traffic in each direction. The tunnel crosses the North Anatolian Fault. The 12 November 1999 the Duzce earthquake (MW = 7.2) caused substantial damage to the tunnel and viaducts, which were under construction at the time of the earthquake. The Gumusova–Gerede section of the Istanbul–Ankara motorway is located in the zone of influence of the North Anatolian Fault. This section of the motorway suffered damage during the Duzce earthquake. Bolu tunnel, which was 20 km from earthquake epicenter, collapsed in the flyshoid parts, with high plasticity and fault gouge. The collapsed part of the tunnel was supported with primary lining and required re-profiling due to excessive deformations under static loading. Investigations showed that the collapsed part of the tunnel extended from the Elmalik portal to the area consisting of the pilot tunnel. Re-excavation of the tunnel in collapsed area would require ground improvement before excavation, slow advance and a high density of tunnel support. Tunnel advance had been carried out along the bypass alignment.

Excavation of the tunnel in the Ankara to Istanbul direction was completed by the beginning of August 2005, using NATM [7].

6.8 Conclusions

Turkey is in a tectonically active region which experiences frequent destructive earthquakes. On the large scale, the tectonics of the region are controlled by the collision of the Arabian Plate and the Eurasian Plate. The Anatolian block is being squeezed to the west. The block is bounded to the north by the North Anatolian Fault and to the southeast by the East Anatolian Fault. Effects of these two faults on TBM performance, including TBMs in the Kargi energy tunnel, Dogancay energy tunnel, Nurdagi railway tunnel and Uluabat energy tunnels. These are explained in detail, giving the causes, the effects and the precautions to be taken in order to eliminate the problems created by two large sets of faults. Some information of the most problematic tunnels (Ayas and Bolu) ever excavated by drill and blast method is also given within this chapter.

References

[1] http://earthquake.usgs.gov (2015).

[2a] Home, L. (2014) *Case Studies On Mechanized Rock Tunneling*. Proceedings of the World Tunnel Congress, Tunnels for Energy, 9–15 May, Iguassu Falls, São Paulo, Brazil.

[2b] Home, L. (2014) *The Kargi Challenge: A study of TBM boring vs. conventional excavation*. Third short course organized by Turkish Tunnelling Society, 28 August 2014, Istanbul, Turkey.

[3] Yurt, M., Ozturk, A., Arslan, A., Nuhoglu, C., Oystein, L., Erdogan, E., Atlar, B., Palakci, Y. and Bilgin, N. (2014) *Factors affecting the performance of a double shield TBM in a very complex geology in Kargi Turkey*. Proceedings of the World Tunnel Congress – Tunnels for a better Life. Foz do Iguaçu, Brazil.

[4] Shaterpour-Mamaghani, A., Tumac, D., Avunduk, E. (2015) Double shield TBM performance analysis in difficult ground conditions: a case study in the Gerede water tunnel, Turkey. *Bull Eng Geo Environ*, **75** (1), 251–262.

[5] Bilgin, N. (2016) An appraisal of TBM performances in Turkey in difficult ground-conditions and some recommendations. *Tunn Undergr Sp Technol* **57**, 265–276.

[6] Bilgin, A., Algan, M. (2012) The performance of a TBM in a squeezing ground at Uluabat, Turkey. *Tunn Undergr Sp Technol,* **32**, 58–65.

[7] Solak, T., Akis, E., Russo, M. (2007) *Detailed seismic analyses for Bolu Tunnel*. The 2nd Symposium on Underground Excavations for Transportation, 15–17 November Istanbul, Turkey.

7 Effect of blocky ground on TBM performance and the mechanism of rock rupture

7.1 Introduction

As explained in previous chapters, Turkey is widely affected by two major fault systems, North Anatolian and East Anatolian Fault. These two fault systems and magmatic inclusions, or 'dykes', fracture the rock making a problematic blocky ground for TBM excavations. This chapter is aimed at explaining the effect of blocky ground on TBM performance and the mechanism of rock rupture in front of a TBM. Two papers recently published on the subject by Delisio et al. [1, 2] will be summarized first, and then a case study from Istanbul metro drivages will be given to clarify the subject.

After Delisio and Zhao [2], the term 'blocky rock conditions' is generally associated with face instabilities in blocky/jointed rock masses. These events are generally promoted by unfavorable rock mass structural conditions, in terms of joint frequency and orientation, and acting stresses. As a result, rock blocks are formed which then detach from the excavation face which becomes 'blocky', with a markedly irregular and uneven profile. This condition can have a major effect on TBM tunneling, leading to a high maintenance frequency and a low TBM advance rates. Based on the TBM performance data recorded during excavation of tunnels in blocky rock conditions, a TBM performance prediction model has been developed.

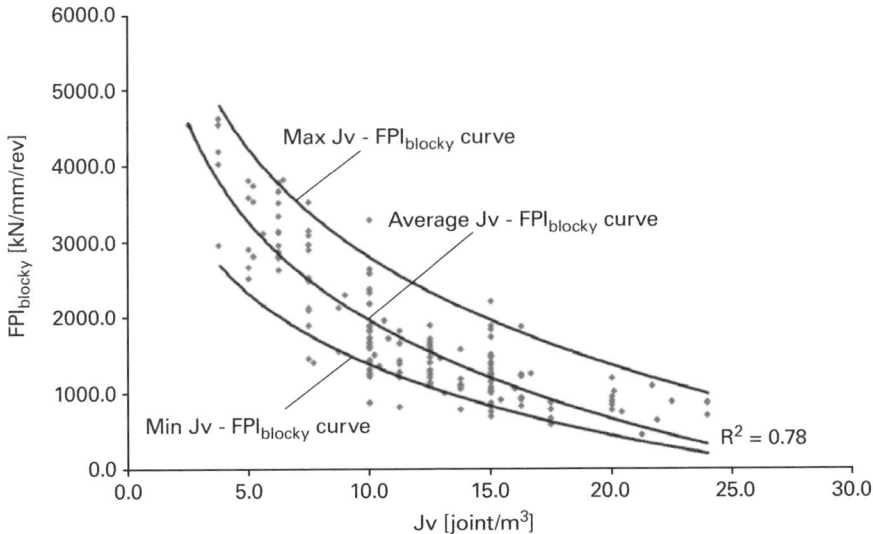

Figure 7.1 Correlation between FPI blocky and volumetric joint count Jv. The trend lines for mean, maximum and minimum FPI blocky values are also shown, after Delisio et al. [1].

TBM Excavation in Difficult Ground Conditions. Case Studies from Turkey. First Edition. Nuh Bilgin, Hanifi Copur, Cemal Balci.
© 2016 Ernst & Sohn GmbH & Co. KG. Published 2016 by Ernst & Sohn GmbH & Co. KG.

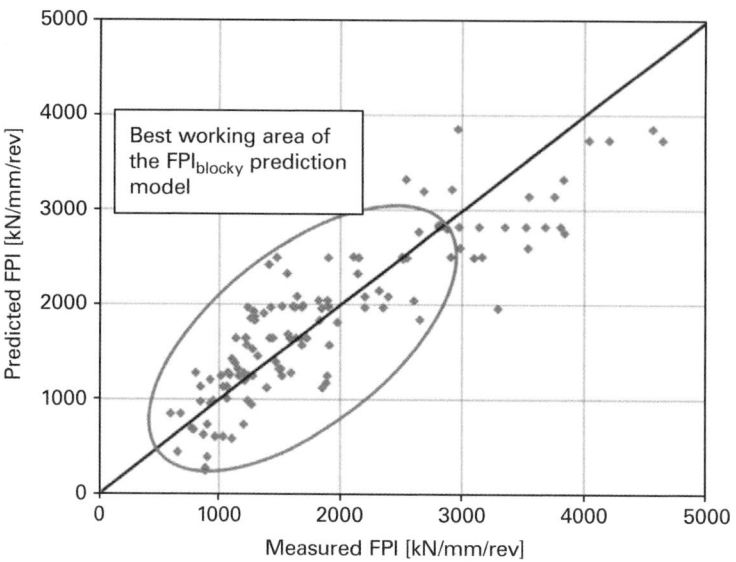

Figure 7.2 Comparison between actual and predicted values of the FPI blocky, after Delisio et al. [1].

The model is based on the field penetration index (FPI) for blocky rock conditions, FPI blocky, which was previously introduced to analyze the TBM performance in blocky ground at the Lötschberg base tunnel, – see Figures 7.1 and 7.2. Through a multi-variate regression analysis, a new expression has been introduced to predict the FPI blocky, based on the volumetric joint count (J_v) and the intact rock uniaxial compressive strength (UCS). An attempt has also been made to quantify the downtimes that can occur in blocky rock conditions and to estimate a reliable value for TBM daily advance. As seen from Figure 7.3, cutter replacement and cutterhead inspection takes 35% of the total shift time.

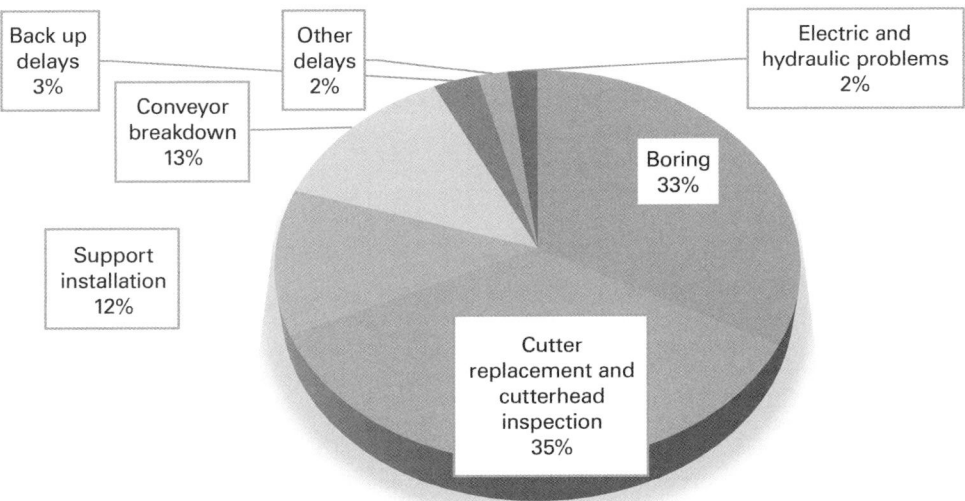

Figure 7.3 Percentage time related to each identified downtime category at the SMTT (Second Manapouri Tailrace Tunnel), after Delisio and Zhao [2].

7.2   Mechanism of rock rupture and face collapse in front of the TBM in the Kozyatagi–Kadikoy metro tunnels in Istanbul

7.2.1   Kozyatagi and Kadikoy metro tunnels and problems related to blocky ground

This chapter concerns the performance of two TBMs working between Kozyatagi and Kadikoy. The project is a part of the metro line starting from Pendik, integrating with the Marmaray project and going as far as Halkali in the south-east of the city. The general route of the metro line is given in Figure 7.4, and the basic technical specifications of the Herrenknecht TBM are outlined in Table.7.1. Sedimentary rock formations of Triassic and Tertiary ages with a complex geology are found in this area. The main geological formation to be excavated is the Kartal formation, with several diabase and andesite dykes approximately every 70 m. The contact zones between dykes and sedimentary rocks are usually highly fractured, causing several problems in front of the cutting head of the TBM [3].

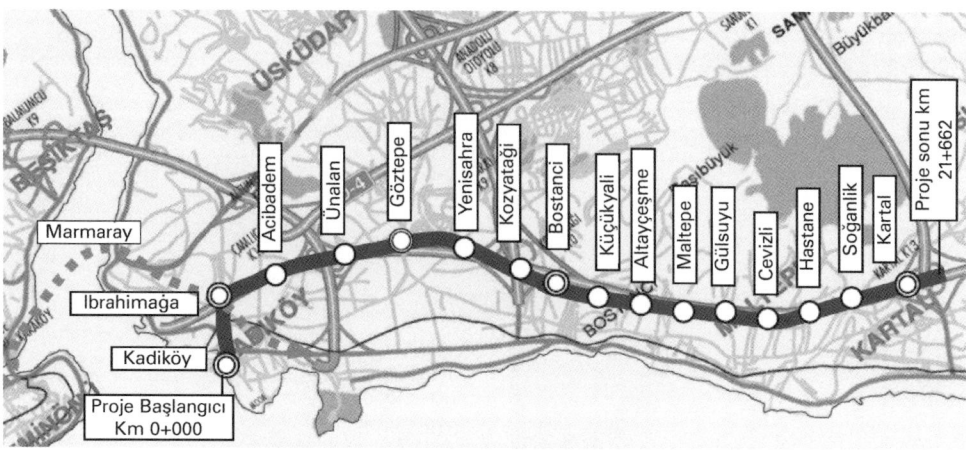

Figure 7.4 The general layout of the Kartal–Kadikoy metro line [3].

Table 7.1 Basic technical specifications of the Herrenknecht TBM [4].

Parameter	Value
Machine diameter	6.57 m
Number of cutters	26 single + 6 double (12) = 38
Max. contact pressure per disk	267 kN
Max. thrust capacity of the main bearing	20,000 kN static
Cutterhead power	1,260 kW (4×315)
Cutterhead rotational speed	1.6–5.5 rpm
Cutterhead torque	5,200 kNm at 1.6 rpm 1,515 kNm at 5.5 rpm
Max. working capacity of thrust cylinders	4,2575 kN at 350 bar

Excavation data was carefully collected during tunnel drivages. At the beginning, it was observed that the contact zones between dykes and the main rock formation were highly fractured, and in some areas big rock blocks having sizes up to $30 \times 40 \times 50$ cm were ripped off by the disc cutters from the fractured zones, passing through the openings of the cutterhead and causing several problems such as collapses of the tunnel face. Typical big blocks are seen in Figures 7.5 and 7.6 and a face collapse in is shown Figure 7.7.

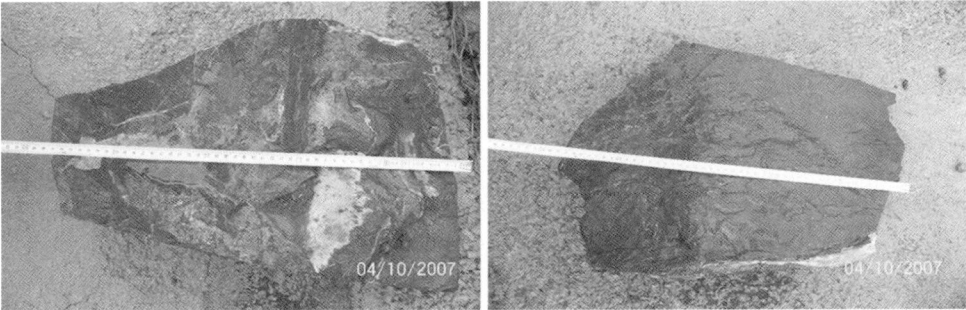

Figure 7.5 Big rock blocks coming from the tunnel face [4].

Figure 7.6 Big rock blocks coming from the tunnel face [4].

Figure 7.7 Collapse in front of cutterhead [3].

The face collapses dramatically decreased the daily advance rates around to 2.5 m/day. After several technical discussions between project management staff and TBM manufacturers it was decided to install some grizzly bars to limit the big rock blocks passing through the openings of the cutterhead. Daily advance rate increased up to 6 m/day after the modification of the cutterhead.

7.2.2   Mechanism of rock rupture and face collapse in front of the TBM

Thrust force, torque and penetration per revolution of the TBM were carefully examined around the collapsed zone within rings 250–270, in order to understand the mechanism of the rupture of big blocks from the tunnel face. Figures 7.8–7.10 show the variation of TBM thrust, torque and penetration in Kartal limestone within the tunnel route [3].

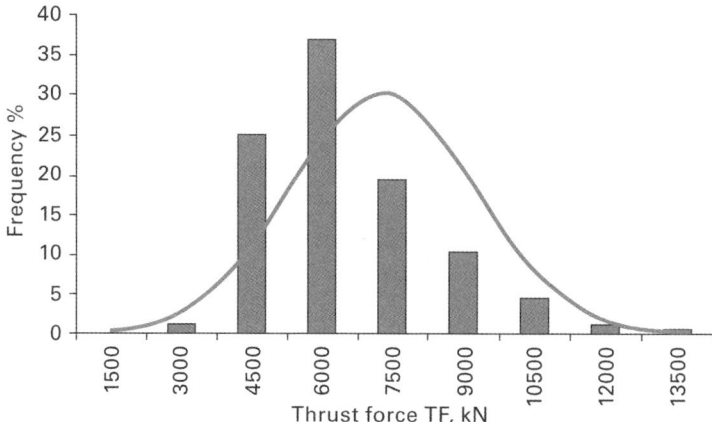

Figure 7.8 The variation of thrust force in open mode in Kartal Formation [3].

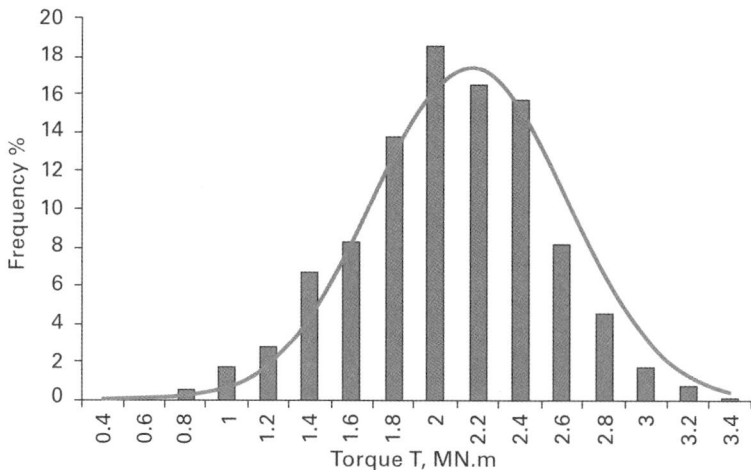

Figure 7.9 The variation of torque in open mode in Kartal Formation [3].

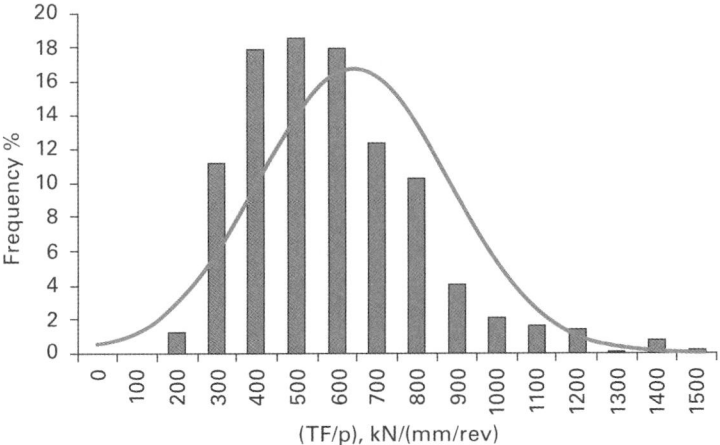

Figure 7.10 The variation of penetration index in open mode in Kartal Formation [3].

One of the collapses occurred in rings 260 and 268. It can be seen from Figure 7.11 that the thrust force of the TBM increases in fractured zones 1 and 2 between ring 250 and 260 until the tunnel face collapses in front of the TBM. The site geological engineer reported that there was a diabase dyke in front of the cutterhead. The thrust force increased in the fractured zone and the discs tried to rip out big blocks from this area. The collapse occurred in ring 260 and this situation continued cyclically, which caused again another collapse in ring 265. This phenomenon is also seen in Figure 7.12, which shows the relationship between ring number and torque. When the machine approached the ring where the collapse occurred, the torque and thrust of the machine increased in the fractured zone and later decreased in the collapse zone [4].

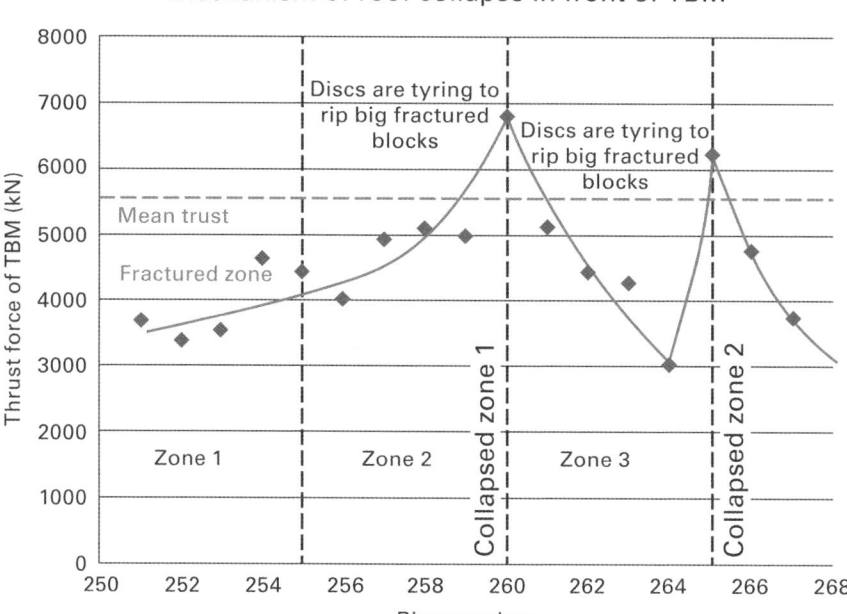

Figure 7.11 The change of TBM thrust force around collapsed zone [4].

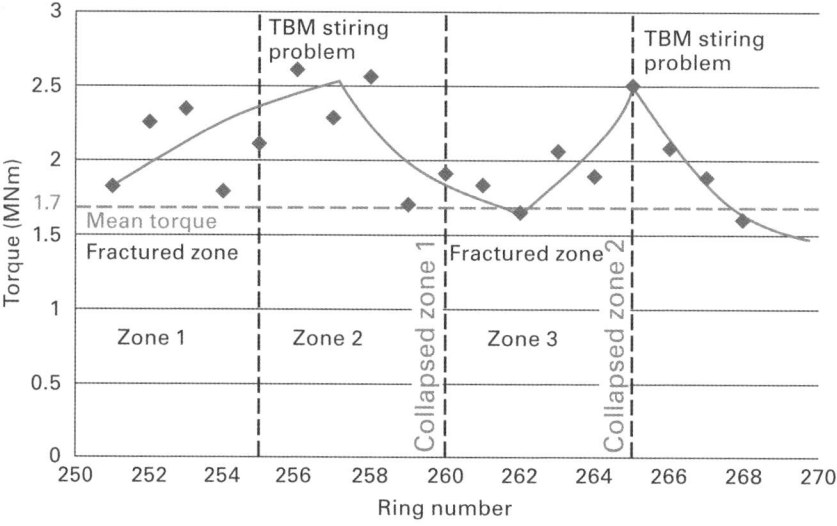

Figure 7.12 The change of TBM torque around collapsed zone [4].

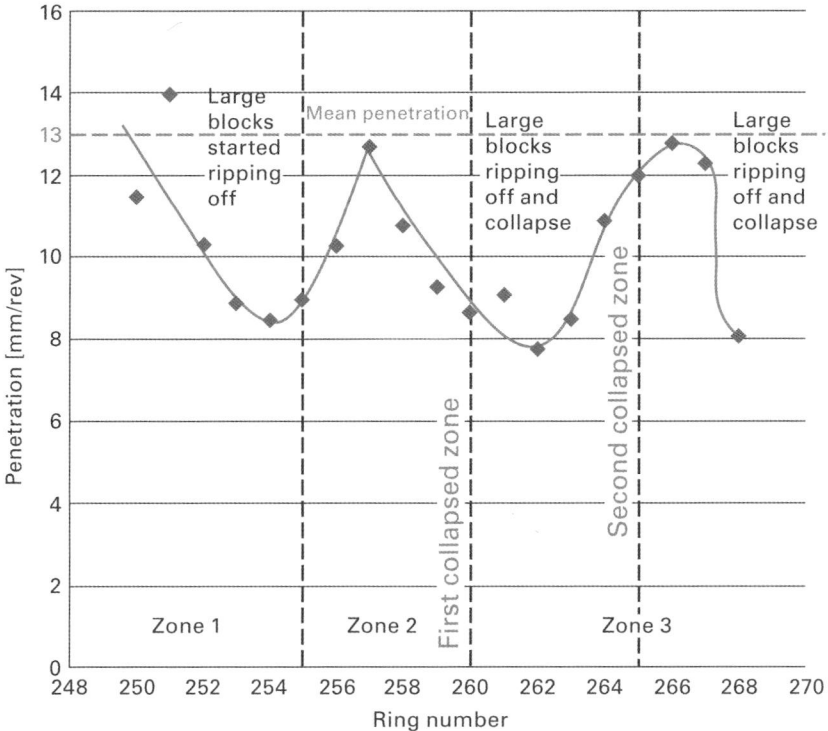

Figure 7.13 The change of TBM penetration around collapsed zone [4].

However, the relationship between ring number and penetration shown in Figure 7.13, around collapse zones, is the inverse of that found in Figures 7.11 and 7.12. Although it is expected that the penetration should be increased with increasing thrust, it is other way around in fractured zones. In fractured zones, high thrust forces are trying to rip off blocky rocks and the big blocks are ripped off with small penetrations. The collapsed areas in front of the cutterhead are shown in Figure 7.7.

The mean value of thrust force, torque and penetration between rings 250 and 270 are calculated from data obtained for the first 400 rings.

Figure 7.14 indicates that the variation of field penetration index within collapsed zone covering rings 250 and 270 is important in explaining the mechanism of rupture of big blocks from the fractured rock mass. Field penetration index is lower than mean penetration index in blocky ground within the same geological formation and has higher values when the blocks are ruptured from the face and face collapse occurs.

Figure 7.14 supports the findings of Delisio et al. [1] and Delisio and Zhao [2], stating that the concept of field penetration index plays an important role in explaining TBM performance prediction in blocky rock conditions and explaining rupture mechanisms of rock blocks from the rock mass.

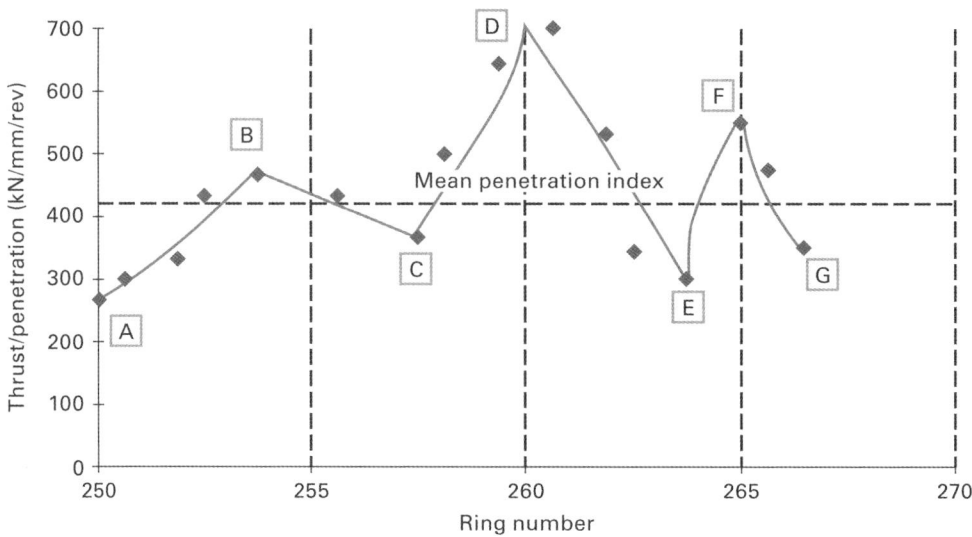

Figure 7.14 The variation of field penetration index within collapsed zone covering rings 250–270.

7.2.3   Other factors affecting the efficiency of tunnel excavation on the Kozyatagi–Kadikoy metro line

Figures 7.15 and 7.16 illustrate the work study carried out for both TBM excavation on both Lines 1 and 2 within a certain period. As seen from these figures, time spent for excavation is 23% and 27% on Lines 1 and 2 and time spent for ring mounting is 23% and 31%. However, one important factor emerging from these figures is that the problems caused by blocky ground were more severe than the problems caused in Second Manapouri Tailrace tunnel as shown in Figure 7.3.

Disc, electricity, water, lubricating grease, and foam chemicals are given in Table 7.2. Disc consumption in Kartal limestone is low as would be expected, with a value of 1,176 m³/disc, but this value is high in Dolayoba formation with a value of 85 m³/disc. This may be explained by the high compressive strength and Cerchar values of this formation with values of 1818 kg/cm² and 3, respectively.

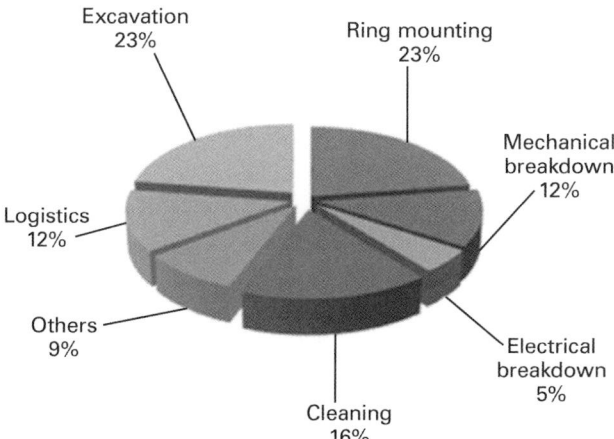

Figure 7.15 TBM Performance in Line 2, during 1 July – 31 December 2008.

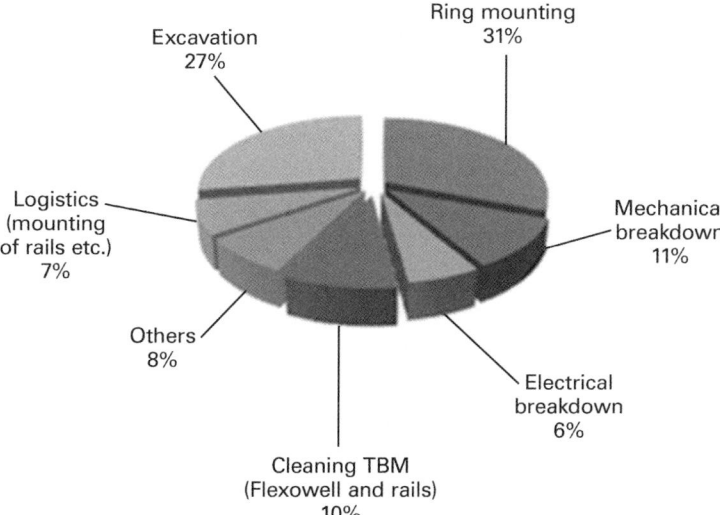

Figure 7.16 TBM Performance in Line1, during 1 July – 31 December 2008.

Table 7.2 Material consumption with TBM excavation in Lines 1 and 2 [3].

Disc consumption in Kartal limestone (m³/disc)	1176
Disc consumption in Dolayoba limestone (m³/disc)	85
Electricity consumption (m³/kW)	22.7
Water consumption (m³/kW)	0.4
Lubricating grease (kg/m³)	0.73
Chemical (SLF30) for foam (lt/m³)	1.4
Chemical (SLF41) for foam (lt/m³)	0.23

7.3   Conclusions

This chapter deals with the performance of two TBM's working in blocky ground on the Kozyatagi–Kadikoy metro line. The project is a part of metro line starting from Pendik, integrating with Marmaray project and going as far as Halkali, south-east, of the city.

Sedimentary rock formations of Triassic and Tertiary ages with a complex geology are found in the area. The main geological formation to be excavated is Kartal formation with diabase and andesite dykes, approximately every 70 m. The contact zone between dykes and sedimentary rocks is usually highly fractured causing several problems in front of the cutting head of the TBM, due to the blocky characteristics of the ground.

Two identical Herrenknecht TBMs were chosen for Lines 1 and 2, and these machines could work in open or closed mode. The TBMs have a diameter of 6.57 m, the total number of discs being 38 and the cutterhead power 1,260 kW. The excavation data was carefully collected during tunnel drives. At the beginning, it was observed that the contact zones between dykes and the main rock formation were highly fractured, and in some area big rock blocks having sizes up to $30 \times 40 \times 50$ cm were ripped off by the disc cutters from the fractured zones, passing through the openings of the cutterhead and causing several problems such as collapses of the tunnel face. The face collapses dramatically decreased daily advance rates around to 2.5 m/day. After several technical discussions between project management staff and TBM manufactures it was decided to install some grizzly bars to limit the big rock blocks passing through the openings of the cutterhead. Daily advance rates increased up to 6 m/day after the modification of the cutterhead.

Thrust force, torque and penetration per revolution of the TBM were carefully examined around the collapsed zone and it is clearly shown that a close look at the change of behavior of the machine including thrust, torque and field penetrations changes of TBM compared to mean values of these parameters provide a good indicator of the forthcoming problems such as face collapses.

Work study carried out showed that time spent for excavation is 23% and 27% in Lines 1 and 2 and time spent for ring mounting is 23% and 31%. However, one important factor emerging from these figures is that the problems caused by blocky ground in the Kadikoy metro tunnel line were more severe than the problems caused in the Second Manapouri Tailrace, as reported by Delisio and Zhao [2].

References

[1] Delisio, A., Zhao, J., Einstein, H.H. (2013) Analysis and prediction of TBM performance in blocky rock conditions at the Lötschberg Base Tunnel. *Tunn. Undergr. Sp. Technol.* **33**, 131–142.

[2] Delisio., A, Zhao, J. (2014) A new model for TBM performance prediction in blocky rock conditions. *Tunn Undergr Sp Technol*, **43**, 440–452.

[3] Yuksel, A. (2013) Mechanical properties and rock fractures affecting the performance of TBMs. Istanbul Technical University, Dissertation.

[4] Bilgin, N., Ozbayir, T, Sozak, T., Eyigun, Y. (2009) *Factors affecting the economy and efficiency of metro drivages with two TBMs in Istanbul in very fractured rock*. World Tunnel Congress, 23–28 May, Budapest, Hungary.

8 Effects of transition zones, dykes, fault zones and rock discontinuities on TBM performance

8.1 Introduction

The geology of Istanbul is complex for tunneling projects. Tectonic activities, faults, diabase, dacite and andesite dykes and several joint sets cause many serious problems during tunnel excavation. The experience gained in different projects in Istanbul has showed that these geological features can cause major problems during TBM excavation resulting in tunnel face collapses, squeezing of the TBM and decreasing daily advance rates. The aim of this chapter is to explain how these geological features affected TBM performance in two projects, Beykoz sewerage project [1, 2, 3] and the Kozyatagi–Kadikoy metro project [4, 5, 6].

8.2 Beykoz sewerage tunnel

8.2.1 Description of the project

The sewerage tunnel excavated between Kavacik and Beykoz in Istanbul is part of an environmental protection project including renewing the inadequate sewerage network around Beykoz and collect the wastewater in a treatment plants and cleaning the polluted water of İstanbul Bosphorus.

The project was owned by Istanbul Water and Sewerage Administration (ISKI) and the contractor was NTF Constructing Company. The ground conditions changed from soft to very hard formations and excessive water ingress was experienced in some areas. The total length of tunnel is 7,190 m, and tunnel excavation started from shaft AT2 on 9 January 2007 and finished during 2009. The general layout of the tunnel is given in Figure 8.1.

TBM Excavation in Difficult Ground Conditions. Case Studies from Turkey. First Edition. Nuh Bilgin, Hanifi Copur, Cemal Balci.
© 2016 Ernst & Sohn GmbH & Co. KG. Published 2016 by Ernst & Sohn GmbH & Co. KG.

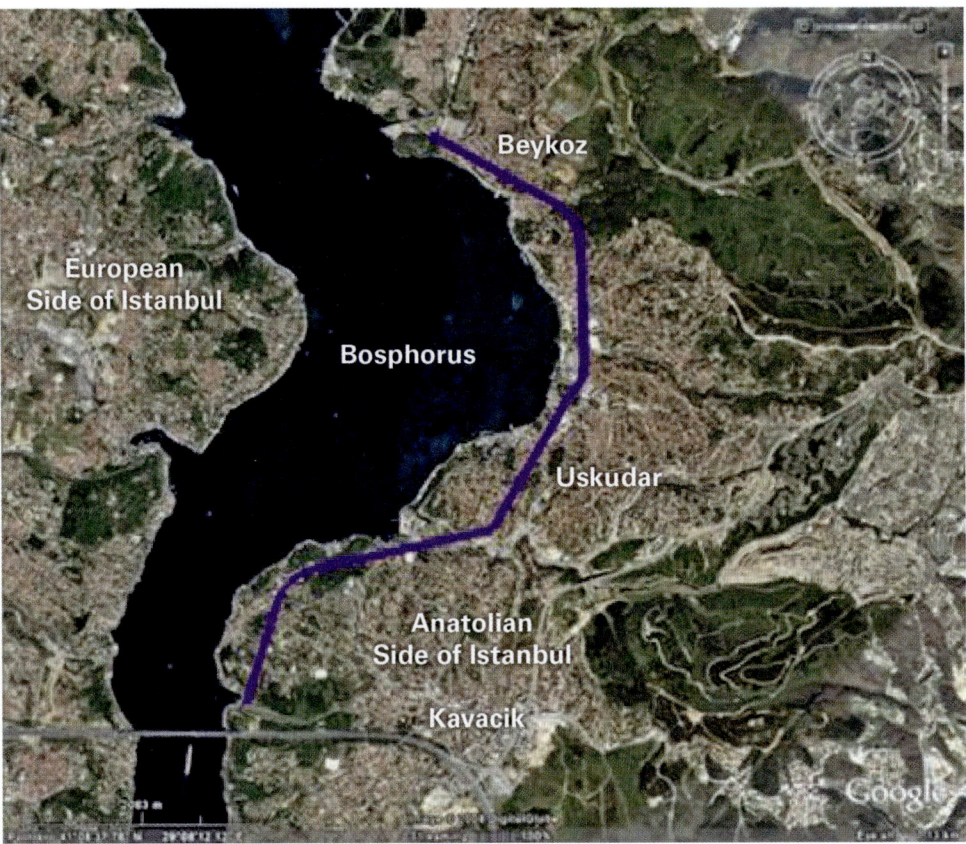

Figure 8.1 The general layout of Beykoz sewerage tunnel.

8.2.2   Geology of the area

The project area is situated in Beykoz in the northern part of the Istanbul Bosphorus. Palaeozoic aged Gozdag, Dolayoba and Kartal formations and alluvium are found in the area. The general stratigraphy of the project area is given in Table 8.1, and the geological profile for the first 4,500 m is given in Figure 8.2, which is where the study presented in this chapter was realized [1, 2, 3].

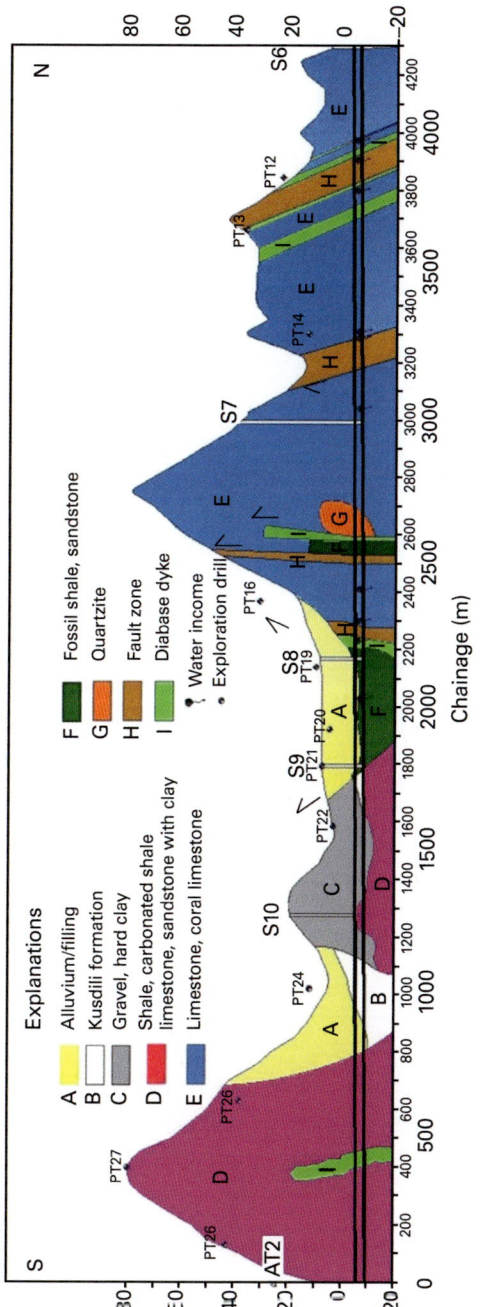

Figure 8.2 Geological profile of the tunnel for the first 4,500 m of the Beykoz Tunnel [2, 3].

Table 8.1 General stratigraphy of the project area [1, 2, 3].

Age	Formation	Lithology
Quaternary	Alluvium	Sands and gravel, hard clay with different thickness
Discordance		
Neogene	Belgrad formation	Sandy with quartz pebbles, hard clay, silty clay
Discordance		
Devonian	Kartal formation	Calcareous shale gray in color, alternating with mudstone, sandstone and limestone in the upper levels
Silurian	Dolayoba formation	Limestone, gray and light blue in color
Ordovician	Gozdag formation	Shale, green and gray in color. Graywacke with interblended mudstone

8.2.3   Description of the TBM

The TBM was a shielded Robbins machine and could work in opened and closed mode. The general view of the TBM is given in Figure 8.3 and characteristics are tabulated in Table 8.2.

Figure 8.3 TBM used in the area [1].

Table 8.2 Technical characteristics of the TBM [1].

Parameter	Value
Excavation diameter	3,175 mm
Disc diameter	336 mm
Number of disc cutters	23
Total power	400 kW (2×200)
Cutting head power	200 kW
Torque	254 kNm, 8.5 rpm
Torque	527 kNm, 4.3 rpm
Maximum thrust force	995 t
Stroke	1,670 mm

8.2.4 Effect of rock formation on chip formation and machine utilization time

The TBM started running on 12 February 2007, and the mean values of TBM advances up to the end of May 2008 are summarized in Table 8.3. TBM operational parameters such as, machine rotational speed, thrust, torque and penetration were carefully collected with the aid of a data acquisition system during the excavation. The muck samples were also collected for strength tests and petrographic and sieve analysis. The muck size in different rock formations are given in Figure 8.4 [2, 3], which clearly shows that the muck size is a good indicator of the main characteristics of the geological formations.

Table 8.3 summarizes the overall mean performances of the TBM in the Beykoz tunnel.

a) Chip formation in massive limestone

b) Chip formation in fractured zone

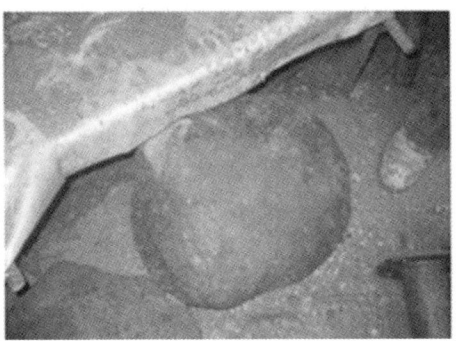

c) Big pebbles in alluvion

d) Muck in mudstone

Figure 8.4 Typical muck size distribution in different geological formations in the Beykoz tunnel [2, 3].

Table 8.3 Mean values of TBM avances.

Best daily advance (m)	23
Best weekly advance (m)	90
Best monthly advance (m)	313
Mean daily advance (m)	6,2
Mean weekly advance (m)	45
Mean monthly advance (m)	180

The effect of geological formations on machine utilization time is given in Table 8.4. As seen from this table, machine utilization time is in the range 41–44 in hard clay and competent Kartal limestone. In alluvium, big pebbles as illustrated in Figure 8.4, decreased the machine utilization time to 35%. In sticky clayey formations the machine utilization time was 17–21%, due to the clogging effect of this formation.

Table 8.4 The effect of geological formation on machine utilization time [2, 3].

Date	Rock formation and major drawbacks	Ring No	Advance (m)	TBM (rpm)	Thrust (bar)	TBM utilization time (%)
13.04.2007	Massif Kartal formation	283–285	13	6–8	110–150	44
12.06.2007	Kartal formation, machine breakdown.	599–609	11	4–6	80–130	20
01.07.2007	Clayey formation, cleaning of the cutterhead	826–631	6	5	70–100	21
03.07.2007	Clayey formation, cleaning of the cutterhead	848–862	15	5–7	60–130	37
04.07.2007	Hard clay	863–883	21	6–7	60–120	41
05.07.2007	Clayey formation, cleaning of the cutterhead	884–893	10	6	60–100	17
24.09.2007	Alluvium with big pebbles	1,389–1,402	14	6–7	80–100	35

8.2.5   Effect of dykes on tunnel face collapse and TBM blockage

The TBM excavated 800 m in Kartal formation, which is mainly composed of shale, carbonated shale, limestone and clayey sandstone. The main problem in this area was an andesite dyke encountered at chainage 0+400 km, which caused stalling of the cutterhead, as seen Figure 8.5. A gallery was opened next to the TBM, to reach and to clean the cutterhead; the operation was completed safely in 20 days [2, 3].

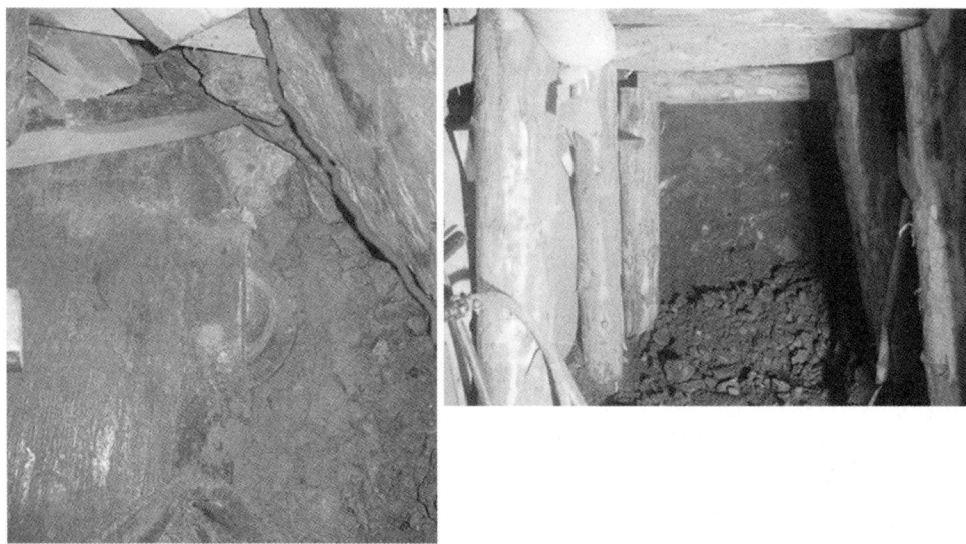

Figure 8.5 Andesite dyke stalling TBM cutterhead and the gallery opened to clean the head in Beykoz tunnel [2, 3].

8.2.6   Effect of transition zones on TBM performance

Transition zones between different geological formations have weak interfaces in most cases, and these zones are always a real problem in tunnel excavation. In most cases, the tunnel face collapses and the TBM gets stuck. Galleries have to be opened to rescue the TBM. The Beykoz sewerage tunnel is a typical example of this.

Figure 8.6 shows a gallery opened at chainage 0+810 km to rescue the TBM in the transition zone between the Kusdili formation and the alluvium fillings, and Figure 8.7 shows the muck transported with wagons in this area. Nine days were needed for rescue operations.

Figure 8.8 shows the transition zone between hard clay and carbonated shale-limestone at chainage 1+208 km, and a rescue gallery was opened for the salvage operations of the trapped TBM. Six days were spent on rescue operations.

Figure 8.6 Rescue gallery opened at chainage 0+810 km in the transition zone between Kusdili formation and alluvium fillings in the Beykoz tunnel [2, 3].

Figure 8.7 Muck in the transition zone between Kusdili formation and alluvium fillings at chainage 0+810 km in the Beykoz tunnel [2, 3].

Figure 8.8 Transition zone between hard clay and carbonated shale-limestone at chainage 1+208 km, and the rescue gallery opened for the salvage of the trapped TBM in the Beykoz tunnel [2, 3].

8.2.7   Effect of fault zones on TBM performance

Fault zones have similar effects on tunneling operations as transition zones, as shown in Figure 8.9 which pictures the gallery opened to rescue the trapped TBM at chainage 3+222 km. Twelve days were spent on TBM salvage operations.

Figure 8.9 Fault zone, and the gallery opened to rescue the trapped TBM at chainage 3+222 km.

8.3   Kartal–Kadikoy metro tunnels, methodology of understanding critical zones

A great deal of effort was spent collecting data concerning the performance of mechanized excavation systems related to geology and rock mass properties on the Kozyatagi–Kadikoy metro tunnels in order to have guidelines for future tunnel projects [4, 5]. The data obtained during the excavation of the twin metro tunnels is evaluated in this respect. This chapter concerns the causes and effects of the TBM face collapses in these tunnels. The project was a part of the metro line starting from Kozyatagi, integrating with the Marmaray project and going as far as Halkali north-east of the city. The length of a single line tunnel is 8,800 m with a diameter of 6.6 m. The general layout of the metro line is given in Figure 8.10 [6].

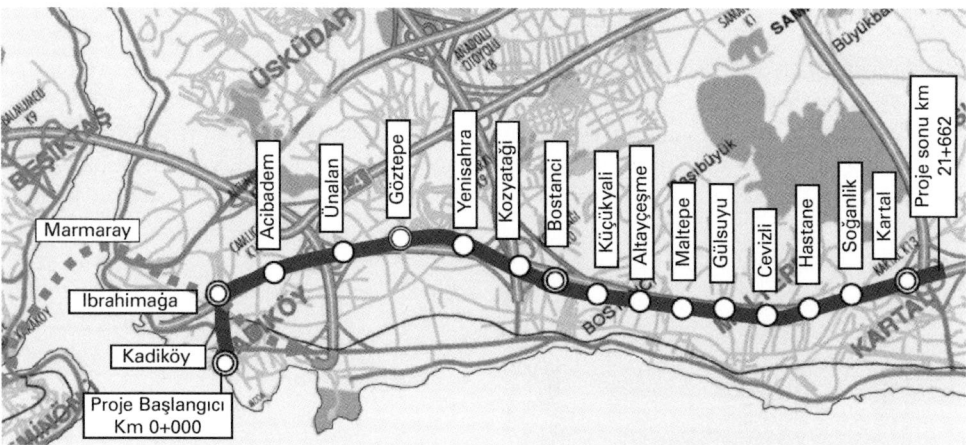

Figure 8.10 The layout of the Bostanci–Kadikoy metro tunnels within the main Kartal–Kadikoy metro line [5].

8.3.1   Geology and physical and mechanical properties of rocks

Sedimentary rock formations of Palaeozoic ages with a complex geology are found in the area. The main geological formations excavated are Kartal formation of Devonian age and Trakya Formation of Carboniferous age with diabase and andesite dykes approximately every 70 m. The contact zone between dykes and sedimentary rocks is usually highly fractured, causing several problems in front of the cutting head of the TBM. Standard physical and mechanical tests were carried out on core samples of NX size. Test results are summarized in Table 8.50, in which UCS is uniaxial compressive strength, E is elastic modulus, RQD is rock quality designation and GSI is geological strength index.

Table 8.5 Some physical and mechanical properties of rock formations [5].

Formation	Rock	UCS MPa±sd	E GPa±sd	RQD±sd	GSI±sd
Kartal Zone A	Limestone shale	32.6±24.8	7.8±5.3	58±13.4	41±13.4
Kartal Zone B	Limestone shale	36.0±18.1	7.3±2.9	38.9+-38.9	35.9±15.2
Trakya	Mudstone siltstone sandstone	52.7±29.3	9.5±5.4	21.2±19.2	22.3±21.2
Dolayoba	Limestone shale	45.2±17.2	18.1±12.1	72.5±72.5	49.1±14.6

8.3.2   Mechanism of face collapse and TBM blockages

During tunnel excavations, the TBM faced severe problems in 11 different areas, as shown in the geological sections given in Figures 8.11 and 8.12. Face collapses started first, and later the cutterhead jammed, which stopped the excavation. A total of 93 days were spent on TBM salvaging. This chapter is aimed at explaining the mechanism of face collapses, in order to help tunnel engineers in future mechanical excavation projects [6].

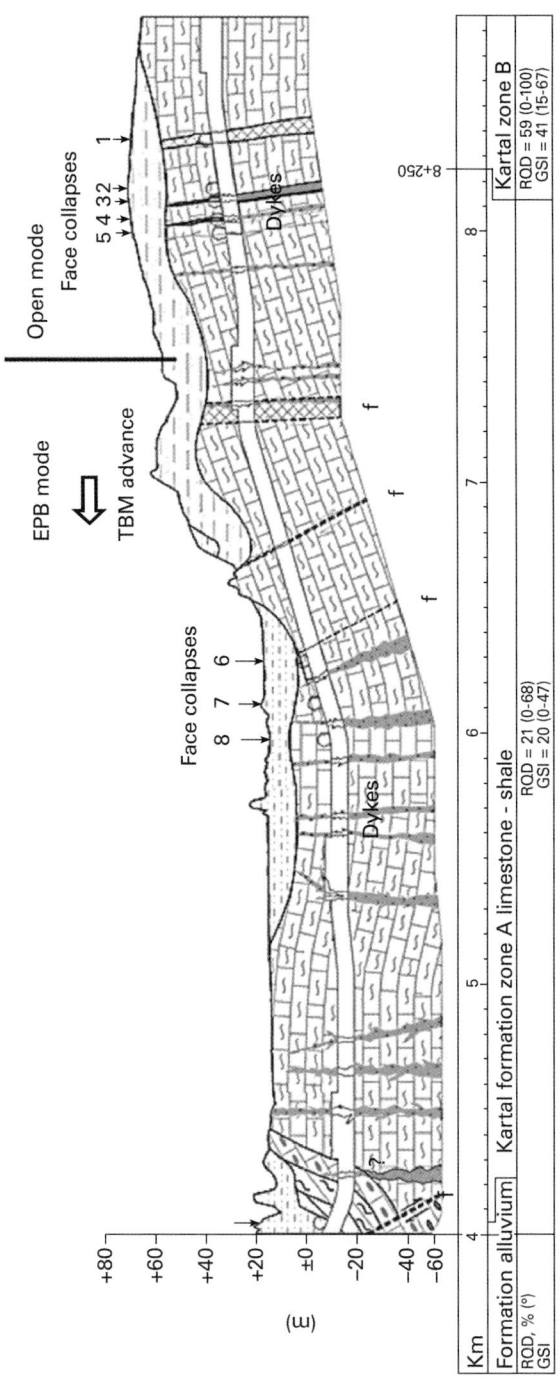

Figure 8.11 TBM blockages between 4+500 and 8+463 km in the Kadikoy–Kozyatagi metro tunnels [5].

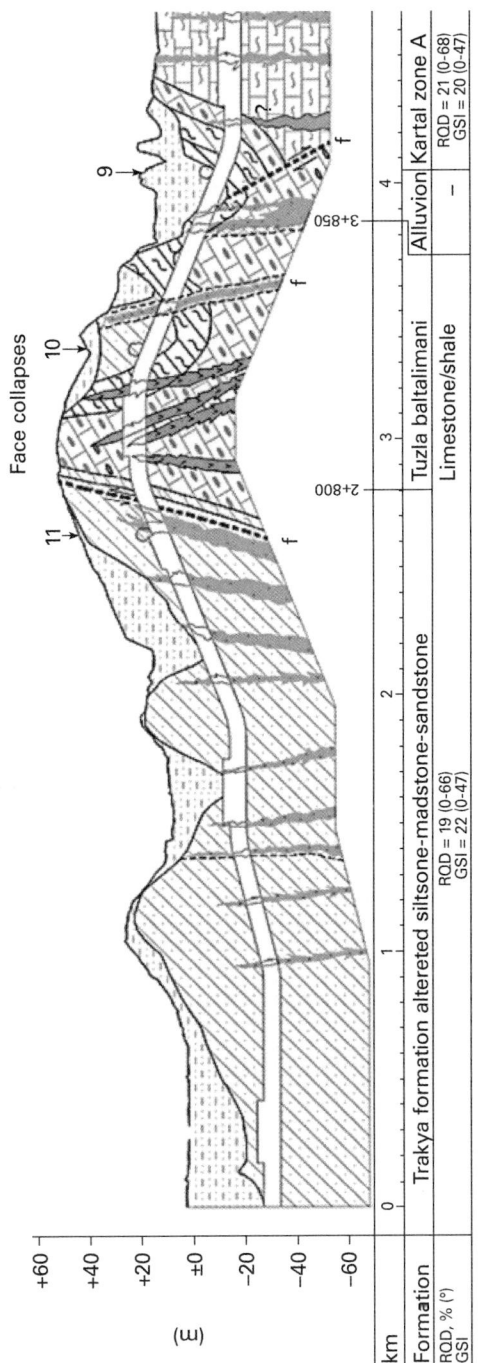

Figure 8.12 TBM blockages between 0+000 and 4+500 km in the Kadikoy–Kozyatagi metro tunnels [5].

The blockages in 11 different areas with tunnel chainage, ring number, time to spent for tunnel salvage, the causes of the problem and remedial works are summarized in Table 8.6 [5]. Note from this table that within the first 200 m of the TBM blocked twice, and it took 41 days to rescue the TBM. Figure 8.13 gives a schematic view of the first TBM blockage, which happened in ring 73, at tunnel chainage of 8+366 km [6]. It can be seen that the face collapse started after ring 54, within the influence zone of the contact area between the main rock and an andesitic dyke. Big blocks, as seen in Figure 8.14, ripped off from the fractured zone, causing the face collapse, 4 m at the top and 1 m at the top end of the cutterhead. To rescue the trapped cutterhead, bentonite injection through 140 mm diameter drill holes were used, and almost 27 days were spent on rescue operations.

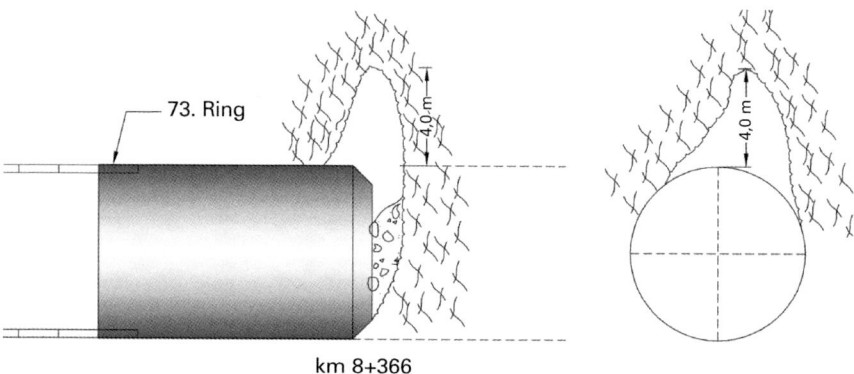

Figure 8.13 The first TBM blockage occurred at chainage 8+366 km, ring 73 [6].

Figure 8.14 Big rock blocks coming from the TBM face at chainage 8+366 km [6].

Table 8.6 Summary of TBM blockages in 11 different areas [5].

Area	Line No	Ring No	Tunnel m	Stoppage	Cause of the problem
1	1	73	8+366	27days	Fault zone
2	2	260–279	8+170	14 days	Contact zone between dyke and Kartal Formation
Opening ratio was reduced by adding grizzly bars on the TBM cutterhead					
3	1	232	8+120	7.5 days	Contact zone between dyke and Kartal Formation
4	1	282	8+046	2 days	Contact zone between dyke and Kartal Formation
5	1	321	7+993	6 days	Contact zone between dyke and Kartal Formation
Screw conveyor was mounted within the cutting chamber and EPB mode was used thereafter					
6	2	1349	6+286	2.5 days	Fault zone
7	2	1461	6+118	3 days	Contact zone between dyke and Kartal Formation
8	1	1483	5+973	1 day	Dyke contact zone
9	1	2272	4+046	2.5 days	Fault zone
10	1	2731	3+356	3 days	Transition zone between Trakya and Baltalimani Formation.
11	1	3021	2+628	24.5 days	Shear zone next to the tunnel connection

After the second TBM blockage, for which salvage operations took 17 days, it was decided to modify the cutterhead by reducing the openings. As seen in Figure 8.15 grizzly bars were added to stop the big rock blocks coming from the openings.

Another three TBM blockages occurred before Yenisahra Station between chainages 7+290 and 7+544 km, where TBMs were advanced in open mode. Thereafter, TBMs were modified to work in EPB modes. The lessons learnt from this project were to work with limited opening ratios and to work in EPB mode in fractured zones when driving tunnels with TBMs in Istanbul. A typical example to a face collapse in faulty zone in a complex geology is given in Figure 8.16.

The last collapse occurred at chainage 6+385 km in the transition zone between alluvium and Kartal formation. A schematic view of the collapse zone is given in Figure 8.17. It caused a major collapse hole in the highway above due to the low shear strength of the geomaterial, as seen in Figure 8.18 [6]. Five injection holes, as seen in Figure 8.17, were drilled into the side of the highway, and later the hole was filled with concrete mixture to stabilize the gauge material.

Figure 8.15 TBM cutterhead before the modification (left), and TBM modified with grizzly bars (right) [6].

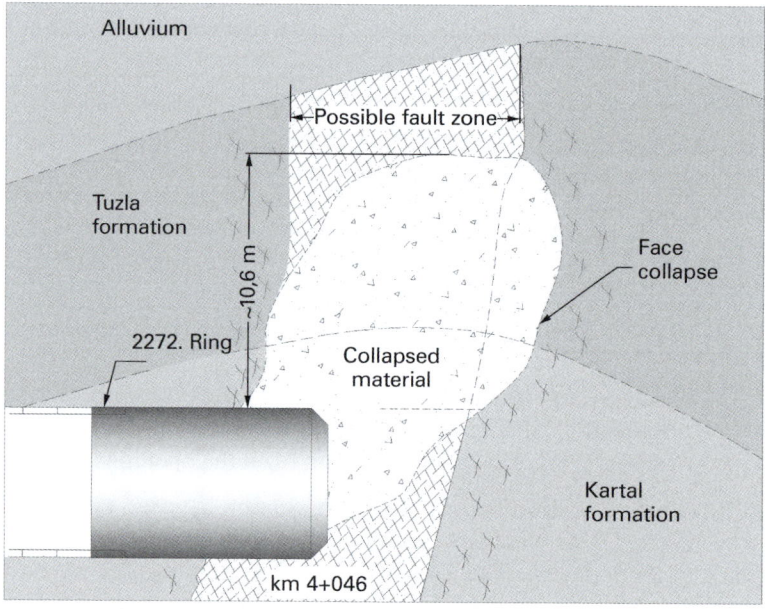

Figure 8.16 Face collapse that occurred within the fault zone in complex geology at chainage 4+046 km (ring 2.272) [6].

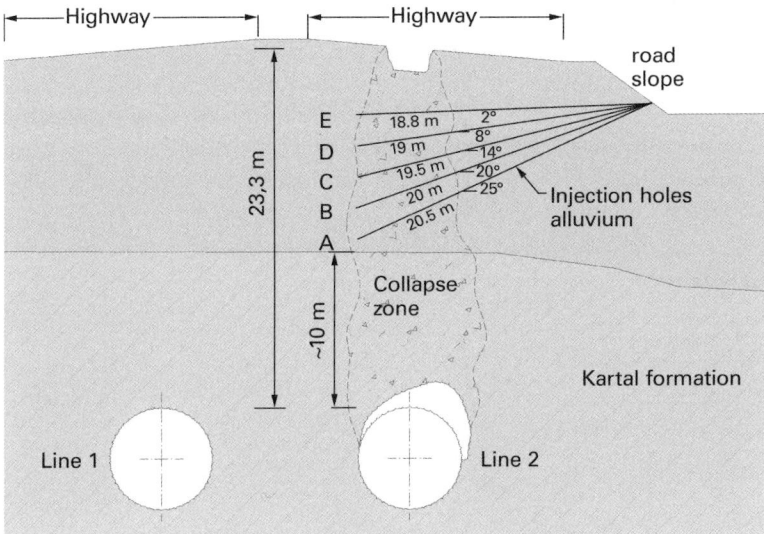

Figure 8.17 Face collapse that occurred in the transition zone between alluvium and Kartal formation at chainage 6+385 km (ring 1.349) [6].

Figure 8.18 The collapse in highway with tunnel chainage 6+385 km (ring 1.349) and concrete mixture filling operation [6].

8.3.3 Change of TBM performance in problematic areas

In this research, excavation taking place 20 rings forward and backward of the point where the collapse occurred is designated 'during collapse' and points within 30 rings backward and forward from these rounds, are defined as 'before collapse' and 'after collapse', respectively. TBM performance records, and the parameters derived from these records, such as total thrust force/penetration ratio (TF/p), torque/penetration ratio (T/p) cutting coefficient (CC), power consumption (P), specific energy (SE), instantaneous cutting rate (*ICR*) and utilization factor (*U*) are examined based on the corresponding collapse regions. It is determined that the TBM excavation parameters fluctuate while approaching the collapse regions, and these parameters show an increasing or decreasing trend inside the 'during collapse' region, as seen in Figures 8.19, 8.20 and 8.21 for thrust/penetration, torque/penetration ratios and specific energy values within these zones. These TBM performance parameters are found to be very significant in predicting the collapse zones [6].

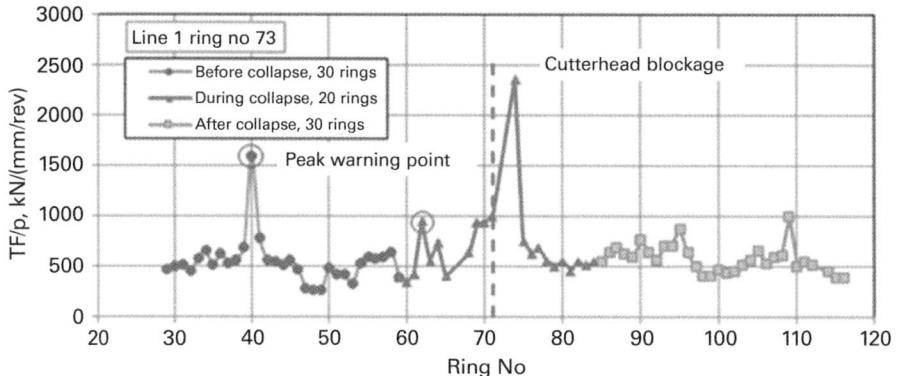

Figure 8.19 The change of TBM thrust force to penetration ratios at 30–115 rings [6].

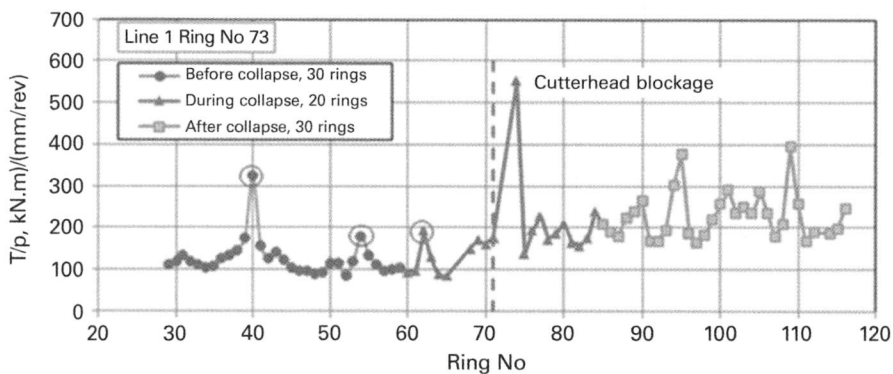

Figure 8.20 The change of TBM torque to penetration ratios at 30–115 rings [6].

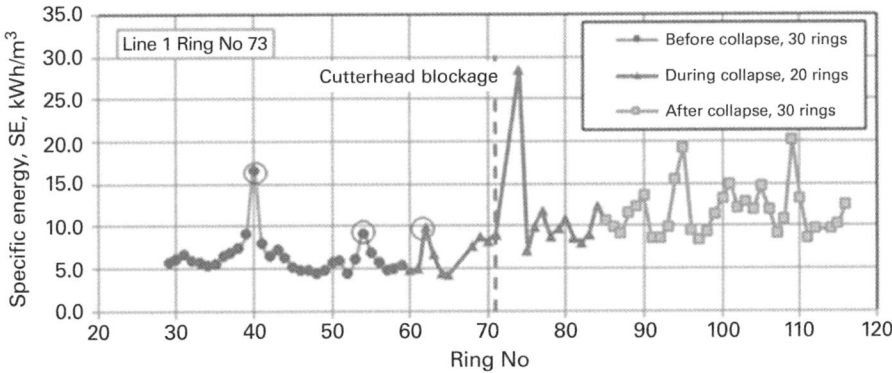

Figure 8.21 The change of specific energy at 30–115 rings [6].

TBM performance parameters related to face collapses and TBM blockage are summarized for 11 different critical zones in Table 8.7. As seen from this table, there is always a critical peak of thrust/penetration, torque penetration ratios and specific energy values 26 rings before TBM blockage. These values may be taking as alerting values for TBM blockage.

Table 8.7 TBM performance parameters related to face collapses and TBM blockage [6].

TBM performance parameters related to face collapses and TBM blockage	Value
Mean thrust/penetration (kN/mm/rev) before the face collapse ±sd	528±146
Peak thrust/penetration (kN/mm/rev) before the face collapse ±sd	1,437±1,203
TBM blockage after peak point, ring number ±sd	26±8
Mean torque/penetration (kNm/mm/rev) before the face collapse ±sd	173±46
Peak torque/penetration (kNm/mm/rev) before the face collapse ±sd	398±117
TBM blockage after peak point, ring number ±sd	26±7
Mean specific energy (kWh/m^3) before the face collapse ± sd	8.4+/2.1
Peak specific energy (kWh/m^3) before the face collapse ± sd	22±8.5
TBM blockage after peak point, ring number ±sd	26±7

8.4   Conclusions

The geology of Istanbul is complex for tunneling projects. Tectonic activities, faults, diabase, dacite and andesite dykes and several joint sets cause many serious problems for tunnel excavation. The experience gained in different projects in Istanbul, especially the Beykoz sewerage project, showed that these geological features cause big problems during TBM excavation resulting in tunnel face collapses, squeezing of the TBM and decreasing daily advance rates.

Eleven different TBM face collapses and blockages which occurred in the very complex geology within the Kadikoy–Kozyatagi metro tunnels were analyzed. The causes and the effects of TBM blockages were explained considering TBM parameters such as opening ratio, working modes and geological parameters. Ninety-three days were spent in rescuing the blocked TBMs. It has been shown that the TBM excavation parameters fluctuate while approaching the collapse regions, and these parameters show an increasing or decreasing trend inside the 'during collapse' region and it is concluded that this trend is a good indicator of potential face collapses, which can serve as a guide to foresee critical areas in front of TBM. There is always a critical peak thrust/penetration, torque penetration ratios and specific energy values at 26 rings before a TBM blockage. These values may be taken as alert values for TBM blockage.

References

[1] Guclucan, Z., Meric, S., Gursoy, C., Algan, M., Bilgin, N., Balci, C., Tumac, D. (2007). *The use of a TBM in difficult ground conditions.* Proceedings of the 2nd Symposium on Underground Excavations for Transportation, Istanbul, Turkey, (in Turkish).

[2] Guclucan, Z., Meric, S., Algan, M., Palakci, Y., Bilgin, N., Bilgin, A.R., Balci, C., Tumac, D. (2008) *The use of a TBM in difficult ground conditions in Beykoz – Kavacik Sewerage Tunnel.* Proceedings of the World Tunnel Congress, Underground Facilities for Better Environment and Safety, Agra, India.

[3] Guclucan, Z., Meric, S., Palakci, Y., Bilgin, N., Balci, C., Copur, H., Namli, M., Bilgin, A.R., Kandemir, E. (2009) *The use of theoretical rock cutting concepts in explaining the cutting performance of a TBM using different cutter types in different rock formations and some recommendations.* Proceedings of World Tunnel Congress, Safe Tunnelling for the City and for the Environment, Budapest, Hungary.

[4] Bilgin, N., Copur, H., Balci, C., Tumac, D., Akgul, M., Yuksel, A. (2008) *The selection of a TBM using full scale laboratory tests and comparison of measured and predicted performance values in Istanbul Kozyatagi-Kadikoy metro tunnels.* In: World Tunnel Congress – Underground Facilities for Better Environment and Safety, eds. Kanjlia, V.K., Ramamurty, T., Wahi, P.P., Gupta, A.C. September 22–24, Agra, India.

[5] Yuksel, A. (2013) Mechanical properties and rock fractures affecting the performance of TBMs. Istanbul Technical University, Dissertation.

[6] Yuksel, A., Arioglu, E., Bilgin, N. (2015) *Mechanism of roof collapses in front of a TBM in a complex geology in Kadikoy-Metro Kozyatagi Metro Tunnels.* International Conference on Tunnel Boring Machines in Difficult Grounds (TBM DiGs), 18–20 November, Singapore.

9   Squeezing grounds and their effects on TBM performance

9.1   Introduction

Squeezing of a TBM or jamming the cutterhead is a nightmare for tunnel engineers, since it affects machine utilization time and realization of the project within the scheduled time. The recovery of the jammed cutterhead often considerably decreases the mean advance rates. Four types of squeezing were encountered in tunnels excavated by TBM in Turkey. The first occurred in weak zones such as graphitic schist with high overburden. A typical example to this is Uluabat tunnel, where the TBM stuck 18 times during the excavation [1, 2], and also the Kosekoy–Bilecik high speed railway tunnel. The second different type of squeezing was due to weak contact zones between the main rock and dykes, as occurred in the Beykoz tunnel [3, 4, 5]. Squeezing of the TBM in the Suruc tunnel occurred due to clayey zones coming in contact with water [6]. However, in some cases like the Kargi tunnel, squeezing of the TBM was the result of either collapsing the blocky ground or squeezing clayey zones [7].

9.2   Basic works carried out on squeezing ground

According to the International Society for Rock Mechanics (ISRM), squeezing is time-dependent large deformation that occurs around the tunnel associated with creep caused by exceeding a limiting shear stress [8].

Aydan et al. [9], taking into account the experience gained in Japan, suggested that squeezing occurs if the ratio of uniaxial compressive strength of intact rock to overburden pressure is less than 2. Hoek and Marinos [10, 11] proposed that the ratio of rock mass uniaxial compressive strength to the in situ stress may be used as an indicator of tunnel squeezing characteristic.

Kovari and Staus [12] found that the low strength and the high deformability of the rock, as well as the presence of pore water pressure, facilitate squeezing and that the following rock types are especially prone to developing large pressure and large deformations: altered gneiss, schist, phyllite, clay, shale, mudstone and tuff, especially if they have a high content of such minerals as micas, chlorite, serpentine and clay.

Panthi and Nilsen [14] combined the semi-analytical method suggested by Hoek and Marinos [10, 11] and the empirical formula proposed by Panthi [13] for Himalayan rock mass conditions. They concluded that squeezing characteristics of the ground may be predicted using some statistical analysis concerning engineering geological properties of rock formations along the tunnel alignment.

Farrokh et al. [15], using convergence-confinement method, introduced a new approach to calculating ground pressure on the TBM shield in the vicinity of the tunnel face in the Ghomroud tunnel in Iran. Based on the results of the analysis, the required TBM thrust was calculated along various geological units. The study concluded that the adopted approach for predicting squeezing characteristic produced realistic results provided that reliable data on rock mass property and ground stress conditions were available.

TBM Excavation in Difficult Ground Conditions. Case Studies from Turkey. First Edition. Nuh Bilgin, Hanifi Copur, Cemal Balci.
© 2016 Ernst & Sohn GmbH & Co. KG. Published 2016 by Ernst & Sohn GmbH & Co. KG.

Farrokh and Rostami [16] emphasized that the existence of weak structures in the rock mass resulted in a high rate of convergence. Results showed that a good correlation existed between TBM operational parameters and tunnel convergence in the Ghomroud tunnel.

Hamidi et al. [17], using FLAC2D software, calculated the maximum radial displacement in determining the squeezing characteristic of the ground in the Nasoud tunnel and suggested some countermeasures to mitigate squeezing problems.

Detailed and comprehensive works carried out by Ramoni and Anagnostou [18, 22] improved our understanding of the effects of squeezing ground characteristics in mechanized tunneling. Specific problems of mechanized tunneling in squeezing ground, the problem of shield jamming and analysis of the effects of countermeasures were first investigated on the basis of the results of numerical computations. They also used the finite element method to develop some dimensionless design nomograms, for gripper, single- and double-shielded TBMs, making a valuable contribution to the decision-making process [19].

Interaction between the shield, ground and tunnel support by means of computational analysis was also investigated by Ramoni and Anagnostou [18, 21]. The suitability of the proposed design nomograms was shown by means of a simplified back analysis of different case studies, including the Uluabat tunnel which is the subject of this chapter. They found good agreement between the results of computational analysis and actual TBM performance [20, 21].

Theoretical considerations, together with numerical investigations carried out by Ramoni and Anagnostou [21], improved the understanding of time dependency of ground control with respect to water-bearing squeezing ground. The governing factor was found to be related to the ratio of gross advance rate to ground permeability.

9.3   Uluabat tunnel

9.3.1   Description of the project

The project area is situated on the southern part of Uluabat–Bursa (Apolyont) Lake. The project consists of the Cinarcik dam at an elevation of 328 m, which is constructed on Orhaneli River. A power tunnel of 11.465 km in length and a pipe conveyance of 1,150 m will deliver the water to the hydroelectric power plant. Tunnel excavation commenced in 2002 from the portal next to Uluabat Lake, using conventional excavation methods with roof bolts, shotcrete, wire mesh and steel arches as the primary tunnel support. Tunneling operations were halted in November 2003 due to extreme roof deformation and floor heave. Figure 9.1 shows a typical deformation curve at one position on the left wall of the tunnel. The deformation increases gradually, and after a certain time it accelerates and then levels off after 1 year, showing a typical creep behavior. However, DSI (the state water authority) and the contractor decided to continue the project with a 5.05 m diameter EPB-TBM. The contractor started the tunnel excavation by TBM from chainage 11+465 km on June 2006, and terminated on March 2010, at chainage 1+792 km, [1, 2].

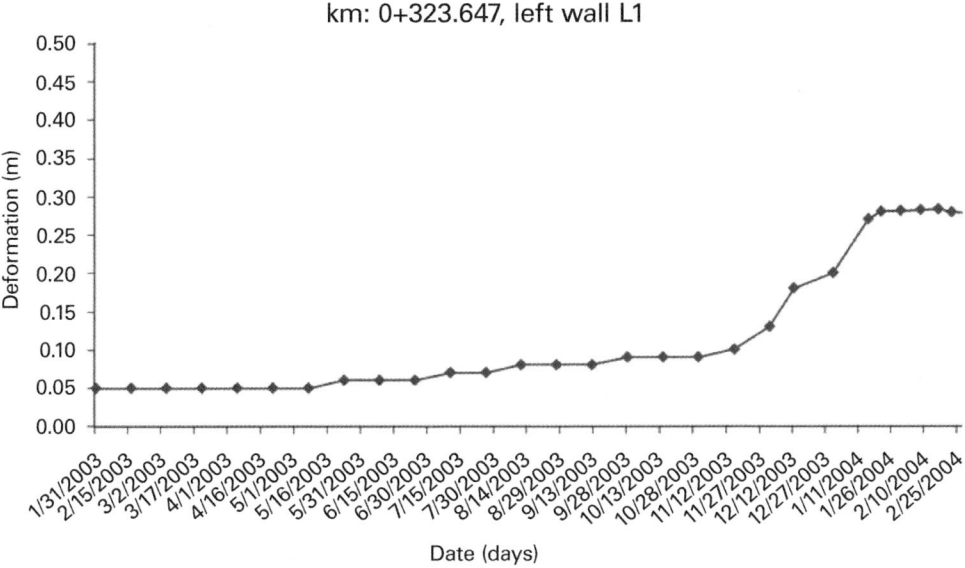

Figure 9.1 Deformation against time on the left wall of the tunnel at km 0+323 [1, 2].

9.3.2   Geology of the project area

The geologic map of the area is given in Figure 9.2, and a general view of geologic formations is seen in Figure 9.3. The tunnel route at chainages 11+465 to 7+750 km and 6+000 to 1+792 km consisted of Karakaya formation of Triassic aged meta-detritic rocks such as fine grained meta-claystone, meta-siltstone, meta-sandstone and graphitic schists. It is important to note that the primary properties of sandstone, claystone and siltstone changed completely due to tectonic deformations causing fragmented meta-detritic formations. Physical and mechanical properties of meta-detritic rocks change between the following values: compressive strength 1–96MPa, elasticity modulus 1.53–39.5 GPa, Poisson ratio 0.1–0.45, density 2.4–2.7. Rock formation may be classified according to RMR as III–V, [1, 2].

Figure 9.2 General view of the ground encountered in Uluabat tunnel [1, 2].

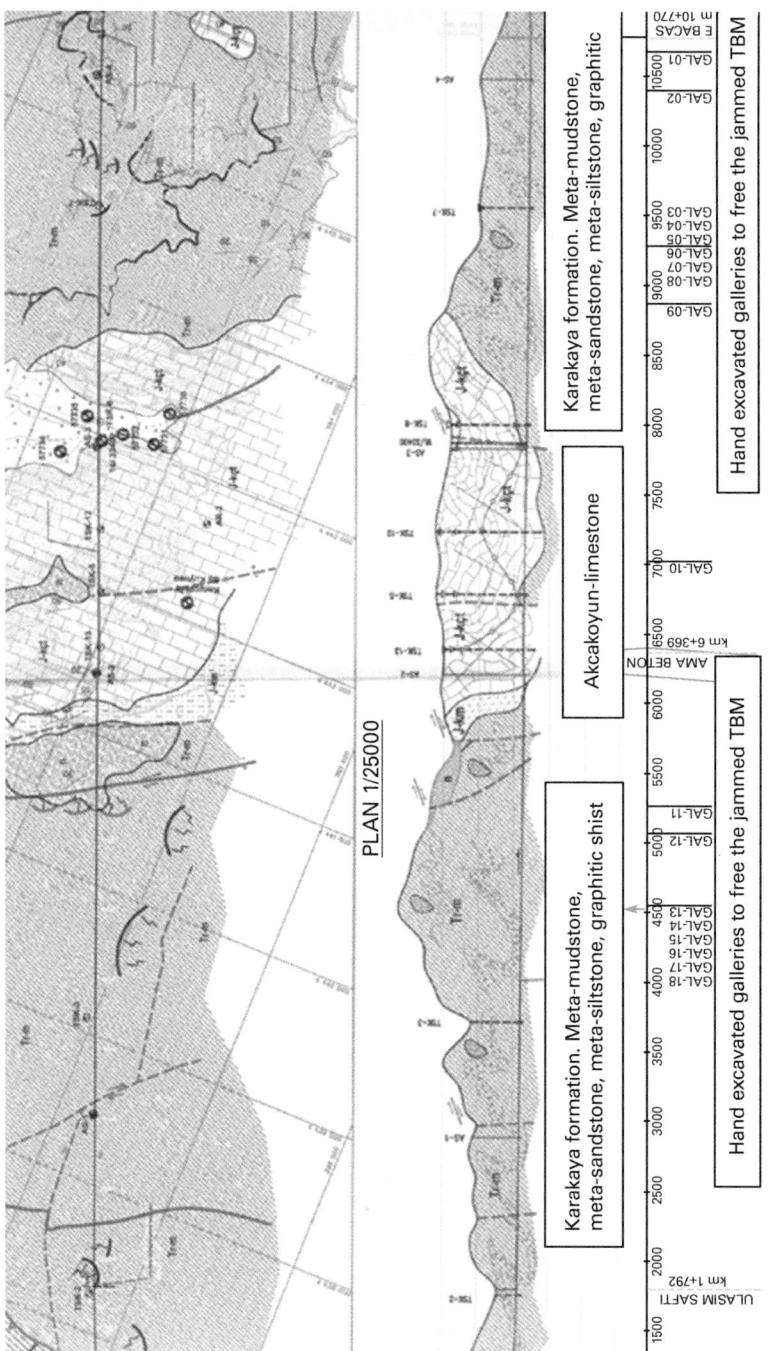

Figure 9.3 The geology of the Uluabat tunnel [1, 2].

9.3.3 Description of the TBM used and the general performance

A single-shielded Herrenknecht EPB-TBM was selected for the project, with specifications as summarized in Table 9.1. An average daily advance rate of 8.6 m was achieved including all stoppages such as TBM standstills and hand mining. The best daily and weekly advance rates were found to be 28.8 m and 198.4 m, respectively. The best monthly advance rate was 583.2 m in February 2007, as shown in Figure 9.4. The breakdown of tunneling activities in February 2007 is illustrated in Figure 9.5. During tunnel excavation, the TBM was stuck a total of 18 times at different tunnel locations, as shown in Figure 9.3. Rescue galleries were driven next to the TBM to free the shield, with a total of 192 days spent on these operations. Another consequence of TBM operation in the squeezing ground was the high rate of disc consumption at 0.034 discs/m^3. This disc consumption is usually 0.001–0.0015 discs/m^3 in non-abrasive, medium strength rock formations, Bilgin et al. [23]. High disc consumption may be explained by the fact that, in squeezing ground conditions, disc cutters are in contact for a longer period of time with excavated material, causing accelerated cutter wear [23]. Another important point is that in weak or soft rock formations, discs fail to turn freely due to lack of friction between the rock and the disc cutter, resulting in uneven disc wear, as seen in Figure 9.6. During tunnel excavation, the TBM became stuck at different places of the tunnel mainly in meta-detritics rock formations due to the highly squeezing characteristics of the Karakaya formation. A general view of the squeezing ground around the TBM shield is shown in Figure 9.7.

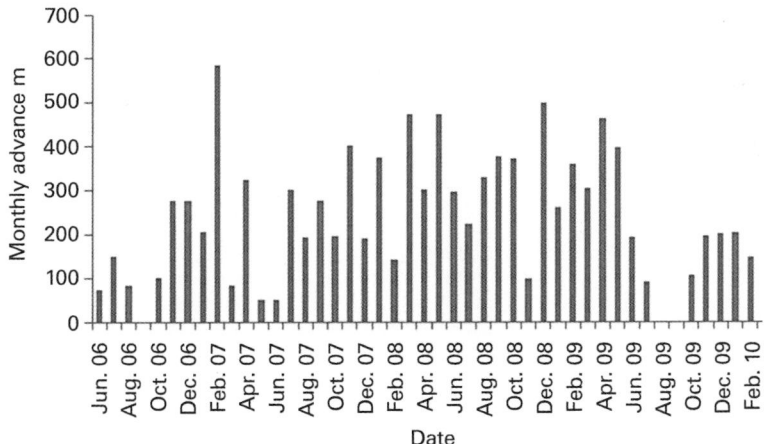

Figure 9.4 Monthly advance rates obtained [24].

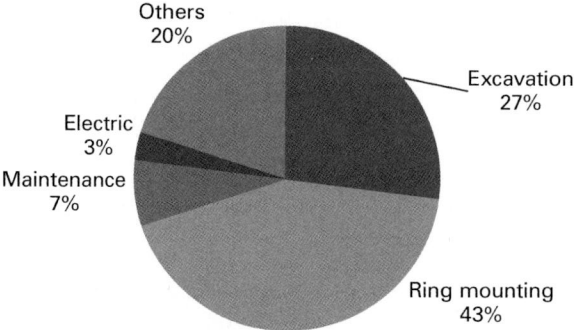

Figure 9.5 Work distribution in February 2007.

Figure 9.6 Uneven disc wear in weak geologic formations in Uluabat tunnel [2].

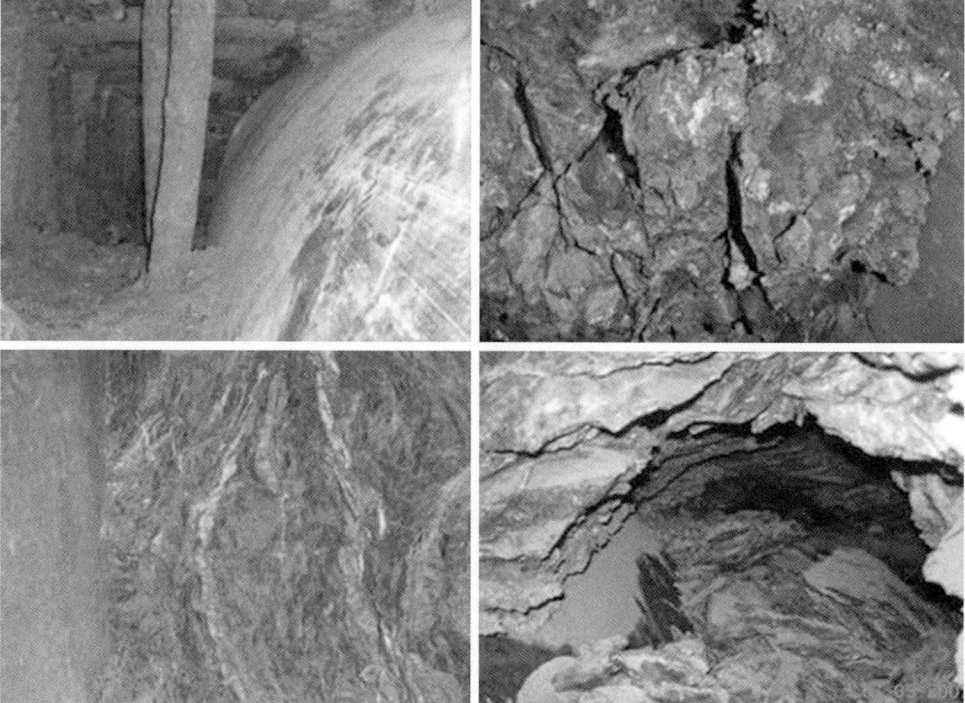

Figure 9.7 General view of the squeezing ground around the TBM shield [1, 2].

Table 9.1 Characteristics of Herrenknecht EPB-TBM

TBM parameter	Value
TBM diameter	5,050 mm
Total power	2,100 kW
Maximum thrust force	29,000 kN
Nominal torque	2,048kNm at 6.25 rpm
Number of discs	34
Disc diameter	17″
Number of scrapers	48
Shield diameter	4,990 mm
Shield weight	69 t
Shield length	12 m
TBM weight	335 t
TBM length	108 m
Length of the band conveyor	65 m
Width of band the conveyor	650 mm
Power of the band conveyor	37 kW
Speed of the band conveyor	0–2.5 m/sec
Capacity of segment erector	1,800 kg

The overall performance of the TBM in 18 critical squeezing zones is summarized in Table 9.2. Thrust forces in squeezing zones 2, 6, 7 and 9 are below the available thrust of 28,000 kN, because there was some weakness in precast segments, and so maximum thrust force was not applied to prevent potential damage to segments. Figure 9.8 is a typical curve showing the increase of the TBM thrust starting at point A, where the squeezing of the ground begins. After an increase of thrust force from 8,276 kN to 20,476 kN (average values within a stroke), the TBM jams at point B after a distance of 24 m (A–A1). Another point of interest in the TBM response is that the TBM torque is also affected, although not as much as the thrust force in the critical zone A–A1, indicating a decrease in thrust force/torque ratio.

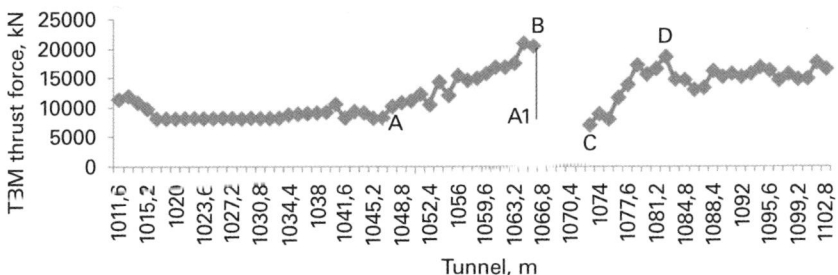

Figure 9.8 The increase of TBM thrust in squeezing zone A–A1, gallery 2 [1, 2].

Table 9.2 The performance of TBM in squeezing zones jamming the shield [2].

Gallery No	Over-burden	Squeezing time for jamming of TBM	Squeezing distance for jamming, A–A1 Figure 9.8	Thrust force where the squeezing is started at point A Figure 9.8	Thrust force at squeezing point B Figure 9.8	Increase in thrust force during the squeezing up to jamming of TBM	Thrust force increasing rate A1–B/A–A1	Squeez-ing index	Torque/thrust before squeezing starts	Torque/thrust just before TBM jamming
—	m	h	m	kN	kN	kN	kN/m	kN/mxh	—	—
1	85	N.A	N.A	N.A	N.A	N.A	N.A	N.A	0.08	0.03
2	125	36.1	24.0	8,276	20,472	12,196	508.2	14.1	0.065	0.02
3	125	32.0	33.6	10,000	26,150	16,150	480.7	15.0	0.065	0.01
4	125	30.2	16.8	9,500	27,000	17,500	1,041.7	34.5	0.05	0.02
5	125	47.9	7.2	15,600	28,600	13,000	1,805.6	37.7	0.04	0.01
6	150	31.0	2.4	20,690	21,670	980	408.3	13.1	0.04	0.008
7	175	24.0	6.0	18,700	21,234	2,535	422.4	17.7	0.05	0.02
8	210	24.1	10.8	18,070	28,600	10,530	975.0	40.5	0.06	0.035
9	225	12.0	13.2	11,700	18,900	7,200	545.5	45.5	0.06	0.03
10	250	12.1	13.2	22,765	28,100	5,335	404.2	33.4	0.05	0.015
11	265	23.8	13.2	14,000	28,000	14,000	1,060.6	44.6	0.05	0.02
12	265	24.0	7.2	14,500	28,500	14,000	1,944.4	81.0	0.05	0.01
13	300	20.1	28.8	5,000	28,000	23,000	798.6	39.7	0.05	0.01
14	300	18.0	7.2	19,950	28,000	8,050	1,118.1	79.9	0.05	0.005
15	265	20.0	9.6	10,700	28,000	17,300	1,802.1	90.1	0.04	0.009
16	265	11.9	12.0	14,500	27,000	12,500	1,041.7	87.5	0.04	0.015
17	250	24.0	26.4	10,000	28,000	18,000	681.8	28.4	0.045	0.008
18	250	24.0	26.4	11,500	28,000	16,500	625.0	26.0	0.04	0.008
Mean	209	24.4	14.3	13,081	24,679	11,599	870.2	42.9	0.05	0.016

As shown in Table 9.2 the average length of squeezing zones (the zone where thrust force starts increasing gradually and ending with squeezing of TBM) is 14.3 m, a little higher than the length of the shield, which is 12 m. This value is in good agreement with calculations given by Ramoni and Anagnostou [18, 22] suggesting that the TBM used at Uluabat would be able to cope with 5–10 m weak zones, and in the case of a weak zone longer than about 10 m, the TBM would be trapped. The mean increase in the thrust force (ΔFT) in critical squeezing zones is 11,599 kN with a mean increase of 870.2 kN/m. However, another important point to note is that squeezing is a large time-dependent deformation. Because of this, the time spent from point A to point A1, or the time between the points where squeezing starts and ends with squeezing of TBM is given in Table 9.3 for each of the 18 jamming locations. The data given in Table 9.2 allows us to calculate a squeezing index for squeezing of the TBM as formulated in Eq. (9.1)

$$SI = \Delta FT / d.t \tag{9.1}$$

where SI is the squeezing index for squeezing of TBM, ΔFT the difference between thrust force where the squeezing of TBM occurred (point B in Figure 9.8) and thrust force where squeezing starts (point A in Figure 9.8), d the distance between where the squeezing starts and ends with jamming (A–A1 in Figure 9.8) and t the time between when the squeezing starts and ends when it is jamming.

The basic idea behind this is to define an index combining squeezing characteristics of the ground with some of the operational parameters of the TBM, to give a quantitative value which will help practicing engineers to take mitigating measures such as injecting bentonite through the shield skin and the ground. This value may also help classify the ground according to susceptibility to squeezing of the TBM. It is clear from Table 9.2 that, for this special case, the squeezing index changes from 13.1 to 90.1. This leads to the conclusion that if careful analysis of tunneling data shows a squeezing index approaching 13.1, a mitigating measure such as annular bentonite injection should be implemented, in order to avoid squeezing of TBM. However, one should be careful about generalizing this index, since absolute values may be different for different tunnel operations using TBMs with different design parameters (single-shield, double-shield, gripper type, different shield length, etc.). Table 9.3 gives rescues galleries with chainage, Q values and a brief description of the zones where the TBM was trapped. An average daily advance rate of 8.6 m was achieved including all stoppages such as TBM standstills and hand mining. The best daily and weekly advance rates were found to be 28.8 m and 198.4 m, respectively. The best monthly advance rate was 583.2 m in February 2007, as shown in Figure 9.4. The breakdown of tunneling activities in general is illustrated in Figure 9.5. Figure 9.9 is the key point in showing that Q values are one of the most important parameters for defining TBM performance in complex geology. Rock mass shows a squeezing characteristic for Q values up to 0.018, where TBM is trapped, and thereafter the TBM machine utilization time increases between Q values of 0.2 and 1.2 and stays constant thereafter [1, 2, 25].

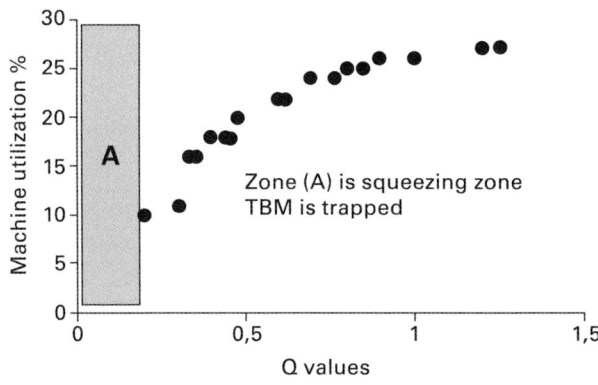

Figure 9.9 The variation of machine utilization with Q values in Uluabat Tunnel [25].

Table 9.3 Squeezing zones and Q values in Uluabat tunnel [25].

Gallery No	Chainage (m)	Q	Lithology
1	785	0.14	Meta-sandstone, meta-siltstone with clay matrix
2	1066	0.16	Meta-sandstone, meta-siltstone with clay matrix
3	2178	0.18	Graphitic shist, meta-siltstone with clay matrix
4	2203	0.014	Wet meta-siltstone, graphitic shist, meta-claystone
5	2218	0.016	Wet meta-siltstone, graphitic shist, meta-claystone
6	2222	0.13	Meta-detritics with clay matrix
7	2245	0.15	Meta-detritics with clay matrix
8	2279	0.14	Meta-detritics with clay matrix
9	2591	0.013	Meta-sandstone, meta-siltstone, graphitic shist
10	4440	0.14	Meta-detritics with clay matrix
11	6205	0.18	Meta-detritics with clay matrix
12	6331	0.13	Wet siltstone with graphitic shist
13	6908	0.2	Meta-detritics with clay matrix
14	6909	0.01	Graphitic shists-meta-mudstone
15	6922	0.014	Graphitic shists-meta-mudstone
16	7078	0.015	Graphitic shists-meta-mudstone
17	7109	0.18	Meta-detritics with clay matrix
18	7142	0.18	Meta-detritics with clay matrix

9.3.4   Effect of TBM waiting time on squeezing

The effect of TBM waiting time on the squeezing is illustrated in Figures 9.10 and 9.11. As seen in Figure 9.10, in very weak rock, the difference between thrust force before stopping and after stopping the TBM increases with time. After 160 min of waiting time, the thrust force increases from 17,500 kN to 28,000 kN, with a thrust force difference of 10,500 kN. Bearing in mind that the maximum available TBM thrust force is 29,000 kN, the TBM shield gets stuck at this point. Figure 9.11 illustrates the effect of waiting time on the increase of TBM thrust force in more competent rock formations between 6,779 and 6,827 m (16.09.2008–21.09.2008). As can be seen from this figure, the effect of waiting time on the increase of thrust force is not as significant as before. Up to 100 min, the difference between thrust force before stopping and after stopping is only 3,600 kN, thereafter it becomes negligible

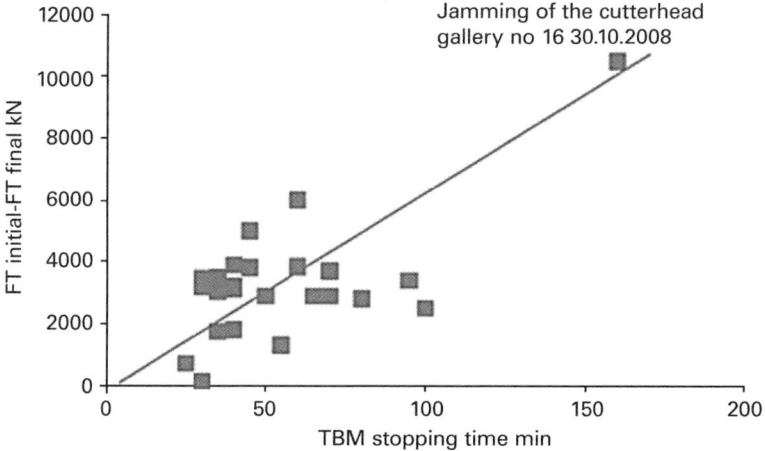

Figure 9.10 The effect of stopping time on the increase of TBM thrust force in very weak formation at 7056–7093.2m (29.10.2008–11.11.2008) [1, 2].

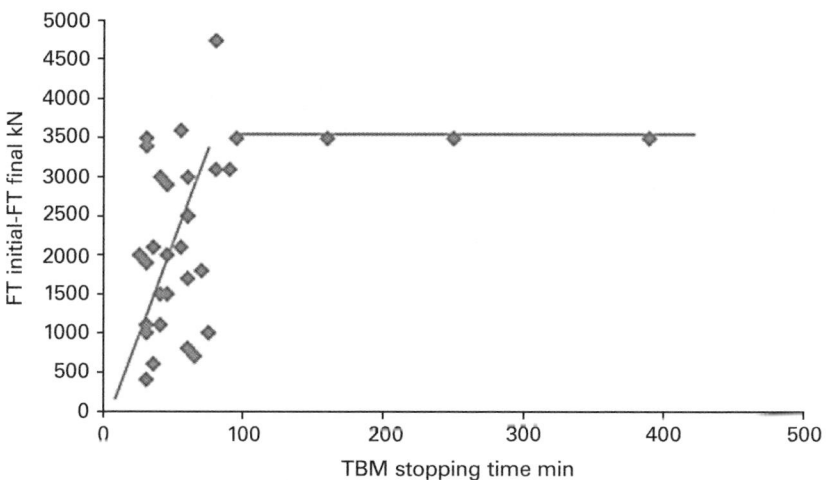

Figure 9.11 The effect of stopping time on the increase of TBM thrust force in more competent rock formation at 6779–6827m (16.09.2008–21.09.2008); jamming of TBM did not occur within these intervals of time [1, 2].

9.3.5   Effect of bentonite application on TBM squeezing

The geology at various chainages presented squeezing or settlement, at extreme rates up to 60 mm per hour. The TBM shield length of 12 m, in the prevailing geology, proved to be a critical factor, since the ground moved so quickly, thereby squeezing the rear edge of the shield body and reducing the advance rates, resulting in a chain of events that eventually stopped the TBM. These values agree well with the results of calculations given by Ramoni and Anagnostou [19, 20]. For a first initiation, the fixed overcut was increased from 35 mm eventually to 95 mm on the radius. However, this was not found to be sufficient in some areas, and bentonite injection was utilized at critical points, and systematically between chainages 7,143 and 7,969.2 m. This was accomplished by pumping a bentonite mixture between the shield skin and the ground, with the aim of reducing the frictional forces. The bentonite mixture was delivered to the shield in a mixing car, to keep it fluid. Injection through 2″ diameter ports, six in total, was done using an electrically operated screw pump. Delivery pressure depended on the proximity of the ground at the injection points, and varied between 0.2 bar and 2.5 bar when ground squeezing had reduced the clearance gap (overcut). Two to three cubic meters of the thin bentonite solution was used for each advance. Steel plates of 20 mm in thickness were welded to the TBM skin surface around injection ports to facilitate the injection. The aim was to produce a groove to permit bentonite injection, because excessive ground squeezing sealed the injection ports so strongly that the bentonite could not be pumped even at pressures in excess of 10 bar. The exception to this statement occurred when the TBM was stopped for a long time, and the ground squeezed down even around the raised plates, thereby sealing the outlet. Figure 9.12 illustrates the change of TBM thrust force within 7,143–7,969.2 m after injection of bentonite through the shield in the ground that had squeezing character-istics. The jamming of the shield was not encountered between these chainages due to the reduction between the shield and the rock which clearly shows the efficiency of the bentonite injections. The geology was the same as in the zones where squeezing of the TBM was encountered. The mean thrust force in this area was 9671 kN with a maximum value 27,463 kN which is close to the maximum available thrust force (Caner, 2010). It is interesting to note that in Figure 9.12, the thrust force values of the TBM at 7,424.4–7,960.8 m of the tunnel are almost 50% less than the thrust force values at 7,143.6–7,424.4 m. This is due to increasing the bentonite injection pressure up to 2.5 bar and welding of steel plates of 20 mm to the TBM skin surface around the injection ports to facilitate the injection.

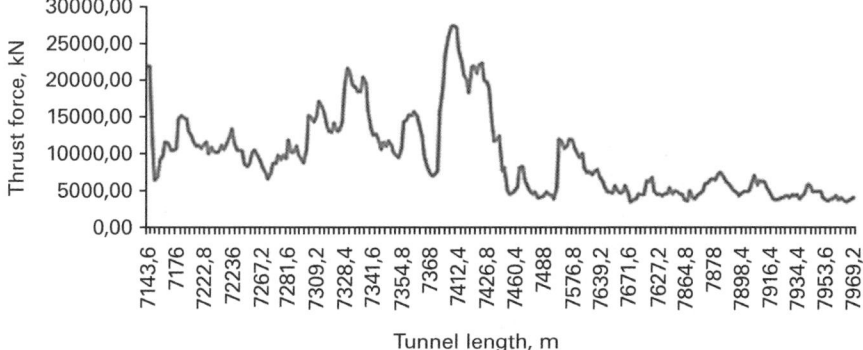

Figure 9.12 The change of TBM thrust force within 7143–7969.2 m after the injection of bentonite through the shield [2].

9.3.6 Conclusions

The Uluabat tunnel case study and the detailed TBM performance analysis clearly show how the squeezing characteristic of the ground can impact the progress rate of mechanized tunneling. A total of 18 rescue galleries were constructed to free the trapped shield of a 5.05 m diameter TBM, and 192 days were spent on TBM rescue operations. Detailed study of the TBM performance data showed that overburden, RMR values, the increase of machine thrust for a given tunnel length and time and the variation in the torque/thrust ratio can be used as a reliable basis for alerting the practicing engineers to implement some mitigating measures, such as using bentonite injection around the TBM shield. The study also shows that waiting time has a major effect on TBM jamming when excavating in weak rocks. TBM waiting time should be minimized with good management practices in critical zones. The Uluabat field results also shows very close agreement with computational results given by Ramoni and Anagnostou [18, 22].

9.4 Kargi Tunnel

9.4.1 Squeezing of the cutterhead and related problems

The main problem was the jamming of the cutterhead due to the squeezing characteristic of the ground. As seen from Figure 9.13, from chainage 11+850 to 9+650 km, 281 days were spent for stoppages, with 68% of the time spent on opening rescue galleries and salvage operations [7].

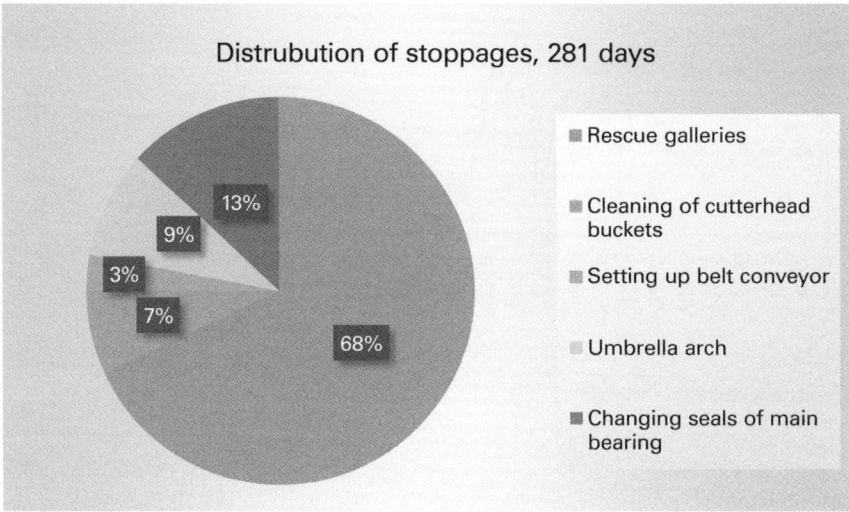

Figure 9.13 Distribution of TBM stoppages in Kargi tunnel [7].

One of the most important factors in TBM squeezing is the existence of swelling minerals like smectite (montmorillonite, vermiculite etc.), anhydrite, and some pyrrhonists in calcareous shale. The type of cation present affects the degree of swelling to a large extent. Na+, for example, will cause a high degree of swelling, while Ca+ will cause a lower one. However, this is a rare case. Under wet tunneling conditions it is easier to detect swelling material. Here, the swelling clay squeezes the gouge material some millimeters out. This feature is often best seen near the lower part of the tunnel wall where the clay can more easily absorb water from the invert [7].

During tunnel advancement in the Kargi tunnel, rock samples were collected in 12 different tunnel excavation areas for detailed geotechnical, mineralogical and petrographical analysis, and the majority of the analysis showed that smectite existed in the rock samples. It is apparent that one of the main reasons for the TBM jamming in the Kargi tunnel was the existence of swelling clay minerals in the tunnel route.

The second reason of TBM squeezing was the weak rock and high overburden conditions. To assess the squeezing potential in rock masses, there are three similar methods including empirical, semi-empirical and theoretical–analytical models. Empirical methods are mainly based on two parameters including rock mass classification systems and tunnel depth [1, 2].

Some of the most frequently referenced investigations commonly used have been carried out by Goel et al. [26]. The semi-empirical methods are focused on squeezing potential by using expected deformation of the rock mass surrounding the tunnel in a hydrostatic stress field. The common starting point of all these methods for quantifying the squeezing potential of rock is the use of the 'competency factor', which is defined as the ratio of uniaxial compressive strength of intact rock mass to overburden/in-situ stress. This competency factor was proposed by Jethwa et al. [27], Aydan et al. [8] and Hoek and Marinos [10, 11]. The theoretical–analytical approaches involve closed form solutions and numerical methods. Determination of the stress–strain behavior of rock mass and its time dependent behavior is the basis for theoretical models.

Based on experiences with tunnels in Japan, Aydan et al. [8] proposed to relate the uniaxial compressive strength of the rock σ_c to the overburden pressure $\gamma*H$, given in equation 9.2.

$$Nc = \sigma_c/\gamma \times H \qquad (9.2)$$

where σ_c is rock uniaxial compressive strength; γ is rock mass unit weight and H is tunnel depth below surface. According to Aydan et al. [8], squeezing conditions will occur if the ratio NC = $\sigma_c / \gamma*H$ is less than 2, where σ_c is rock uniaxial compressive strength; γ is rock mass unit weight and H is tunnel depth below surface. According to Aydan et al. [8], squeezing conditions will occur if the ratio NC / $\gamma*H$ is less than 2.0.

Table 9.4 is a summary of the ground characteristics where jamming of the TBM occurred in the Kargi tunnel. In most of the locations, for example cases 3, 4 and 5 where the jamming of the TBM occurred, this ratio is less than 2, suggesting that TBM jamming is a result of the swelling clay and weak character of the ground.

Table 9.4 Characteristics of seven areas where jamming of the TBM occurred [7].

Bypass	UCS MPa	γ g/cm^3	N_c	Ground characteristic
1	20	2.39	16.7	Fault zone
2	15	2.09	4.8	Swelling clay, ground is wet
3	5	1.99	1.3	Swelling clay, very weak rock
4	5–23	2.30	1–1.5	Swelling clay, very weak rock
5		2.30	1–1.5	Fault, swelling clay, weal rock
6	5–23	2.30	4.6	Fault, swelling clay, ground is wet
7	5–23 5.23	2.30	2.2–10	Weathered dyke

The ratio of thrust force to torque and time dependency of these parameters are good indicators for estimating TBM jamming, as explained by Algan, et al. [1, 2]. It is also important to follow theoretical concepts given by Ramoni and Anagnostou [18, 21] to understand the mechanism of TBM jamming.

If this ratio given in Equation (9.1) is less than 2, this points out that torque values are extremely high, due to the swelling characteristics of the ground. If this ratio is higher than 2, this points out that there are excessive face collapses in front of TBM, necessitating higher thrust values. In most cases, TBM jamming occurs due to swelling of the ground and excessive face collapses, as indicated in Table 9.4.

The variation of thrust and torque values in seven different cases are seen in Figures 9.14–9.20 [7].

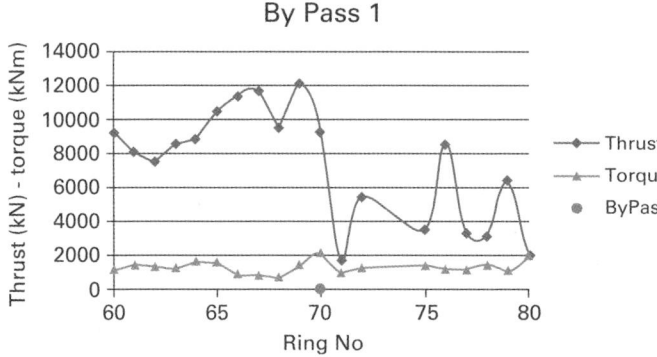

Figure 9.14 The variation of thrust and torque values in the jamming area no (1).

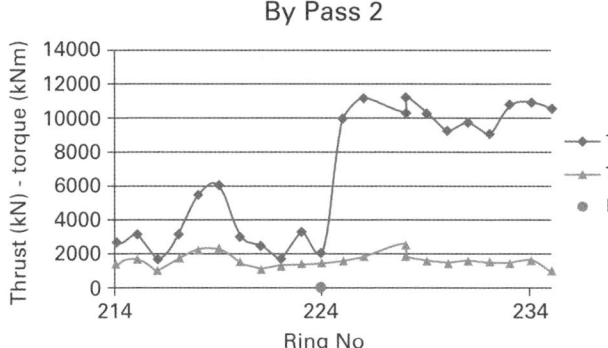

Figure 9.15 The variation of thrust and torque values in the jamming area no (2).

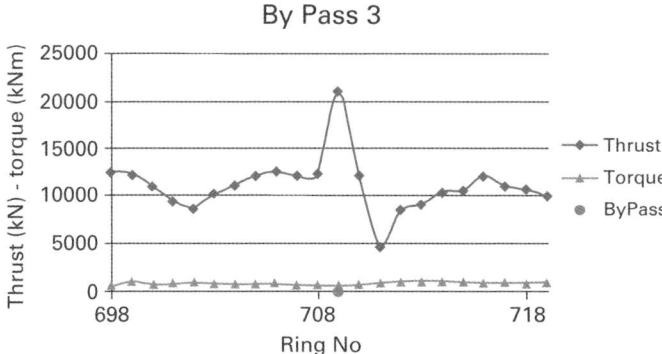

Figure 9.16 The variation of thrust and torque values in the jamming area no (3).

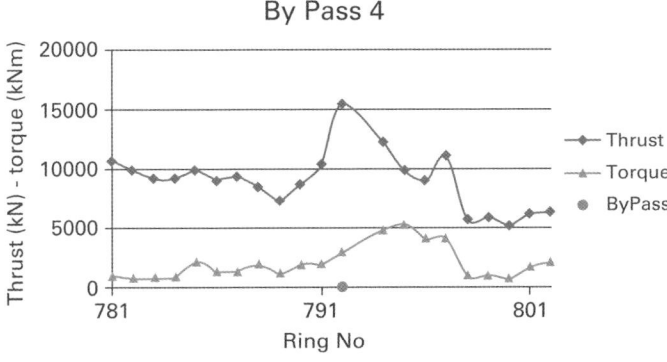

Figure 9.17 The variation of thrust and torque values in the jamming area no (4).

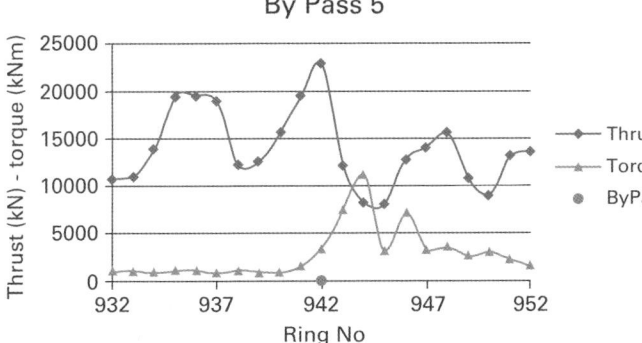

Figure 9.18 The variation of thrust and torque values in the jamming area no (5).

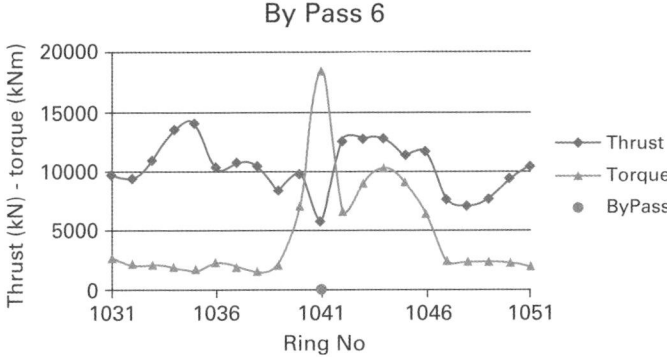

Figure 9.19 The variation of thrust and torque values in the jamming area no (6).

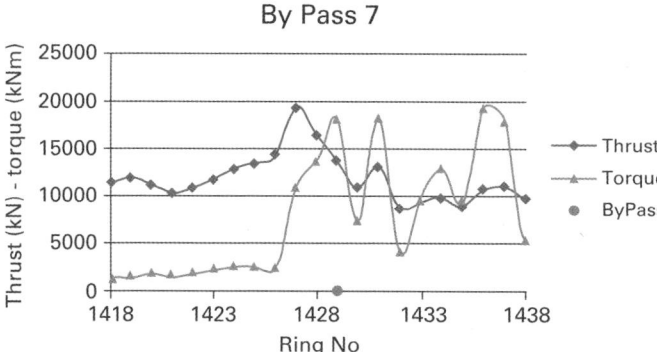

Figure 9.20 The variation of thrust and torque values in the jamming area no (7).

From Table 9.4 and Figures 9.14–9.20 it is apparent that the causes of TBM jamming differ from case to case. Competency factor, Nc, as defined by Aydan at al. [9], plays an important role in determining the squeezing zones, but the existence of swelling clay and the humidity or water also play an important role in determining squeezing zones, as pointed out in Table 9.4 for cases (2) and (6), where severe jamming problem occurred even when the competency factor was high [7].

9.4.2   Effect of Q values on squeezing of TBM

The chainage of the galleries opened to rescue the trapped TBM and associated faults are given in Table 9.5. This table clearly shows that TBM cutterhead blockage is associated to fault zones and Q values, as noticed in the Uluabat formation. However, the direct effect of Q values on machine utilization was not apparent in the previous example. Apart from the geological factors, the main factors improving TBM utilization time and daily advance rates were: machine modifications such as increasing cutterhead drive torque capability by more than 50% and increasing the thrust by hydraulic power pack cylinder jacks, providing overbore capabilities, shield lubrication, probe drilling of 4 km around and umbrella arches at nine points. In most cases if the geological parameters are not in favor of using a TBM, the performance of conventional excavation with D&B becomes competitive with mechanical excavation [25].

Table 9.5 The chainage of the galleries, Q values and associated faults [25].

Gallery No	Chainage km	Q	Affected by faults at chainages
1	11+756	0.016	Major East Anatolian Fault at 11+874
2	10+796	0.02	10+900
3	10+671	0.18	?
4	10+478	0.014	10+478
5	10+445	0.18	10+478
6	10+295	0.012	10+295
7	9+712	0.016	9+712

9.4.3 Discussions and conclusions on TBM swelling in Kargi project

The main objective of this section was first to investigate the causes and effects of the TBM jamming, resulting in 192 days spent opening rescue galleries to free the TBM and 25 days carrying out umbrella arch operations to ease TBM advancement. The TBM was found to be affected by the existence of smectite minerals (swelling clays) in a fault or shear zones causing squeezing of the machine. The second important point in TBM jamming is that the ground competency factor as defined by Aydan et al. [8] as the ratio of intact rock compressive strength over the product of rock density by overburden. This ratio being less than 2 in most cases (bypass 3, 4, 5 and 7) points to the fact that the ground has squeezing characteristic due to the weak characteristic of the rock formation, where in these areas, faults do not exist. In most cases, the swelling characteristic of the ground was combined with the weak characteristic of the ground, exacerbating the situation.

The following measures were taken in order to solve the problems associated with the squeezing characteristic of the ground: the volume of the excavated material for each ring was kept close to the theoretical excavated volume, continuous probe drilling was applied, minimum at two different points due to the heterogeneous characteristics of the ground, umbrella arches were carried out, especially in blocky ground, and in such cases different ground improvement techniques with different chemicals such as resins were used as well as a skin lubrication system. After these remedial measures TBM stoppages were kept to a minimum. Due to a careful side observation, analyzing the TBM data and remediation program, daily advance rates of TBM increased considerably, reaching a mean daily advance rate of 20 m in September and October 2013 [7].

References

[1] Algan, M., Palakci, Y. and Bilgin, N. (2011) Performance analysis of a TBM in Uluabat power tunnel within a squeezing ground. World Tunnel Congress, Helsinki.

[2] Bilgin N., Algan M. (2012) The performance of a TBM in a squeezing ground at Uluabat, Turkey. *Tunn Undergr Sp Technol*, **32**, 58–65.

[3] Guclucan, Z., Meric, S., Gursoy, C., Algan, M., Bilgin, N., Balci, C., Tumac, D. (2007) The use of a TBM in difficult ground conditions. Proceedings of the 2nd Symposium on Underground Excavations for Transportation, Istanbul, Turkey. (in Turkish).

[4] Guclucan, Z., Meric, S., Algan, M., Palakci, Y., Bilgin, N., Bilgin, A.R., Balci, C., Tumac, D. (2008) The use of a TBM in difficult ground conditions in Beykoz–Kavacik Sewerage Tunnel. Proceedings of the World Tunnel Congress, Underground Facilities for Better Environment and Safety, Agra, India.

[5] Guclucan, Z., Meric, S., Palakci, Y., Bilgin, N., Balci, C., Copur, H., Namli, M., Bilgin, A.R., Kandemir, E. (2009) The use of theoretical rock cutting concepts in explaining the cutting performance of a TBM using different cutter types in different rock formations and some recommendations. Proceedings of World Tunnel Congress, Safe Tunnelling for the City and for the Environment, Budapest, Hungary.

[6] Ilci, N., Temel, M., Sezgin, S., Akpinar, T., Guarasio, S., Polat, C., Bilgin N. (2013) Clogging and squeezing effect of marl-clayey limestone on the performance of a hard rock TBM in Suruc Tunnel, Turkey. World Tunnel Congress, Geneva, Switzerland.

[7] Yurt, M., Ozturk, A., Arslan, A., Nuhoglu, C., Oystein, L., Erdogan, E., Atlar, B., Palakci, Y., & Bilgin, N. (2014) Factors affecting the performance of a double shield TBM in a very complex geology in Kargi Turkey. Proceedings of the World Tunnel Congress 2014 – Tunnels for a better Life. Foz do Iguaçu, Brazil.

[8] Barla, G. (2001) Tunnelling Under Squeezing Rock Conditions. Euro summer-School in Tunnel Mechanics, Innsbruck. Logos Verlag, Berlin, Germany.

[9] Aydan, O., Akagit, T., Kawamoto, T. (1993) The squeezing potential of rocks around tunnels. *Rock Mechanics and Rock Engineering*, **2**, 137–163.

[10] Hoek, E., Marinos, P. (2000a) Predicting tunnel squeezing problems in weak heterogeneous rock masses. *Tunnels and Tunnelling International Part 1* (November), 45–51.

[11] Hoek, E., Marinos, P. (2000b) Predicting tunnel squeezing problems in weak heterogeneous rock masses. *Tunnels and Tunnelling International Part 2* (December), 33–36.

[12] Kovari, K., Staus, J. (1996) Basic considerations on tunnelling in squeezing ground. *Rock Mechanics and Rock Engineering* **29** (4), 203–210.

[13] Panthi, K.K. (2006) Engineering geological uncertainties related to tunnelling in Himalayan rock mass conditions. Norwegian University of Science and Technology (NTNU), Dissertation.

[14] Panthi, K.K., Nilsen, B. (2007) Uncertainty analysis of tunnel squeezing for two tunnel cases from Nepal Himalaya. *Int J Rock Mech Min Sci*, **44**, 67–76.

[15] Farrokh, E., Mortavazi, A., Shamsi, G. (2006) Evaluation of ground convergence and squeezing potential in the TBM driven Ghomroud tunnel project. *Tunn Undergr Sp Technol*, **21**, 504–510.

[16] Farrokh, E., Rostami, J. (2009) Effect of adverse geological conditions on TBM operation in Ghomroud tunnel conveyance project. *Tunn Undergr Sp Technol*, **24**, 436–446.

[17] Hamidi, J.K., Bejari, H., Shahriar, K., Rezai, B. (2008) Assessment of ground squeezing and ground pressure imposed on TBM shield. In: Proceedings of the 12th International Conference of International Association for Computer Methods and Advances in Geomechanics, Gao, India.

[18] Ramoni, M., Anagnostou, G. (2006) On the feasibility of TBM drive in squeezing ground. ITA World Tunnel Congress, Safety in Underground Space, 22–27 April, COEX, Seoul, South Korea.

[19] Ramoni, M., Anagnostou, G. (2010a) Tunnel boring machines under squeezing conditions. *Tunn Undergr Sp Technol*, **25**, 139–157.

[20] Ramoni, M., Anagnostou, G. (2010b) Thrust force requirements in squeezing ground. *Tunn Undergr Sp Technol*, **25**, 433–456.

[21] Ramoni, M., Anagnostou, G. (2011a) The interaction between shield ground and tunnel support in TBM tunnelling through squeezing ground. *Rock Mechanics and Rock Engineering*, **44**, 37–61.

[22] Ramoni, M., Anagnostou, G. (2011b) The effect of consolidation on TBM shield loading in water-bearing squeezing ground. *Rock Mechanics and Rock Engineering*, **44**, 63–83.

[23] Bilgin, N., Copur, H., Balci, C. (2012) Effect of replacing disc cutters with chisel tools on performance of a TBM in difficult ground conditions. *Tunn Undergr Sp Technol*, **27**, 41–45.

[24] Caner, E. (2010) Performance Analysis of Full Face Tunnel boring machine in squeezing ground. Uluabat power tunnel example. Istanbul Technical University, Dissertation.

[25] Barton., N., Bilgin, N. (2016) Fast or slow progress with TBM in ideal or faulted conditions. EUROCK 2016 Cappadocia, Turkey.

[26] Goel, R.K., Jethwa, J.L. and Paithakan, A.G. (1995) Tunnelling through the young Himalayas – a case history of the Maneri–Uttarkashi power tunnel. Eng. Geol. 39 (1–2), 31–44.

[27] Jethwa, J.L. and Singh, B (1984) Estimation of ultimate rock pressure for tunnel linings under squeezing rock conditions – a new approach. In: Brown, E.T., Hudson, J.A. (Eds.), Design and performance of underground excavations, ISRM Symposium, Cambridge, London.

Recommendations in Geotechnical Engineering

Ed.: Deutsche Gesellschaft
für Geotechnik e.V.
**Recommendations on
Excavations**
3. Edition 2013. 324 pages.
€ 79,–*
ISBN 978-3-433-03036-3
Also available as **ebook**

For the new 3rd edition, all the recommendations have been completely revised and brought into line with the new generation of codes (EC 7 and DIN 1054), which will become valid soon. The book thus supersedes the 2nd edition from 2008.

Ed.: Deutsche Gesellschaft
für Geotechnik e.V.
**Recommendations on
Piling (EA Pfähle)**
2013. 496 pages.
€ 119,–*
ISBN 978-3-433-03018-9
Also available as **ebook**

This handbook provides a complete overview of pile systems and their application and production. It shows their analysis based on the new safety concept providing numerous examples for single piles, pile grids and groups. These recommendations are considered rules of engineering.

Ed.: Deutsche Gesellschaft
für Geotechnik e.V.
**Recommendations for
Design and Analysis of
Earth Structures using
Geosynthetic Reinforce-
ments – EBGEO**
2011. 316 pages.
€ 99,–*
ISBN 978-3-433-02983-1
Also available as **ebook**

The Recommendations deal with analysis principles and the applications of geosynthetics used for reinforcement purposes in a range of foundation systems, ground improvement measures, highways engineering projects, in slopes and retaining structures, and in landfill engineering.

Ed.: HTG
**Recommendations of the
Committee for Water-
front Structures Har-
bours and Waterways
EAU 2012**
2015. 676 pages.
€ 129,–*
ISBN 978-3-433-03110-0
Also available as **ebook**

The "EAU 2012" takes into account the new generation of the Eurocodes. The recommendations apply to the planning, design, specification, tender procedure, construction and monitoring, as well as the handover of and cost accounting for port and waterway systems.

DGGT
Deutsche Gesellschaft
für Geotechnik e. V.
German Geotechnical Society

**Order online:
www.ernst-und-sohn.de**

Ernst & Sohn
Verlag für Architektur und technische
Wissenschaften GmbH & Co. KG

Customer Service: Wiley-VCH
Boschstraße 12
D-69469 Weinheim

Tel. +49 (0)6201 606-400
Fax +49 (0)6201 606-184
service@wiley-vch.de

* € Prices are valid in Germany, exclusively, and subject to alterations. Prices incl. VAT. excl. shipping. 1036336_dp

10 Clogging of the TBM cutterhead

10.1 Introduction

Tunneling through clay soils or rocks with a TBM can result in serious clogging of the cutters, which in turn can lead to delays in job termination time. The clogging of a TBM, as is encountered in clay-containing ground, has extensive consequences for the construction process and can severely affect performance of the machine, by increasing torque, thrust and specific energy and lowering advance rates owing to extra cleaning efforts. So it is essential to understand the reasons for and the effects of clogging including mitigation programs to combat the negative effects of the clogging. This chapter is written with the intention of clarifying the subject by giving three examples from tunneling projects in Turkey.

10.2 What is clogging of a TBM cutterhead and what are the clogging materials?

The clogging potential of a clay formation (clay soils or clay rich rocks) was defined as the interaction of four single effect mechanisms by Thewes and Burger [1] as: adhesion of clay particles on a component surface, bridging of clay particles over openings in the path of the spoil transport, cohesion of clay particles, sticking to each other and the low tendency of a clay towards dissolving in water. In fine-grained soil or rock, and particularly in combination with water inflow, the excavated material often sticks to the cutting tools or the conveying system, which can cause major difficulties in the construction process, which can in turn cause difficulties in its excavation, transport and reuse or dumping. Figure 10.1 shows a typical example of clogging of the disc cutters in the Mahmutbey–Mecidiyekoy metro tunnels in Istanbul.

Figure 10.1 Typical example of clogging of discs in the Mahmutbey–Mecidiyekoy metro tunnel in Istanbul.

10.3 Testing clogging effects of the ground

The clogging potential of a soil depends on the clay content and the type of clay minerals present, as was shown by Thewes and Burger [1]. Cohesive soils therefore have to be evaluated according to different properties. The clay content, which is the quantity of the clay grain fraction (< 0.002 mm), is not necessarily identical to the content of clay minerals. Increased adhesion, and thereby clogging, is to be expected above a content of more than 10% of swelling clay minerals. The plastic properties of cohesive soils are defined by their natural water content, the liquid limit and the plastic limit. At the plastic limit, the soil changes from very stiff to stiff consistency, resulting in plastic behavior. Then, at the liquid limit, it changes to liquid consistency, which has no further cohesive bonding. The tendency of a soil to clog can essentially be determined by evaluating the combination of its plasticity index and its consistency index. In order to evaluate ground for slurry shield tunneling, Hollmann and Thewes [2, 3] developed a diagram for the assessment of the clogging potential of cohesive soils, Figure 10.2, as a universal classification for the evaluation of possible critical conversion of soils. As a conclusion, it should be stated that to evaluate the clogging potential of cohesive soil during EPB tunneling it is necessary to analyze in-situ soil moisture content in relation to Atterberg Limits, including the adhesion limit and to analyze the relationship between soil consistency and soil plasticity.

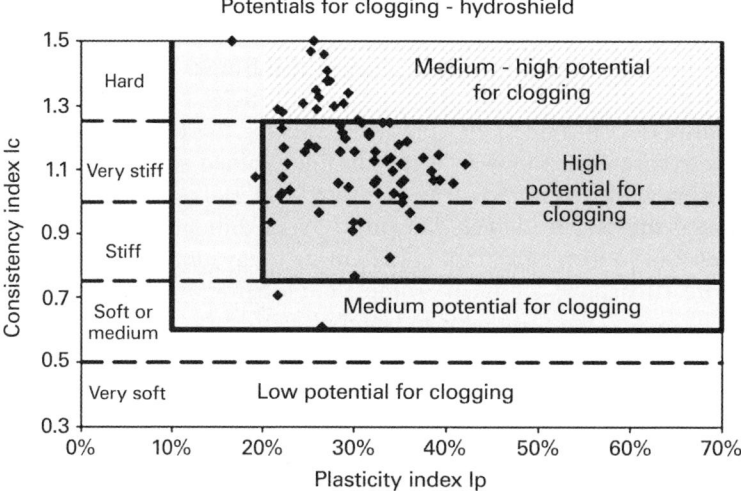

Figure 10.2 Clogging potential for slurry-supported shield drives, modified classification diagram according to Thewes with value pairs from face samples from a hydroshield drive with massive clogging problems, Hollman and Thewes [2, 3].

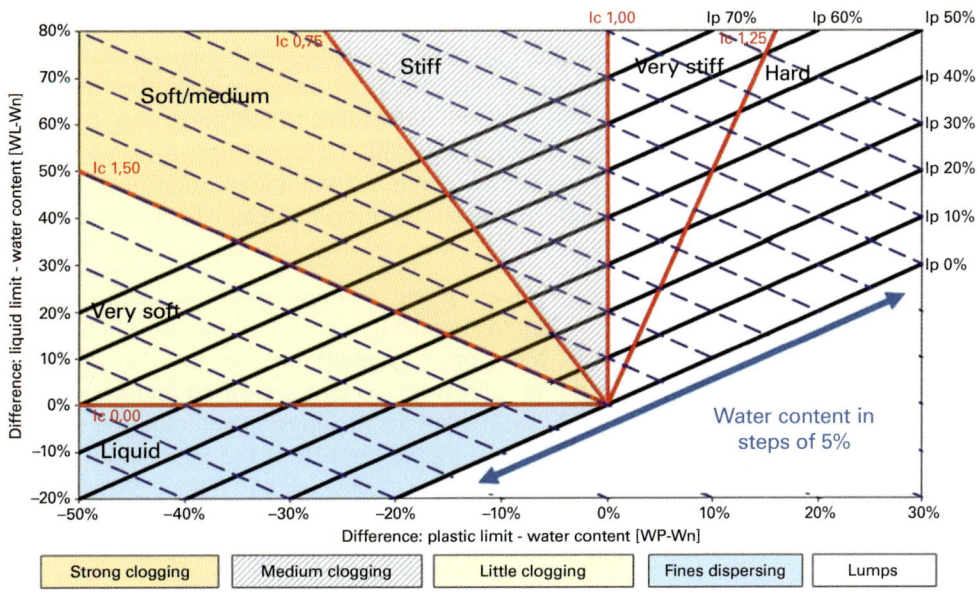

Figure 10.3 Universal classification diagram according to Hollmann and Thewes for the evaluation of possible critical conversion of soils [2, 3].

10.4   Mitigation programs to eliminate clogging

According to Langmaack and Ibarra [4] using anti-clay additives is one of the solutions for reducing adhesion and transport problems and increasing the TBM speed, as suggested in Figure 10.4.

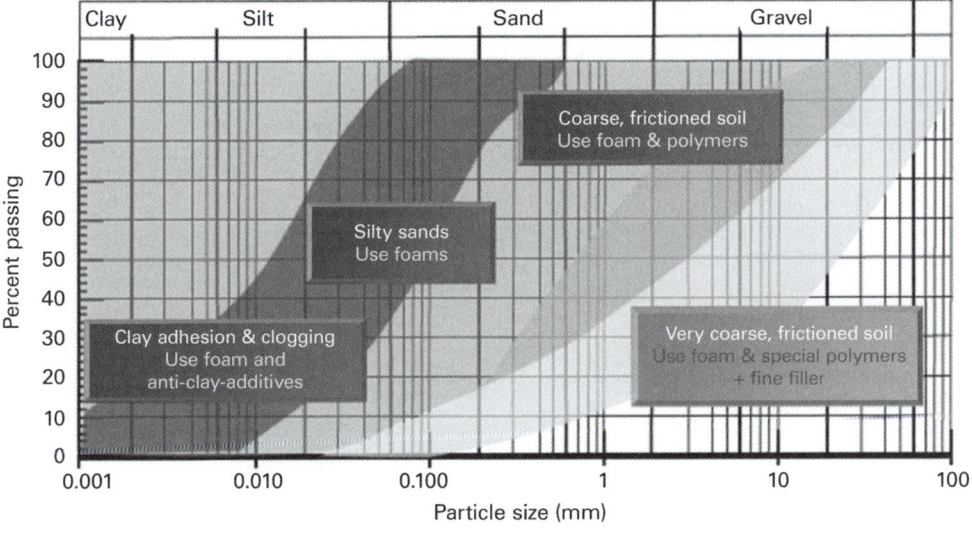

Figure 10.4 Anti-clay additive limits to stop clogging of the muck, after Langmaack and Ibarra [4].

Thewes [5] suggest that the principal aims in the design of shield machines, to prevent clogging, are to generate large soil chips to reduce the adhesion-prone surface of the excavated clay in relation to its volume, to avoid narrow passages and other obstructions for the transport of the clay chip from the tunnel face to the slurry line, to avoid clay accumulation and to minimize the time spent by the clay chips in the chamber, by increasing the ratio of the suspension flow rate to the volume of excavated soil and to avoid clay agglomerations through increased agitation in areas which are prone to material settlement.

Ball et al. [6] concluded that for soils that are prone to clogging (1) the adhesion limit test does not have enough empirical data to support its use in estimating soil clogging in an EPB-TBM, (2) the addition of water prior to addition of an anti-clay agent or foam substantially increased efficacy of additives, (3) addition of an anti-clay agent and a foam additive resulted in significant reductions in adhesion and required torque, while creating a uniform consistency, (4) the higher the FIR of the foam additive, the more likely it is that the pressure change at the screw conveyor will induce effervescing of the material and could promote soil bypassing through a stationary conveyor which may be undesirable during mining operations, (5) a FIR of 15–30% seems to be the most effective to establish a good mix in the EPB-TBM chamber.

10.5 Clogging of TBMs in Turkish projects

10.5.1 Suruc Project

The Suruc tunnel, with a length of 17,185 m is the longest tunnel ever excavated in Turkey. It is aimed to irrigate the Eastern Anatolian part of Turkey, as seen in Figure 10.5, Ilci et al. [7].

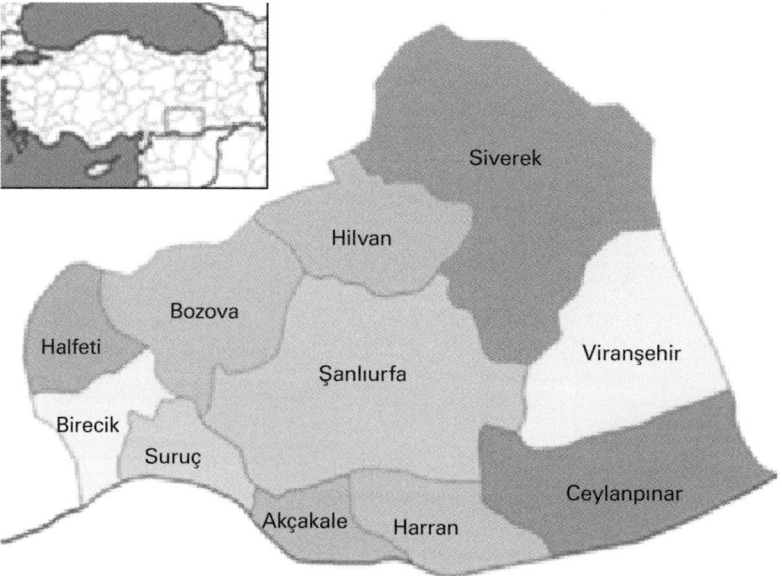

Figure 10.5 Location of Suruc Tunnel [7].

The expected water flow rate is 90 m^3/s and it will irrigate an area of 94,814 ha. The excavation of the 7.88 m diameter tunnel started with a double-shield hard rock TBM in September 2010, and 11,508 m of tunnel had been excavated by 10 October 2012 and the excavation finished in 2013. The power of the cutterhead was 4,500 kW with a maximum torque of 4,000 kNm and maximum thrust of 20,000 kN. The best daily advance was 38.4 m (26 April 2012), the best weekly advance was 206.8 m (16–22 April), the best monthly advance was 757.1 m, the mean daily advance rate including main stoppages was 15.7 m, and the mean daily advance excluding main stoppages was 17.3 m. Although the mean advance rate was one of the highest within those obtained with mechanized tunneling in Turkey, the clogging effect of clayey formation on the cutterhead and on muck transport systems decreased the excavation efficiency. The marl, in combination with water, had a sticky character that clogged the cutterhead. XRD analysis showed that clay minerals such as kaolinite, smectite and illite caused clogging of the cutterhead. Clogging can be defined as excavated material adhering to the steel surfaces, conveyor system or cutting tools of a TBM. This problem causes a lot of difficulties, such as blocking of excavation tools, breakdown of conveyor systems and high energy demand caused by clogging of the TBM cutterhead. It is clearly seen from Figure 10.7 that the muddy muck is responsible of clogging the cutterhead of the TBM. However, it is important to notice that excess water of 20 m^3/s, as seen in Figure 10.8 was encountered within 5,248–5,275 m of the tunnel. After contact injections, water ingress decreased to a considerable extent [7].

General views of the rock in the tunnel, muddy characteristic of the rock and the excessive water within the tunnel are seen in Figures 10.6, 10.7 and 10.8. A close-up view of clogging muck is given in Figure 10.9 [7].

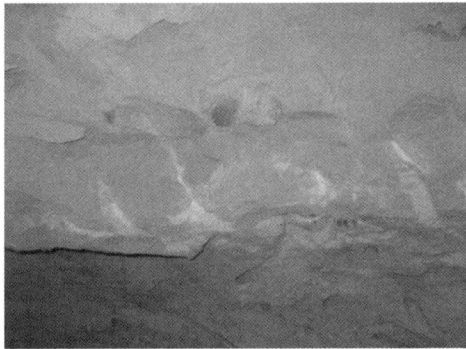

Figure 10.6 General view of the rock in the tunnel [7].

Figure 10.7 Muddy characteristic of the rock [7].

Figure 10.8 Excessive water within the tunnel [7].

Figure 10.9 Clogging muck [7].

The main rock formation in the area is Gaziantep formation of Eocene-Oligocene aged karstic chalky-clayey limestone. Clay bands sometimes have a thickness of 50–75 cm. Groundwater was encountered in the majority of drill holes. Mechanical properties of the geological formations are given in Table 10.1.

Table 10.1 Physical and mechanical properties of rock formations [7].

	Chainage [km]	UCS [MPa]	Es [MPa]	C [MPa]	φ [°]	γ [g/cm^3]	RQD [%]	Permeability Lugeon
Karstic chalky limestone	30+000 to 34+500	10–44	8,648–15,126	4.51–5.17	27.7–30.8	2.33–2.42	15–70	2.17–7.8
Karstic clayey limestone	34+500 to 36+300	4.1–7	—	—	—	—	25–45	5.41–7.8
Claystone marl	36+300 to 46+470	2.0–13.5	963–5,692	3.1–3.95	14–29.9	1.93–2.49	15–85	0.13–0.71

The effect of clogging of the cutterhead on thrust force, torque, net excavation rate, depth of cut per revolution and specific energy is dramatic, as seen in Figures 10.10, 10.11 and 10.12. As seen from Figure 10.10, in cases where the cutterhead is cleaned with anti-clogging agents, thrust force decreased from 4 MN to 2 MN and torque values decreased from 3,500 kNm to 1,000 kNm [7]. In general, the ratio of thrust force to torque is 2.3, but the situation is different when clogging of the cutterhead occurs. As seen in Figure 10.10, the ratio of thrust force to torque becomes to 1.3 where the clogging occurs at chainages 43+780 to 43+800 km. Then, after cleaning the cutterhead, the ratio rises again to 2.3–2.5, showing that this ratio is a good indicator of clogging of the cutterhead. Another very important point in the clogging phenomenon is that mean penetration decreases from 12 mm/rev to 4 mm/rev and mean specific energy increases from 4 kWh/m^3 to 24 kWh/m^3. In other words, clogging of the cutterhead causes the energy to excavate the rock to be six times more than a clean head situation [7].

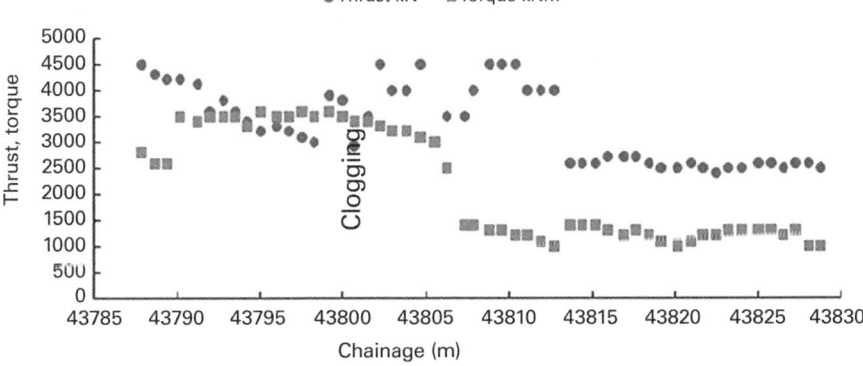

Figure 10.10 Effect of cutterhead clogging on TBM thrust and torque [7].

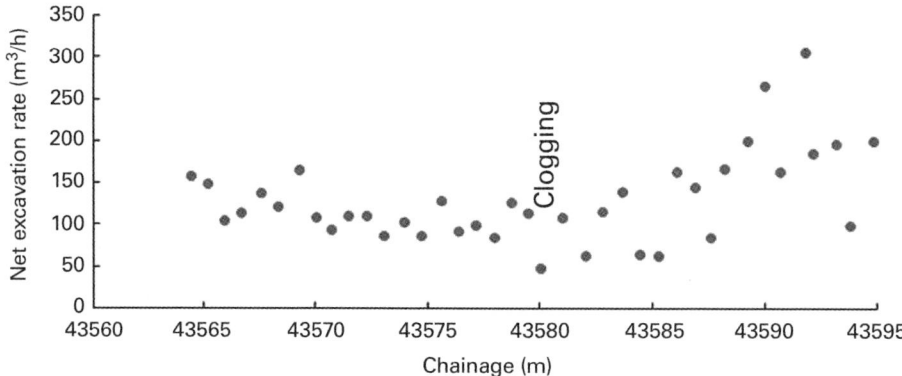

Figure 10.11 Effect of clogging on net excavation rate [7].

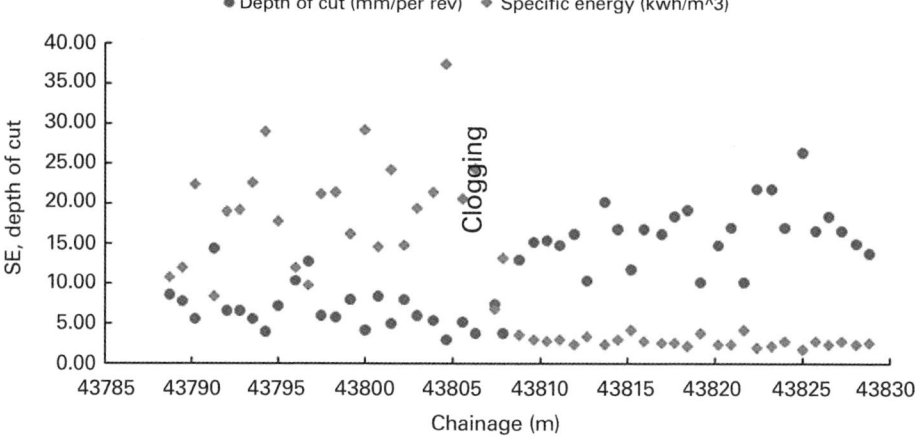

Figure 10.12 Effect of clogging on depth of cut (penetration per revolution) and specific energy [7].

10.5.2 Selimpasa sewerage tunnel in Istanbul

Selimpasa wastewater tunnel project is constructed along the alignment of Selimpasa–Kumburgaz–Guzelce, located in the European part of Istanbul on the coast of the Sea of Marmara, to transport the wastewater to the Selimpasa wastewater treatment plant. The project owner, Istanbul Water and Sewage Administration (ISKI), granted the project involving around 11.7 km of tunnel length (2,000 mm completed diameter) to Ozka-Kalyon JV, as their first tunneling contract.

The project is constructed in the Oligocene aged deltaic Gurpinar (Danisment) formation consisting mostly of sandstones and shales. It also includes thin coal bands, tuffs intercalations, gray carbonated clays and claystones (the most common lithology along the tunnel excavation level). Kusdili formation consisting of late Quaternary aged sand, gravel and fossiliferous sand is unconformable in the Gurpinar formation.

The units described as beach dunes, talus, alluviums and artificial fill are the other young units encountered in the region.

An EPB-TBM manufactured on the basis of customer needs by Herrenknecht AG having an excavation diameter of 2966 mm was used for excavation. The TBM was transported to the site on 13 July in 2009. Precast reinforced concrete rings, each with six segments, having a length of 950 mm, inner diameter of 2,300 mm and outer diameter of 2,700 mm, are used for lining. A set of four wagons each with 1 m^3 capacity were used for transportation of muck from the rear of the TBM to the shaft bottom, pulled by a battery-driven locomotive. The working pattern was 7 days/week, 2 shifts/day and 12 hours/shift. Usually, ten personnel worked at the face in each shift.

10.5.2.1 Soil conditioning experiments, field studies and results

All of the soil conditioning test results and field studies mentioned in this section are summarized based on an MSc dissertation performed by Aksu [8] at the Istanbul Technical University and on Copur et al. [9]. Experiments include ground and foam-ground mixture characterization tests. Experimental procedures were summarized in Copur et al. [9] and will not be given here.

Average natural water content (W_n) of the claystone sample is around 20.6%. Size distribution of the sample is presented in Figure 10.13. As seen, the sample includes around 60% of granular material. Liquid and plastic limits of the claystone are found to be 67.6% and 27.9%, respectively. Plasticity and consistency indices are estimated to be 39.7% and 1.18%, respectively. The place of the claystone sample on the plasticity chart is presented in Figure 10.14. It is seen that the claystone sample contains high plasticity clay.

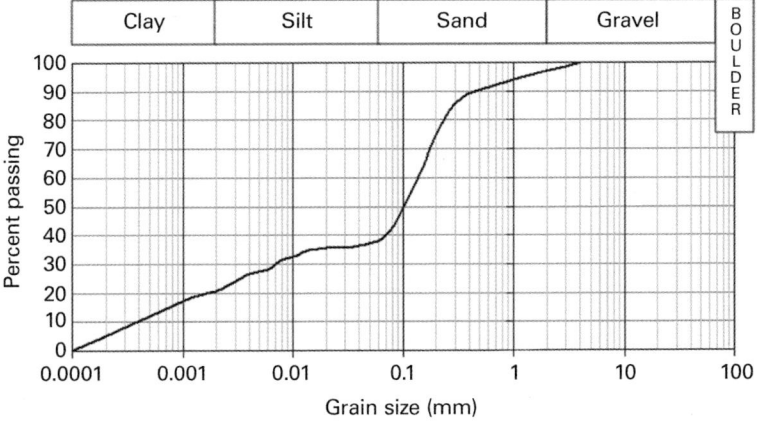

Figure 10.13 Size distribution of the claystone sample [9].

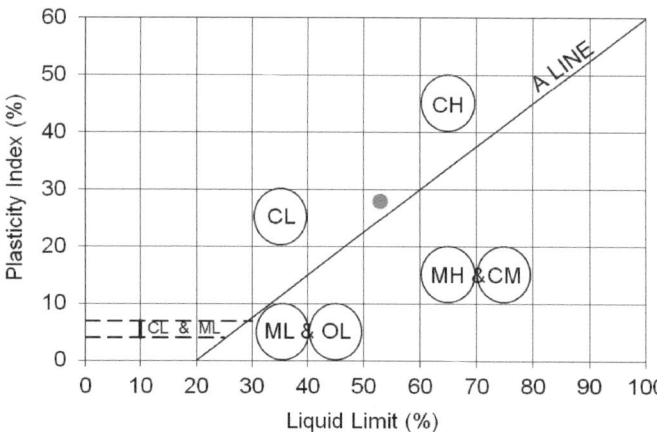

Figure 10.14 Place of the claystone sample on the plasticity chart given by Thewes [10].

The foaming agent, "Exfoam-1" selected by the contractor company, is also used in the foam-ground mixture tests by taking C_f of 1% and FER of 10.

Cone penetrometer tests are first performed by using only water–claystone mixture (no foam) at different water contents (25, 35, 40, 45, 55, 60, 65 and 75%). The size of the claystone sample used for these tests is below 0.425 mm. It is observed that the liquid limit value of 67.6% yields 17 mm of cone penetration. Then, the tests are repeated after adding foam at constant C_f of 1.0% and FER of 10 with varying levels of FIR (5, 10, 20 and 40%) and water content (35 and 45%). It is observed that 17 mm of cone penetration is obtained at 45% water content with 20% FIR, which is lower than the FIR value (30% FIR) at 35% water content. This indicates that the conditioned ground behaves like a liquid over the FIR value of 20%, and plastic under the FIR value of 20%. Variation of penetration depth of the cone penetrometer with FIR at 45% water content is presented in Figure 10.15 [9].

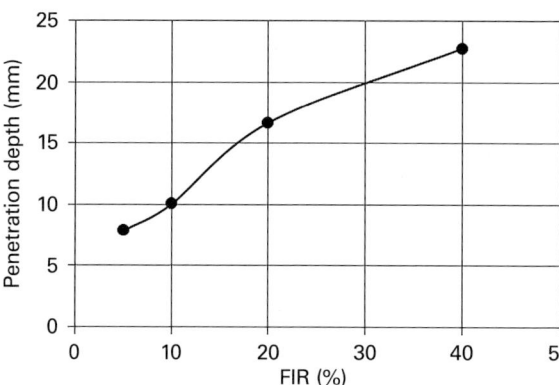

Figure 10.15 Variation of penetration depth with FIR at water content of 45% [9].

The mixing tests with power measurement are performed at constant C_f and FER with varying levels of FIR and water content. The size of claystone sample used for these tests is below 2 mm. The tests are first performed by using only water–claystone mixture at different water contents. It is observed that the sample starts sticking to the mixing apparatus at 20% water content, and that net power consumption steadily increases from 20% up to 35% water content, and then, sharply decreases at 40% water content. Variation of net power consumption with water content is presented in Figure 10.16 [9].

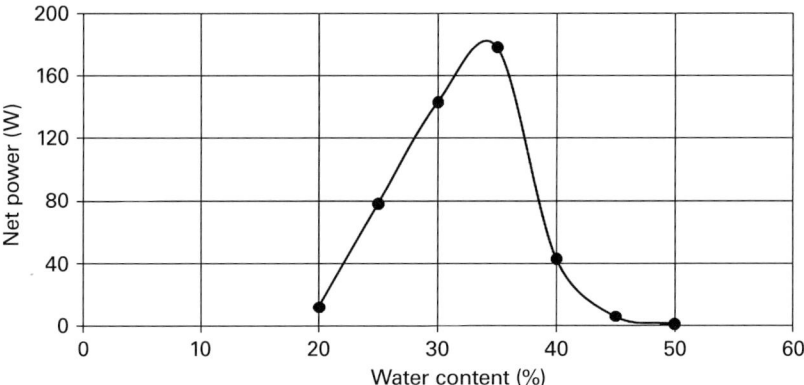

Figure 10.16 Variation of net power consumption with water content in mixing tests [9].

The tests are then repeated by adding foam at constant C_f of 1.0% and FER of ten with different levels of FIR (5 and 10%) and water contents (35 and 45%). It is observed that increasing the FIR value from 5 to 10% decreases the net power consumption. Also, increasing the water content from 35 to 45% decreases the net power consumption. It is also observed that increasing the amount of water being added to the sample, without any foam, decreases the power consumption while increasing visually the liquidity of the mixture (from paste consistency) making transportation of muck more difficult.

The place of the claystone sample on the clogging risk chart is presented in Figure 10.17. It is seen that the sample has a high clogging risk. An anti-clay polymer, Exfoam-Anticlay is used for mixing tests with power measurements for varying levels of anti-clay addition (1, 2, 3, 4, 5 and 6%) to the claystone sample at 45% water content. It is observed that addition of 5% anti-clay agent to the sample totally eliminates the sticking problem and minimizes power consumption. Addition of anti-clay agent more than 5% does not affect the power consumption, as seen in Figure 10.18 [9].

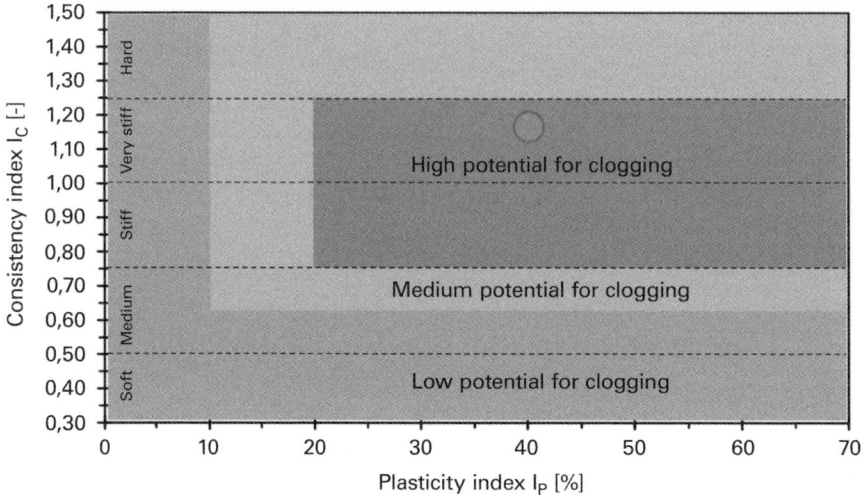

Figure 10.17 Place of the claystone sample on the clogging risk chart given by Thewes [10].

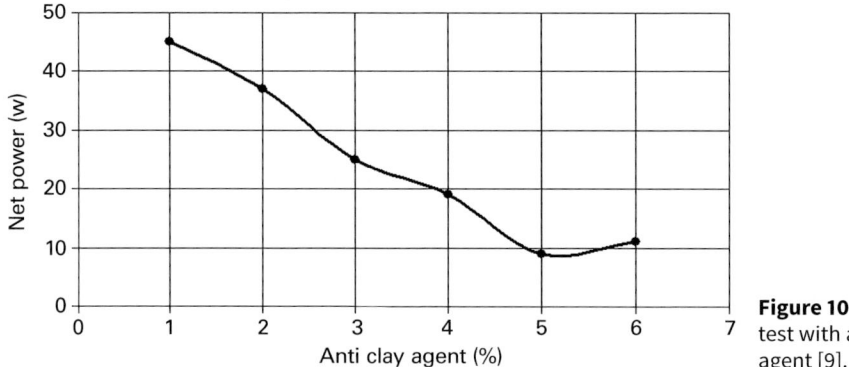

Figure 10.18 Mixing test with anti-clay agent [9].

The same anti-clay agent is also used for adhesion plate tests, which are first performed at constant water content of 45% with no addition of anti-clay agent. It is observed that the claystone sample sticks tightly to the plate and the plate does not slide at all. Then, anti-clay agent up to a maximum of 5% is added to the claystone sample at constant water content of 45%. The plate slides at the first second of the tests, which means that there is almost no sticking problem.

The results indicate that the optimum conditioning is obtained at C_f of 1.0%, FER of 10, water content of 40–45% (minus natural water content) and FIR of 20% for the claystone sample. This result is in agreement with the general suggestion of Thewes et al. [10, 11] for cohesive grounds. It is also indicated that 5% anti-clay agent can be used for reducing the sticking problem and power consumption with or without using foam, depending on the ground behavior. It should also be noted that foaming agent at the correct dosage also reduces the sticking problem, but not totally.

The suggested foaming design was validated by field application and tests in the Selimpasa wastewater tunnel project. Since a methane explosion occurred in the tunnel and data logging/recording unit of the EPB-TBM was damaged, the power, torque, thrust and instantaneous penetration rate measurements could not be performed during the preparation of this study; only general performance values were later obtained from the representatives of the contractor, Copur et al. [12]. The tunnel crew completed excavation of the tunnel without any problem, at foam feeding rates of 250–280 l/min. The EPB-TBM achieved 21.2 m/day average daily advance rate between 6 May 2010 and 19 May 2012, with a machine utilization time ranging usually between 25 and 35%. The best monthly advance rate was 809.4 m in April 2011, [12]. It should be noted that the excavation chamber was open without any pressure since the face was stable.

In average conditions, the TBM reached an average net advance rate of 50 mm/min at around 3 rpm. This advance rate requires around 177 l/min water addition, at C_f of 1.0%, FER of 10 and water content of 45% and FIR of 20%. Some of this water requirement comes from the natural water content (20.6%) of the claystone. Since the water pumping capacity of the TBM was 90 l/min (meaning ~13% water addition to the muck), the TBM operator increased the FIR from the suggested 20% to 60% (which means an additional 6% water, total water content around 40%) for easing transportation of the muck. The contractor company also used an anti-clay agent in addition to the foaming agent when sticking problems occurred.

10.5.3 Zeytinburnu Ayvalidere-2 wastewater tunnel project

Zeytinburnu Ayvalidere-2 tunnel project is a wastewater project being excavated on the European side of Istanbul with the specific goal of transporting wastewater from Zeytinburnu to the Yenikapi treatment plant through the Bayrampasa, Bagcilar and Gungoren alignment.

The general geology of the tunnel alignment is mostly Cekmece formation, Bakirkoy division which consists of grayish beige, low plasticity, consisting of gravels, carbonate lumps, iron oxides, sandy and silty rigid clay. The other part of the project consists of off-white, degree of weathering: W2 W3, porous, calcareous surface layer, low–mid strength, and clay intercalated limestone. The studied area in this section consists generally of low plasticity carbonated hard clay consisting of silt in gray color.

All of the soil conditioning test results and field studies mentioned in this section are summarized from Copur et al. [13] and Tolouei et al. [14]. Experiments include ground, foam and foam–ground mixture characterization tests. Experimental procedures were summarized in Tolouei et al. [14] and will not be given here.

10.5.3.1 Soil conditioning experiments, field studies and results

Average natural water content (W_n) of the sample is determined as being around 33%. The grain size distribution of the sample is presented in Figure 10.19. As can be seen, the sample incudes 80% fine material (clay + silt). Plastic and liquid limit values of the soil sample are 29.7% and 56.2%, respectively. Plasticity and consistency indexes are calculated to be 26.5 and 0.87, respectively. The place of the sample on a plasticity chart is presented in Figure 10.20, and also the place of the sample on the newly developed clogging risk chart and consistency chart given by Hollmann and Thewes [3] are presented in Figure 10.21, which shows that the sample has a medium–high potential for clogging and requires at least 12% of water addition for a better consistency.

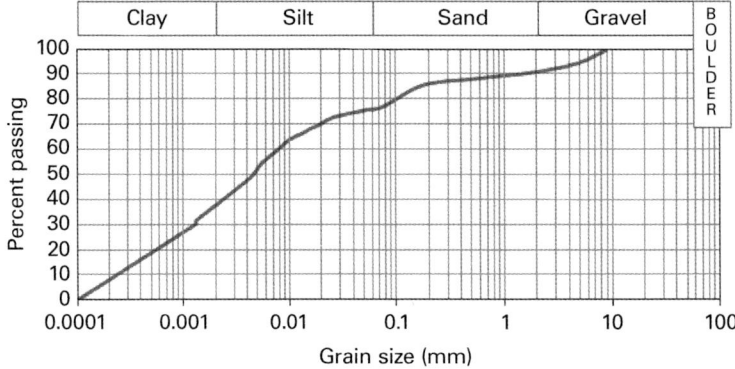

Figure 10.19 Grain size distribution of soil sample [13].

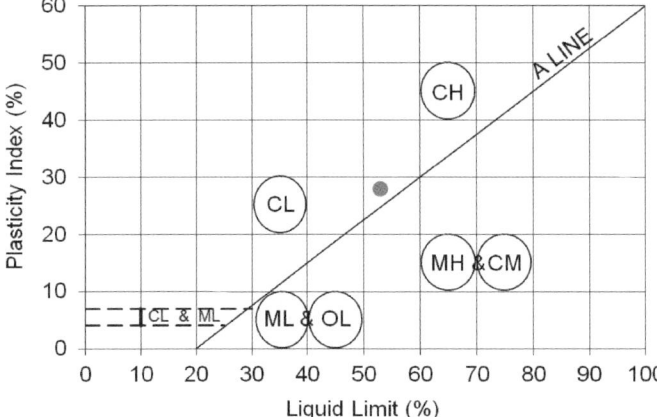

Figure 10.20 Place of the sample on the plasticity chart [13].

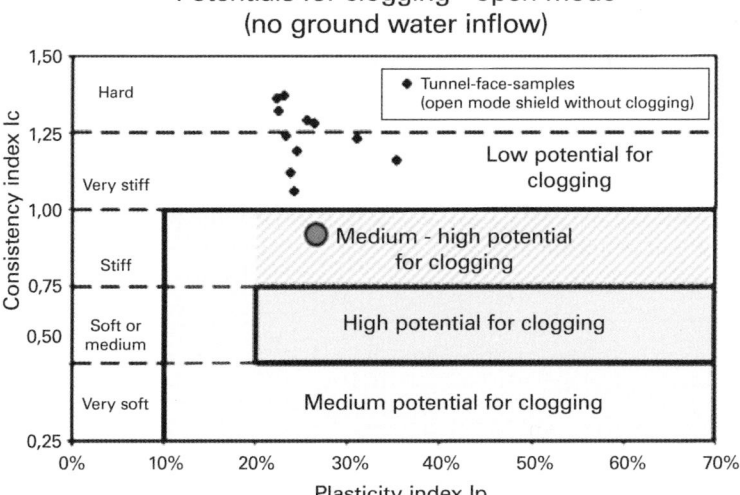

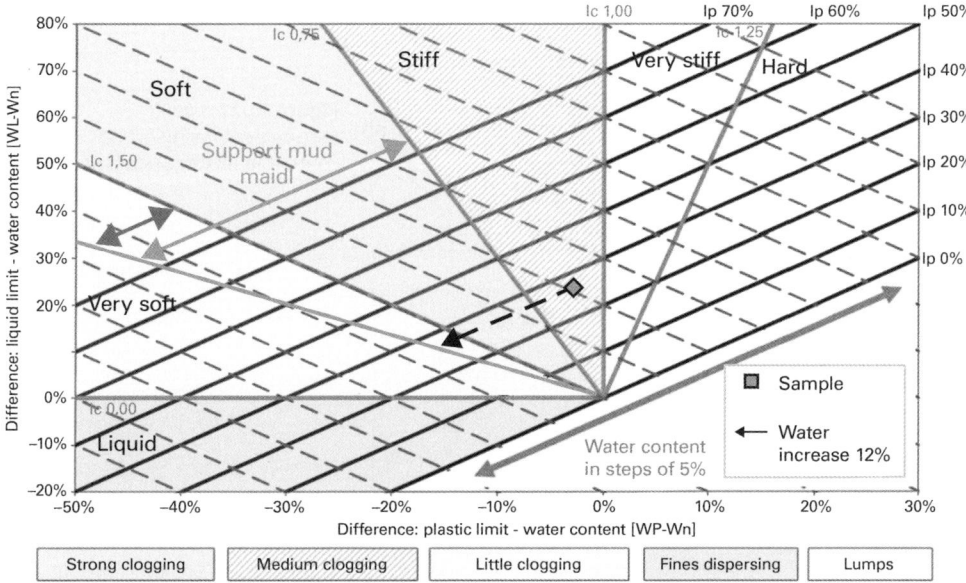

Figure 10.21 Place of the sample on the clogging risk chart (left) and the consistency chart (right) for open mode EPB applications given by Holmann and Thewes [13].

Conditioning tests are started, using just distilled water to observe behavior of the soil sample without any foaming agent. The range of water content is considered from 33 to 54% at which the power requirement became around half of the starting point of the test. After mixing the sample with different amount of water (33, 35, 40, 42, 45, 47, 50, 52 and 54%), the mixing, cone penetration and flow table tests were performed in turn. It was observed that power requirement (net, reduced by free rotational power) decreases from 92 to 35 W (Figure 10.22) and sticking amount (to the stirring blade) decreases from 2,042 to 295 grams (Figure 10.23) by adding and mixing only water to the soil sample. Water content of over 50% would reduce the power consumption (Figure 10.22). Water content of over 50% would reduce the sticking problem (Figure 10.23). On the other hand, the water content between 33 and 54% makes the penetration amount change from 5.5 to 22.6 mm (Figure 10.24), which indicates that an acceptable consistency can be obtained with 42–52% water content. It should also be mentioned that flow table values could only be obtained at 50% or more water content. However, it is known that using only water usually does not solve the problem of consistency and adhesion in soil conditioning.

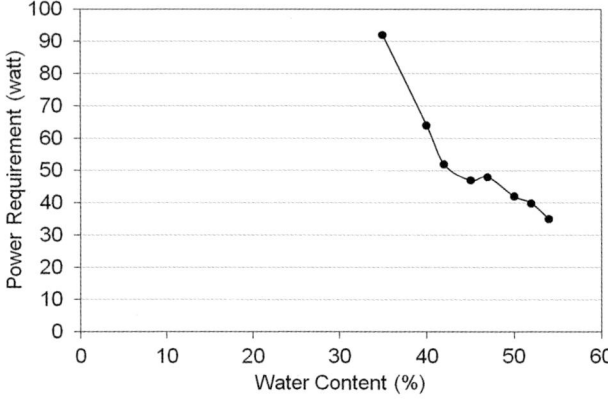

Figure 10.22 Power requirement versus water content in mixing tests [13].

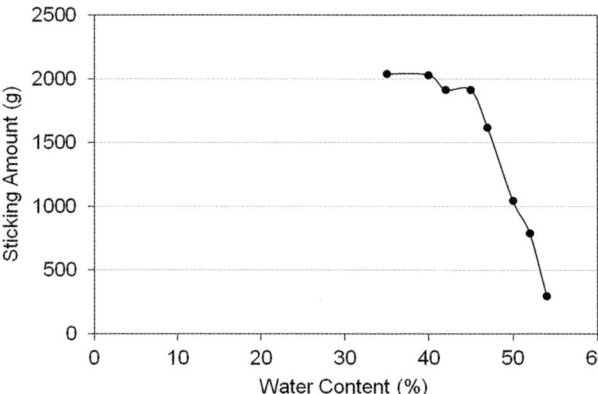

Figure 10.23 Sticking amount (adhesion to mixing blade) versus water content in mixing tests.

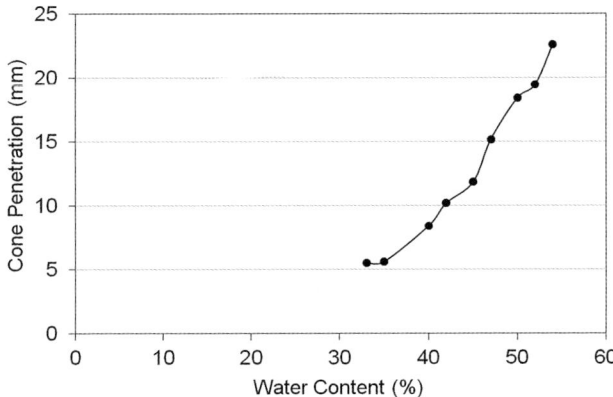

Figure 10.24 Cone penetration versus water content.

Soil conditioning tests are also performed by using two different additives: anti-clay Foam 1 and anti-clay Foam 2 (at 33% of water content, which is the natural water content of the soil sample) produced by different companies, and Foam 2 plus water (total 45% water, including natural water content) to analyze the effect of different additives on soil conditioning. C_f and FER are selected as 3% and 16, respectively, with varying FIR values (between 50% and 400%). The following results can be inferred: for Foam 1, by increasing FIR from 50 to 400% the power requirement decreases from 87 to 44 W (Figure 10.25), the sticking amount decreases from 1958 to 1570 g (Figure 10.26), the cone penetration increases from 5.5 to 10.5 mm (Figure 10.27). For Foam 2, by increasing FIR from 50 to 300% power requirement decreases from 67.5 to 42 watts (Figure 10.25), sticking decreases from 1,802 to 1,600 g (Figure 10.26), the cone penetration increases from 7.5 to 13.1 mm (Figure 10.27). For Foam 2 plus water (additional 12% to natural water content), by increasing FIR from 50 to 300%, the power requirement decreases from 45 to 30 W (Figure 10.25), sticking decreases from 1937 to 650 g (Figure 10.10) and penetration increases from 13.5 to 22.4 mm (Figure 10.27).

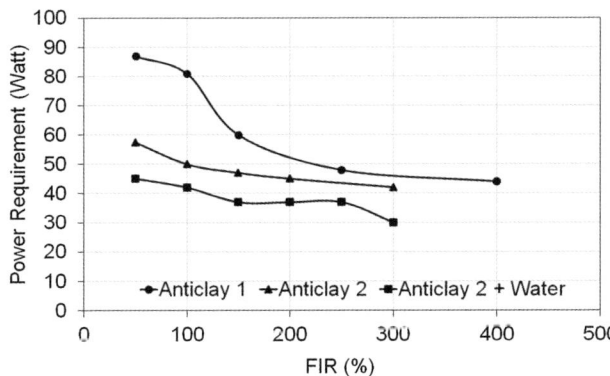

Figure 10.25 Power requirement for three foam types in mixing tests.

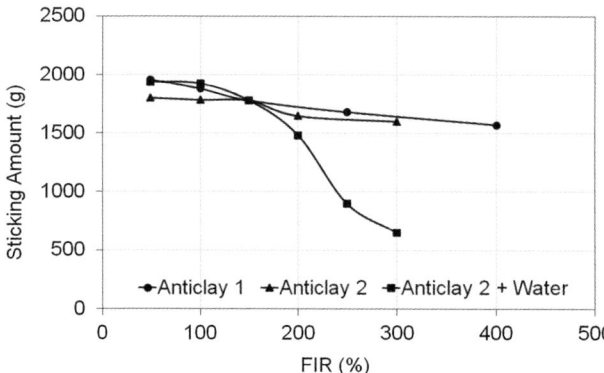

Figure 10.26 Sticking amount versus FIR in mixing tests.

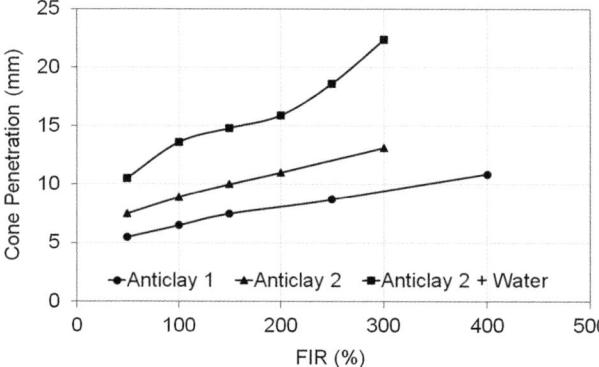

Figure 10.27 Penetration versus FIR in mixing tests.

These results indicate that Anticlay 2 gives better results than Anticlay 1. When power consumption, cone penetration and sticking values are considered together, a better conditioning and consistency is obtained with Anticlay 2 plus 12% additional water to the natural water content and at 250% or more FIR values. However, the experimental studies also indicate that this conditioning might be slightly improved in order to minimize the amount of anti-clay foaming agent used. Therefore, the experimental studies will be continued to reduce the sticking problem further, maybe by adding more water to the soil, instead of increasing the FIR (adding more anti-clay agent) since its cost is relatively higher.

In order to validate the experimental studies, some field trials were performed. The cutterhead torque was considered to be high during excavation of the Zeytinburnu Ayvalidere-2 wastewater tunnel and increased more after 150 m of excavation from the shaft due to clogging on cutterhead, although the cutterhead was half filled (not full) with muck since the face was stable. As a result, the operation stopped for remediation actions.

First, the grizzly bars on the cutterhead were uninstalled to reduce sticking and cutterhead torque, and a new conditioning design was applied according to the experimental results obtained in this study. After negotiating with the representatives of the contractor, C_f of 1.5%, FER of 15–16, FIR of 150–200% and total water content of 50% (around 17% additional water to the natural water content of 33%) was defined as the target conditioning and applied in the field with the EPB-TBM.

It was observed that clogging decreased significantly (with very little clogging) and hence the penetration rate of the EPB-TBM increased from 24 to 32 mm/rev. However, additional field trials are on the way to improve performance of the EPB-TBM further.

10.6 Conclusions

The clogging of a TBM, as encountered in clay-containing ground, has extensive consequences for the construction process and can severely affect performance of the machine by increasing torque, thrust and specific energy and lowering advance rates due to the extra cleaning efforts. So it is essential to understand the reasons for and the effects of clogging including mitigation programs to combat the negative effects of the clogging. This chapter is written with the intention of clarifying the subject by giving three examples from tunneling projects in Turkey.

The Suruc tunnel with a length of 17,185 m, is the longest water tunnel ever excavated in Turkey. The excavation with a hard rock double-shield TBM was one of the most efficient applications in Turkey in difficult conditions with a mean daily advance rate of 17.5 m, including main stoppages. Observations made during the execution showed that clogging of the cutterhead had an major adverse effect on thrust force, torque, depth of cut per revolution, specific energy and net excavation rate. Generally, the ratio of thrust force to torque decreased from 2.3 to 1.3. Specific energy increased from $4\,\text{kWh/m}^3$ to $24\,\text{kWh/m}^3$. In other words, clogging of the cutterhead caused the energy to excavate a unit volume of ground to be six times higher than with a clean cutterhead. Anti-clogging agents were therefore used to prevent the cutterhead clogging.

Although the ground was stable in both Selimpasa and Zeytinburnu Ayvalidere 2 wastewater tunnels, the ground was quite sticky, in both cases reducing excavation rates and causing stoppages due to cleaning up operations and muck transportation problems (sliding on the tail belt conveyor). Experimental studies performed in the soil conditioning laboratory indicated that regular application of foam selected by the contractor was adequate to solve the sticking and clogging problems in Selimpasa, while an anti-clay agent selected by the contractor was used in Zeytinburnu. The representatives in both cases applied the laboratory results in the field. The field measurements validated the experimental studies and the net advance rates of the EPB-TBMs increased at least 1.3 to 1.5 times, and the stoppages due to clogging problems reduced to normal ranges.

References

[1] Thewes, M., Burger, W. (2004) Clogging risks for TBM drives in clay. *Tunnels and tunneling Int.*, 6, 28–31.

[2] Hollmann, F., Thewes, M. (2012) Evaluation of the tendency of clogging and separation of fines on shield drives. *Geomechanics and Tunnelling*, 5, 574–580.

[3] Hollmann, F.S., Thewes, M. (2013) Assessment method for clay clogging and disintegration of fines in mechanised tunnelling. *Tunn Undergr Sp Technol*, 37, 96–106.

[4] Langmaack, L., Ibarra, J. (2011) *Speciality Chemicals for Tunnel Boring Machines*, Tunneling Lectures.

[5] Thewes, M., Burger, W. (2005) *Clogging of TBM drives in clay – identification and mitigation of risks*. WTC, Istanbul Underground Space Use: Analysis of the Past and Lessons for the Future – Erdem & Solak (eds) Taylor & Francis Group, London.

[6] Ball, R.P.A., Young, D.J., Isaacson, J., Champa, Gause, J. (2009) *Research in soil conditioning for EPB tunnelling through difficult soils*, RETC, USA.

[7] Ilci, N., Temel, M., Sezgin, S., Akpinar, T., Guarasio, S., Polat, C., Bilgin, N. (2013) *Clogging and squeezing effect of marl-clayey limestone on the performance of a hard rock TBM in Suruc Tunnel, Turkey*, World Tunnel Congress 2013 Geneva Underground – the way to the future!, Geneva, Switzerland.

[8] Aksu, F. (2010) The Effects of Soil Conditioning on Earth Pressure Balanced (EPB) TBM Performance. Istanbul Technical University, Dissertation. (in Turkish with Extended English Abstract).

[9].Copur, H., Aksu, F., Levent, L., Cinar, M. (2013) *Studies by a mobile laboratory on soil conditioning for EPB tunneling*. World Tunnel Congress, Geneva Underground – the way to the future. Geneva, Switzerland.

[10] Thewes, M. (2007*) Mechanized urban tunnelling – Machine technology*. The 3rd Training Course – Tunnelling in Urban Area, Proceedings of the World Tunnel Congress, Prague, Czech Republic.

[11] Thewes, M., Budach, C., Galli, M. (2010) *Laboratory Tests with various conditioned Soils for Tunnelling with Earth Pressure Balance Shield Machines*. The 4th BASF TBM Conference, London, England.

[12] Copur, H., Cinar, M., Okten, G., Bilgin, N. (2012) A case study on the methane explosion in the excavation chamber of an EPB-TBM and lessons learned including some recent accidents. *Tunn Undergr Sp Technol*, 27 (1), 159–167.

[13] Copur, H., Bilgin, N., Balci, C., Tumac, D., 2016. *Prediction and Optimization of Excavation Performance of Earth Pressure Balance Tunnel Boring Machines (EPB TBMs)*. Report submitted to The Scientific and Technological Research Council of Turkey, Project No: 213M487. (in Turkish with English Abstract).

[14] Tolouei, S., Avunduk, E., Copur, H., Tumac, D., Balci, C., Bilgin, N., Cinar, M., Kahya, S. (2016) *Foam optimization and relationship between foam use and EPB TBM performance*, Proceedings of the 13th International Conference on Underground Construction Prague 2016, 23–25 May, Prague, Czech Republic.

11 Effect of high strength rocks on TBM performance

11.1 Introduction

The strength of rock mass is an important parameter in determining the thrust–torque values and the type of cutters for a TBM. Large diameter CCS type disc cutters with larger tip widths are preferred in high strength and abrasive rocks. Sometimes in complex geologies, a rock formation having unexpectedly high strength characteristics may suddenly appear in the tunnel route, limiting the penetration of the cutters, thus decreasing the advance rates in most cases to undesired values. The Beykoz sewerage tunnel in Istanbul is a typical example to this, a sudden mass of quartzite, a few hundred meters in length, affected the cutter penetration to an undesirable, even to practically zero level. In that case CCS disc cutters were replaced with V-type disc cutters, because, as explained in earlier chapters, for a limited thrust force the penetration is higher in V-type disc cutters compared with other types of disc cutter [1, 2, 3, 4].

Sometimes in very high strength rock formations, as encountered in the Nurdagi railway tunnel, it is essential to carry out full-scale laboratory cutting tests to predict the behavior of the cutters to obtain the optimum design parameters of a TBM [5].

In the light of the information given above, the in-situ observation obtained with a TBM in the Beykoz tunnel will be given first, and then full-scale laboratory cutting tests will be detailed for the Nurdagi tunnel.

11.2 Beykoz sewerage tunnel, replacing CCS disc cutters with V-type disc cutters to overcome undesirable limits of penetration for a maximum limit of TBM thrust

The sewerage tunnel excavated between Kavacik and Beykoz in the northern part of the Istanbul Bosphorus is part of an environmental protection project concerned with renewing the inadequate sewerage network around Beykoz and collecting the wastewater in a treatment plant and cleaning the polluted water of the Istanbul Bosphorus.

The ground conditions change from soft to hard formations, and excessive water ingress was expected in some areas. The total length of the tunnel is 7,253 m, tunnel excavation started from shaft AT2 on 9 January 2007 in open mode, and 4,267 m of the tunnel had been excavated by 26 January 2009. The general layout of the tunnel and previous works carried out are given in Guclucan et al. [1, 2, 3].

Figure 11.1 shows the relationship between penetration index in kN/mm of penetration and torque penetration index in kNm/mm. These relationships are only valid for disc cutters used in chainages 0–800 and 2,300–2,900 m and may be used to explain the performance of the TBM in relevant geological conditions.

TBM Excavation in Difficult Ground Conditions. Case Studies from Turkey. First Edition. Nuh Bilgin, Hanifi Copur, Cemal Balci.
© 2016 Ernst & Sohn GmbH & Co. KG. Published 2016 by Ernst & Sohn GmbH & Co. KG.

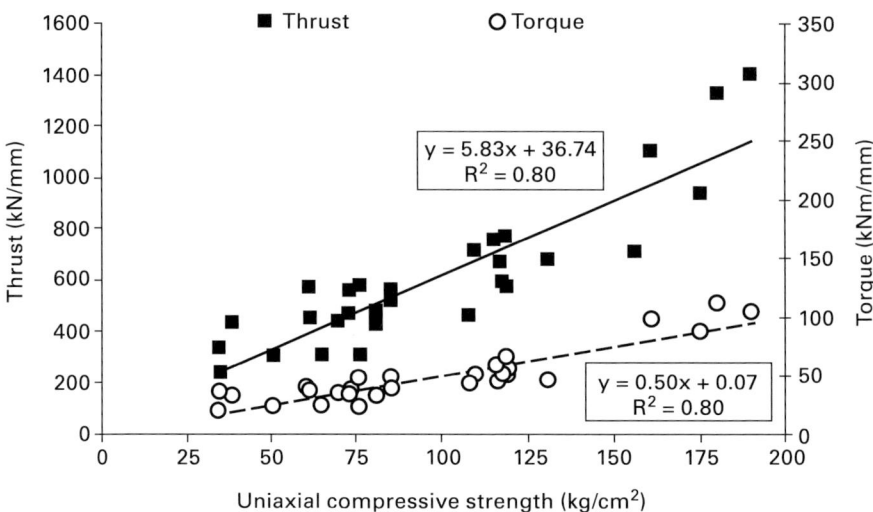

Figure 11.1 The change of TBM penetration index and torque per unit penetration for disc cutters [4].

Quartzite veins of 50 m in thickness and having compressive strength of 110–225 MPa were faced after chainage 2,550 m, which caused very low disc penetration of around 1 mm/rev with excessive dust problems. It was therefore decided to change all discs with 76° edge angle V-type disc cutters, except the eight central CCS type disc cutters. Penetration index in kN/mm (the ratio of disc thrust to the penetration per revolution) is seen for both CCS disc cutters and V-type disc cutters in Figure 11.2. It is clear from this figure that for a unit penetration, V-type disc cutters work with less thrust compared to CCS disc cutters. However, this advantage disappears when the abrasivity of the rocks is considered. Therefore, the V-type disc cutter was changed to CCS disc cutters when passing to limestone formation.

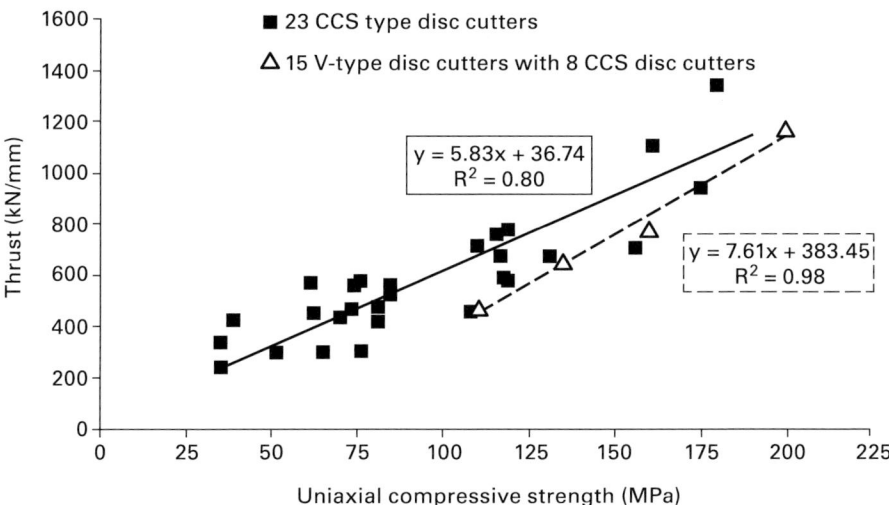

Figure 11.2 The variation of penetration index with compressive strength for CCS and V-type disc cutters [4].

11.3   Nurdagi tunnel, full-scale cutting tests to obtain optimum TBM design parameters in very high strength and abrasive rock formation

The geology and outline of the Nurdagi tunnel is summarized in Chapter 6. Cuttability of the two block samples (meta-sandstone and meta-mudstone) representing the tunnel route are determined by full-scale linear cutting tests in ITU Laboratories, using a disc cutter aimed at selecting a suitable TBM, defining its basic specifications and predicting its excavation performance for the Bahce–Nurdagi tunnel crossing. Also, some tests were performed to determine some of the physical and mechanical properties of the block samples by using chunk samples obtained from the block samples. The full-scale linear cutting tests were performed by using a constant cross-section disc cutter with a diameter of 13″ (330 mm) and a tip width of 1.2 cm at 80 mm of cutter (line) spacing and different depths of cut (3, 5, 7 mm). The normal and rolling forces acting on the disc cutter and specific energy values were measured, optimum cutting geometry is identified and size distribution and coarseness index values of the chip samples were identified for defining the efficiency of the excavation process.

Based on experimental results, optimum cutting geometry (optimum line spacing to depth of cut (s/d) ratio) was determined for the TBM to be used for excavation of the tunnel, its preliminary net cutting rate was predicted by using specific energy method, and some nomograms were developed for predicting its daily advance rates.

Uniaxial compressive strengths of the meta-sandstone and meta-mudstone samples were found to be 223 MPa and 118 MPa, and their Cerchar abrasivity index values were found to be 3.87 (high abrasive) and 1.84 (low abrasive), respectively.

Optimum cutting geometry was obtained at cutter (line) spacing to depth of cut (s/d) ratio of 16, and at this geometry, the optimum specific energy was found to be

11.08 kWh/m^3 for meta-sandstone sample. The thrust force that should be applied to the disc cutter at optimum cutting geometry was found to be 198.65 kN.

Optimum cutting geometry was obtained at s/d ratio of 16, and at this geometry, the optimum specific energy was found to be 1.59 kWh/m^3 for the meta-mudstone sample. The thrust force that should be applied to the disc cutter at optimum cutting geometry was found to be 38.6 kN.

Some nomograms were developed aimed at predicting preliminary net excavation (cutting) rates and daily advance rates for the meta-sandstone and meta-mudstone samples by using the specific energy method with some assumptions (especially for hard rock TBMs). Based on selected TBM and tunnel conditions, a preliminary performance can be estimated. It was also seen that the excavation rates in meta-mudstone could be higher by as much as twice that in meta-sandstone.

Instantaneous (net) cutting rate of a mechanical miner is that achieved only during excavation, excluding stoppages. Net cutting rates of mechanical miners can be predicted using the model given below (Eq. 11.1), which is based on linear cutting tests [6, 7].

$$ICR = k \cdot P_{net}/SE_{opt} \qquad\qquad (11.1)$$

where ICR is the instantaneous (net) cutting rate (m^3/h), P_{net} is the cutterhead power of the mechanical miner while cutting at optimum conditions (kW), SE_{opt} is optimum specific energy obtained from full-scale linear cutting tests (kWh/m^3) and k is the energy transfer coefficient (which can be taken as 0.8–0.9 for hard rock TBMs).

Optimum specific energy values are 11.08 kWh/m^3 for meta-sandstone and 1.59 kWh/m^3 for meta-mudstone. It is assumed that any variation of disc cutter diameter would not affect these values. Assuming k value of 0.85, the relationship between ICR and P_{net} shown in Figure 11.3 can be obtained. ICR values for meta-mudstone are divided by 2 in this graph (worst case), since a TBM would excavate in both parallel and perpendicular to bedding planes in reality. By using this graph, ICR can be estimated by using installed cutterhead power reduced by an efficiency factor (or estimating deterministically). However, estimated ICR values should be corrected for fractures-joints or for RQD, which increase ICR. Net cutting rate (as IPR in m/h) can be estimated by dividing ICR by tunnel cross-section area (Figure 11.3).

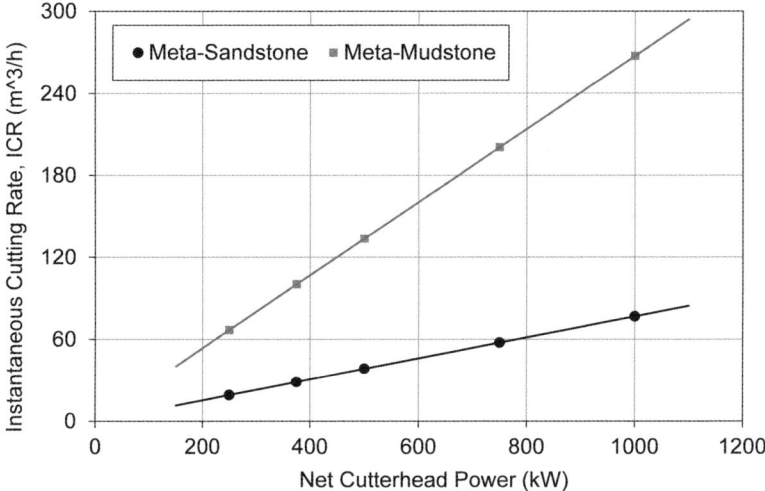

Figure 11.3 Relationship between instantaneous cutting rate and net cutterhead power for Nurdagi project based on laboratory rock cutting tests [5].

A suitable machine utilization factor (AR) should be estimated to predict the daily advance rates. The machine utilization factor determines the percentage of the time used just for excavation (excluding stoppage time) over a whole shift time (stoppages + excavation time); it varies from project to project. By assuming a net cutterhead power of 750 kW for meta-sandstone and 375 kW for meta-mudstone, a machine utilization factor of 50%, and 24 hours of daily working time, the variation of daily advance rate with tunnel diameter is obtained, as given in Figure 11.4.

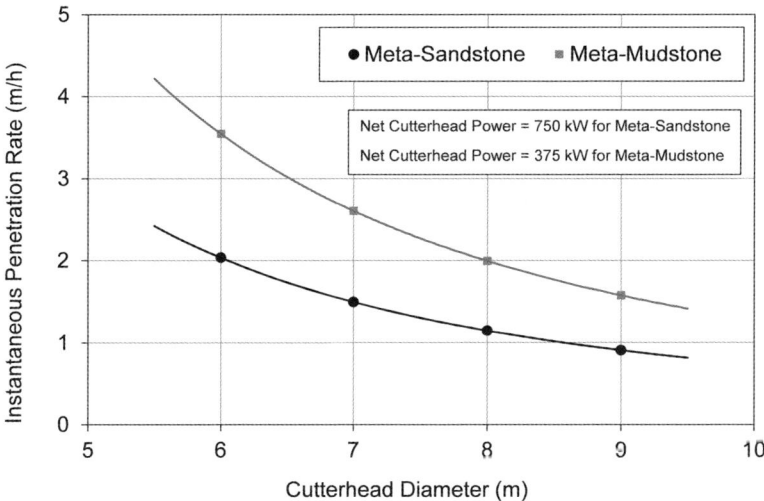

Figure 11.4 Relationships between instantaneous cutting rate and net cutterhead power for Nurdagi project based on laboratory rock cutting tests [5].

A single-shield TBM has been designed and manufactured based on site investigations and full-scale laboratory cutting test results. The TBM is now in the process of excavating the tunnel. Performance of the TBM is being collected currently; predicted and actual performance values will be the subject of a coming research topic.

11.4 Beylerbeyi–Kucuksu wastewater tunnel, TBM performance in high strength rock formation

This section is summarized based on [8]. The Beylerbeyi–Kucuksu wastewater tunnel aligned along the Istanbul Strait having a length of 4,344 m and finished diameter of 2.2 m was constructed by Unal Akpinar Construction Inc. for the owner Istanbul Water and Sewage Authority (ISKI).

Palaeozoic aged geological units are found between Beylerbeyi and Kucuksu. Lithology basically includes limestone, laminated mudstone with interbedded siltstone, sandstone, carbonated-laminated-fossiliferous shale interbedded with limestone. These units are cut sometimes by andesite and diabase dykes having thicknesses of up to several meters and significantly fractured contact zones and high strength. Physical and mechanical properties of the formations along the alignment are summarized in Table 11.1. As seen, the general strength of the formations is not very high, but from time to time very hard and massive rock zones were encountered.

Table 11.1 Some of the properties of the rock masses [9].

Formation	Dolayoba formation (limestone)	Kartal formation (shale)	Tuzla formation (nodular limestone)	Dykes
UCS (MPa)	34.0 ± 14.0	32.4 ± 18.0	34.0 ± 13.5	76.0 ± 26.6
BTS (MPa)	6.3 ± 3.0	5.2 ± 2.4	5.5 ± 2.8	7.8 ± 2.7
E_S (GPa)	5.2 ± 3.5	5.3 ± 3.8	4.26 ± 3	4.55 ± 1.62
v_S (—)	0.27 ± 0.15	0.32 ± 0.13	0.28 ± 0.14	0.26 ± 0.11
RQD (%)	70–90	25–50	60–80	25–70
RMR	66	45	50	59
Q	8	1	2	10

UCS: Uniaxial compressive strength, BTS: Brazilian (indirect) tensile strength, E_S: Static elasticity modulus, v_S: Static Poisson's ratio, RQD: Rock quality designation, RMR: Rock mass rating class, Q: Q class.

The EPB-TBM used for excavation is a new machine manufactured by Herrenknecht and designed to work in closed and open modes and for mixed ground conditions. Its basic characteristics are summarized in Table 11.2. A photograph of the cutterhead after the second revision during the construction is presented in Figure 11.5.

Figure 11.5 General view of cutterhead of EPB-TBM [8].

Table 11.2 Technical features of Herrenknecht (M1801M) EPB-TBM.

Total length (TBM + backup)	62 m
Excavation diameter	3,251 mm
Shield diameter	3,195 mm
Number of push cylinders	12
Stroke of push cylinders	1,700 mm
Total thrust capacity	10,688 kN
Maximum torque (continuous)	620 kNm
Maximum torque (intermittent)	780 kNm
Variable rotational speed	0–9.2 rpm
Installed power	400 kW
Cutterhead power	315 kW
Number of disc cutters	6 single + 9 double
Diameter of disc cutters	12″ (305 mm)
Number of scrapers + buckets	18 scrapers+6 buckets
Nominal diameter of screw conveyor	500 mm
Rotational speed of screw conveyor	0–27 rpm

Monthly advance rates and general performance nomograms for formations having certain characteristics are presented in Figures 11.6–11.11, considered as helpful for decision-makers. Advance per revolution in fractured/jointed limestone was almost 60% higher than massive limestone, while thrust force was almost 30% higher in the massive zone. While 29% of TBM utilization time was achieved, the TBM-related delays took almost 32% of the whole delays. Cutterhead maintenance and revisions had the highest share of around 25% of the TBM related stoppages.

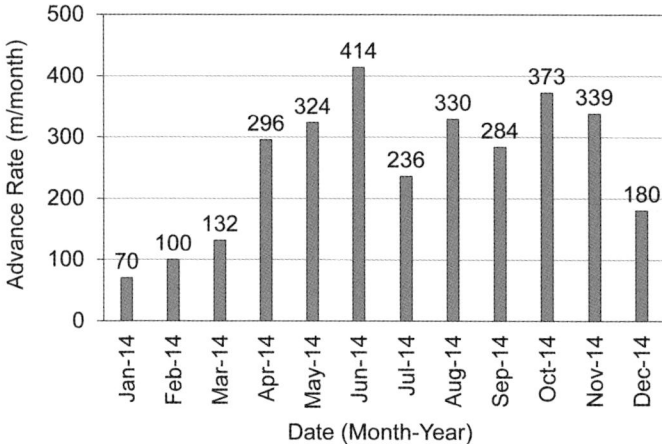

Figure 11.6 Monthly advance rates in Beylerbeyi–Kucuksu wastewater tunnel [8].

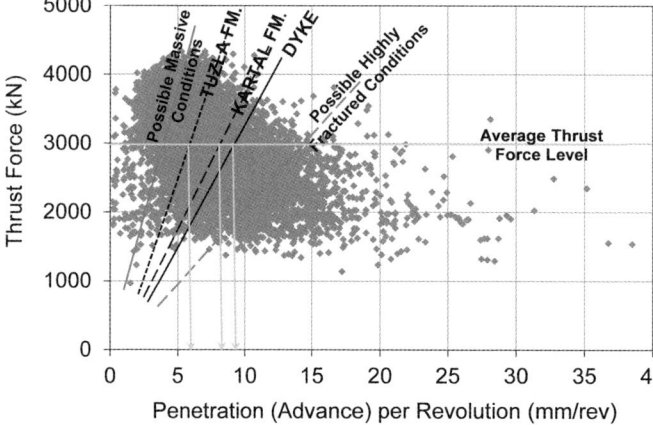

Figure 11.7 Relationship between advance per revolution and thrust force in the Beylerbeyi–Kucuksu wastewater tunnel [8].

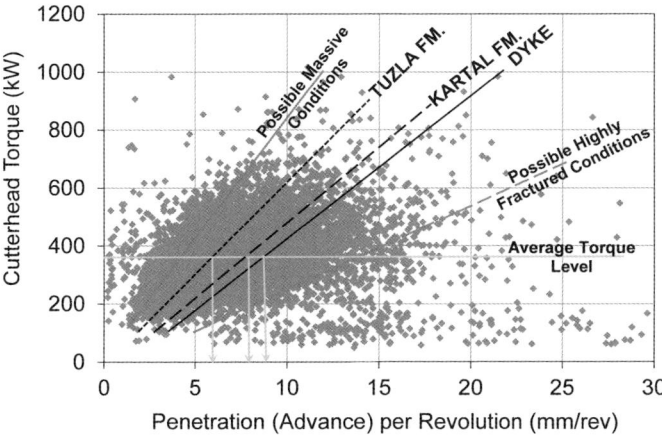

Figure 11.8 Relationship between advance per revolution and cutterhead torque in the Beylerbeyi–Kucuksu wastewater tunnel [8].

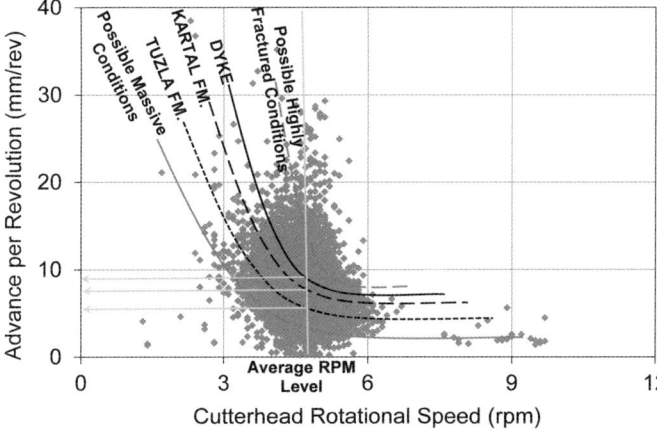

Figure 11.9 Relationship between cutterhead speed and advance per revolution in the Beylerbeyi–Kucuksu wastewater tunnel [8].

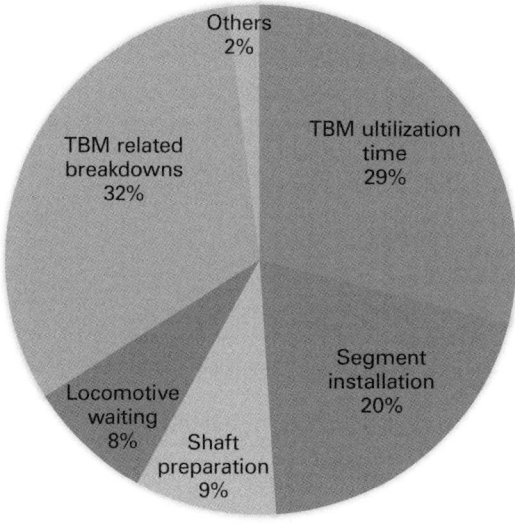

Figure 11.10 Job time utilization distributions in the Beylerbeyi–Kucuksu wastewater tunnel [8].

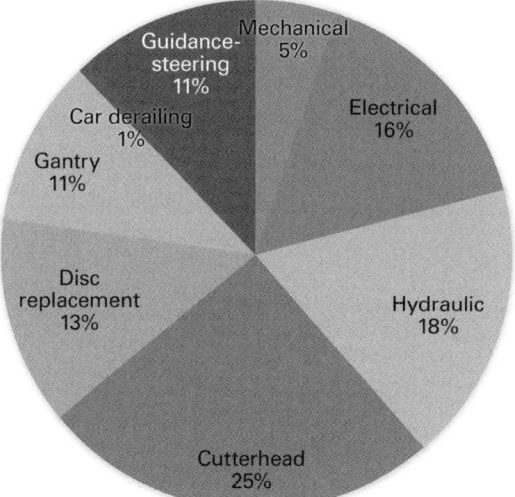

Figure 11.11 Distribution of TBM related stoppages in the Beylerbeyi–Kucuksu wastewater tunnel [8].

Average daily advance rate of 9.3 m/day, monthly advance rate of 256.5 m/month, and penetration per revolution of 7.1 mm/rev were achieved. Average thrust force was 2994 kN reciprocating to average 173 kWh of cutterhead power and 356 kNm cutterhead torque at an average 4.7 rpm rotational speed. Average cutter life was realized around 158 m³/cutter ring (20 m/cutter ring), which was quite low. It was seen that advance per revolution in fractured or jointed limestone was almost 60% higher than the massive limestone, while the thrust force was almost 30% higher in the massive zone. While 29% of TBM utilization time was achieved, TBM related delays took almost

32% of job time. Cutterhead maintenance and revisions had the highest share of TBM related stoppages at 25%.

This performance could be considered quite low, mostly because the ground conditions encountered are classed as difficult ground conditions for the EPB-TBM used in the Beylerbeyi–Kucuksu wastewater tunnel, since it is considered that a larger diameter TBM equipped with larger diameter single disc cutters would excavate these rocks without any significant problem. Since the EPB-TBM used in the Beylerbeyi–Kucuksu wastewater tunnel had a small diameter and were designed for mixed ground conditions, it had a limited thrust and power for the rocks excavated. The disc diameter of 12″ limited the applied load/thrust on the cutters and thus the penetration rate. It is also known that applied thrust on double discs is split into two halves, reducing its penetration capability into rock face and increasing tool consumption rates.

11.5 Tuzla-Akfirat wastewater tunnel, TBM performance in high strength rocks

The detailed information about Tuzla-Akfirat wastewater tunnel is given in Chapter 12 of this book. Very hard and abrasive rock zones up to 200 MPa uniaxial compressive strength were encountered in this tunnel [10]. This project saw uniaxial compressive strength values of over 50 MPa, which can be considered as difficult ground conditions for an EPB-TBM with diameter of 3,151 m designed for mixed ground conditions and equipped with small diameter double disc cutters. The relationship between advance per revolution and uniaxial compressive strength of the rocks encountered along the tunnel alignment is presented in Figure 11.12. It is seen that the advance rate of the TBM decreases below 2 mm/rev when excavating rocks having compressive strength values over 50 MPa.

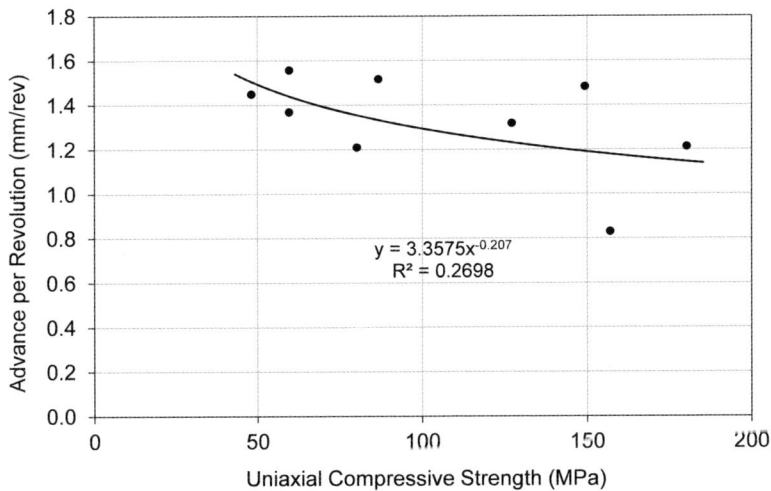

Figure 11.12 Relationship between uniaxial compressive strength of the rocks and advance per revolution in the Tuzla-Akfirat wastewater tunnel [10].

11.6   Conclusions

Very hard rock formations affect tremendously the performance of both hard rock TBMs and soft ground TBMs excavating hard rocks, decreasing advance rates, increasing dust problems and causing long stoppages. This effect becomes more serious in the case of soft ground TBMs (especially EPB-TBMs mentioned in this section) with small diameter and low capability. If the selected cutters have small diameter and multi-row cutters in one cutter bearing, the difficulties in the excavation operation increases significantly.

Quartzite veins having compressive strength of 110–225 MPa encountered in the Beykoz sewerage tunnel caused very low disc penetration of around 1 mm/rev with excessive dust problems. Replacing CCS cutters with 76° edge angle V-type disc cutters improved the penetration and increased the cutter life.

In the Tuzla-Akfirat, Beykoz and Beylerbeyi-Kucuksu wastewater tunnel projects, the EPB-TBMs used were of small diameter, designed for mixed ground conditions, causing low capability. The penetration rates of these TBMs were usually lower than 2 mm/rev for excavation of rocks having over 50 MPa compressive strength. Machine utilization times were quite low and cutter consumption rates were quite high in these projects.

References

[1] Guclucan, Z., Meric, S., Gursoy, C., Algan, M., Bilgin, N., Balci, C., Tumac, D. (2007) The use of a TBM in difficult ground conditions. Proceedings of the 2nd Symposium on Underground Excavations for Transportation, Istanbul, Turkey (in Turkish).

[2] Guclucan, Z., Meric, S., Algan, M., Palakci, Y., Bilgin, N., Bilgin, A.R., Balci, C., Tumac, D. (2008) The use of a TBM in difficult ground conditions in Beykoz – Kavacik Sewerage Tunnel. Proceedings of the World Tunnel Congress, Underground Facilities for Better Environment and Safety, Agra, India.

[3] Guclucan, Z., Meric, S., Palakci, Y., Bilgin, N., Balci, C., Copur, H., Namli, M., Bilgin, A.R., Kandemir, E. (2009) The use of theoretical rock cutting concepts in explaining the cutting performance of a TBM using different cutter types in different rock formations and some recommendations. Proceedings of World Tunnel Congress, Safe Tunnelling for the City and for the Environment, Budapest, Hungary.

[4] Bilgin, N., Copur, H., Balci, C. (2012) Effect of replacing disc cutters with chisel tools on performance of a TBM in difficult ground conditions. *Tunn Undergr Sp Technol,* **27**, 41–51.

[5] Copur, H., Balci, C., Tumac, D., Avunduk, E., Comakli, R. (2013) Full-scale linear cutting tests with disc cutter on block samples from "Bahce-Nurdag tunnel crossing" *Project in the Context of Bahce-Nurdag (Fevzipasa) Variant Infrastructure Jobs*, ITU investigation Report, p 83.

[6] Rostami, J., Ozdemir, L., Neil, D.M. (1994) Performance prediction: a key issue in mechanical hard rock mining. *Mining Engineering*, **11**, 1263–1267.

[7] Copur, H., Tuncdemir, H., Bilgin, N., Dincer, T. (2001) Specific energy as a criterion for the use of rapid excavation systems in Turkish mines. *The Institution of Mining and Metallurgy, Transactions Section – A Mining Technology*, **110**, A149–157.

[8] Gencay, Y., Akgul, M., Bilgin, A.R., Copur, H., Bilgin, N. (2015) *Performance of EPB-TBM in Beylerbeyi–Kucuksu Wastewater Tunnel in Istanbul.* WTC2015 World Tunnel Congress, May 22–28, Dubrovnik, Croatia.

[9] Vardar, M., Bilgin, N., Akkok, R. (2010) Re-evaluation of Beylerbeyi–Kucuksu wastewater tunnel in terms of rock cutting mechanics and engineering geological properties. *Report*. Istanbul Technical University.

[10] Akyuz, S., Copur, H. (2015) Investigations into rock and soil related problems encountered during construction of Tuzla wastewater basin (Omerli Dam) collectors and networks. *Unpublished Report submitted to Nas-Akad JV* (in Turkish).

12 Effect of high abrasivity on TBM performance

12.1 Introduction

Cutter consumption is one of the important cost items in mechanized tunneling due to replacement costs, cutting efficiency (reduction with worn tools), and also time and workmanship spent on replacement. Since the cutters carry the cutting force of the excavation machine to the rock to be cut, they generally define the economical limits of the excavation system. Cutter consumption has become more important, especially because TBMs have been used very widely and usually equipped with disc cutters. Predicting cutter cost is also a very important task, especially at the feasibility stage of a tunneling project.

For various reasons cutters wearout, get blunt and are broken during excavation. Because of blunting and non-functioning of the cutters, the cutting forces increase dramatically and thus also the specific energy to cut the rock. And since penetration of the cutters into the rock becomes more difficult, the cutting depth decreases and excavation machine vibration increases drastically. In addition to cutter consumption cost, blunting causes excessive dust occurrence, sparks, and reduction on machine utilization time due to cutter replacement, and these all increase the cost of excavation.

Basic parameters affecting cutter life, replacement and consumption rates for TBMs can be classified into three general groups, as summarized in Table 12.1: Geological and geotechnical parameters, excavation machine (TBM) related parameters and operational parameters [1–2]. The geological and geotechnical parameters are the dominant ones. The machine-related parameters should be arranged in accordance with the expected geological and geotechnical conditions for the best TBM excavation performance and minimum cost.

Table 12.1 Some parameters affecting cutter life, replacement and consumption rates for TBMs (revised from [1, 2]).

Mechanical (TBM related) parameters
General excavation machine properties Excavation machine type (hard rock TBM, EPB, SPB, double-shield, single shield, etc.) Excavation machine weight and dimensions Thrust capacity
Cutterhead properties Cutterhead type, stability, opening ratio Cutterhead power (torque) and RPM capacities Cutter allocation (positioning, spacing design)
Cutting tool (cutter) properties Cutter type (roller type, such as single disc, multi-raw disc or strawberry; drag type such as conical, radial, ripper, chisel, etc.) Cutter dimensions (diameter, tip width, etc. for roller cutters; tip diameter, blunting radius, etc. for drag type cutters) Metallurgical properties of cutters

TBM Excavation in Difficult Ground Conditions. Case Studies from Turkey. First Edition. Nuh Bilgin, Hanifi Copur, Cemal Balci.
© 2016 Ernst & Sohn GmbH & Co. KG. Published 2016 by Ernst & Sohn GmbH & Co. KG.

Geological and geotechnical parameters
Rock mass properties   Rock quality designation (RQD)   Bedding, foliation, fault zones, shear zones   Joint sets (orientation, spacing, filling, etc.)   Hydrogeology (water table, water ingress)   Adverse geology (squeezing, swelling, blocky grounds, etc.)
Physical and mechanical properties   Abrasivity and texture (mineral/quartz content, grain size, interlocking, etc.)   Cuttability (cutter forces, specific energy, optimum cutting geometry)   Strength (compressive strength, tensile strength, elasticity, etc.)   Others (brittleness, water content, etc.)
Operational parameters
Applied thrust force, torque, penetration rate   Applied face pressure   Foam or slurry use (for EPB or SPB TBMs)   Labor, operator and material quality

It is generally difficult to assess the effect of any single parameter given in Table 12.1 on cutter life in TBM applications since usually several parameters come into effect together. However, it is known that abrasivity of rocks is an important indicator for cutter wear rates. It is also known that wear is not the only reason for cutter damage, blunting and consumption. There are different causes or mechanisms of cutter failure, as shown in Table 12.2 [3, 4]. Frictional and abrasive wear is the most common failure mechanism in practice. In most cases, all mechanisms operate at the same time with various degrees of intensity.

There are different types of cutter damage as summarized in Table 12.3 for disc cutters [5]. Examples of different disc cutter damage types encountered in the Uskudar–Umraniye–Cekmekoy–Sancaktepe metro tunnel in Istanbul are presented in Figure 12.1 [6].

Figure 12.1 Disc cutter damage type examples in the Uskudar–Umraniye–Cekmekoy–Sancaktepe metro tunnel in Istanbul [6].

Table 12.2 Major cutter failure mechanisms [3, 4].

Major failure type	Failure mechanism
Wear	Frictional and abrasive wear
Fracture	Impact and vibration damage
Wear + fracture	Thermal fatigue
Fracture	Stress fatigue

Table 12.3 Basic disc cutter damage types [5].

Abrasive ring wear
Plastic deformation
Ring chipping
Secondary ring wear
Ring breakage
Cutter blockage
Hub breakage

Abrasive wear can occur tribologically as a result of two-body or three-body frictional contacts of materials [7]. When the mineral grains of the excavated rock protrude solidly on the rock surface, two-body friction is dominant, but when the mineral grains come out and stay free in between the contacting surfaces of the cutting tool and rock, three-body friction is dominant. It can be deduced from the practical experience of the authors of this book that wear of roller cutters such as disc cutters can occur as a combination of two- and three-body frictional contacts, when considering the cutting mechanism of a roller cutter generating a large crushed zone under the tip of the cutter, although one or the other might be more dominant depending on the characteristics of the rock being excavated and the type and geometry of the cutter used.

Cutter wear and damage is the most important cost item for a TBM-excavated tunnel. However, the abrasive effects of the ground being either rock or soil is not limited to the cutters; abrasivity of the ground usually also affects other metal parts of the TBM cutterhead, such as the front plate, cutting wheel (especially the side surfaces), buckets, screw conveyors (especially in EPB-TBMs), transportation pipes in SPB (slurry) TBMs, and so on. Also, soil might be more destructive in terms of abrasivity, especially in face-pressurized excavation systems due to sanding effect.

General disc cutter usage based on their positions on cutterhead is presented in Figure 12.2 [8]. As seen, replacement number increases from the center of the cutterhead towards the periphery area (gauge cutters).

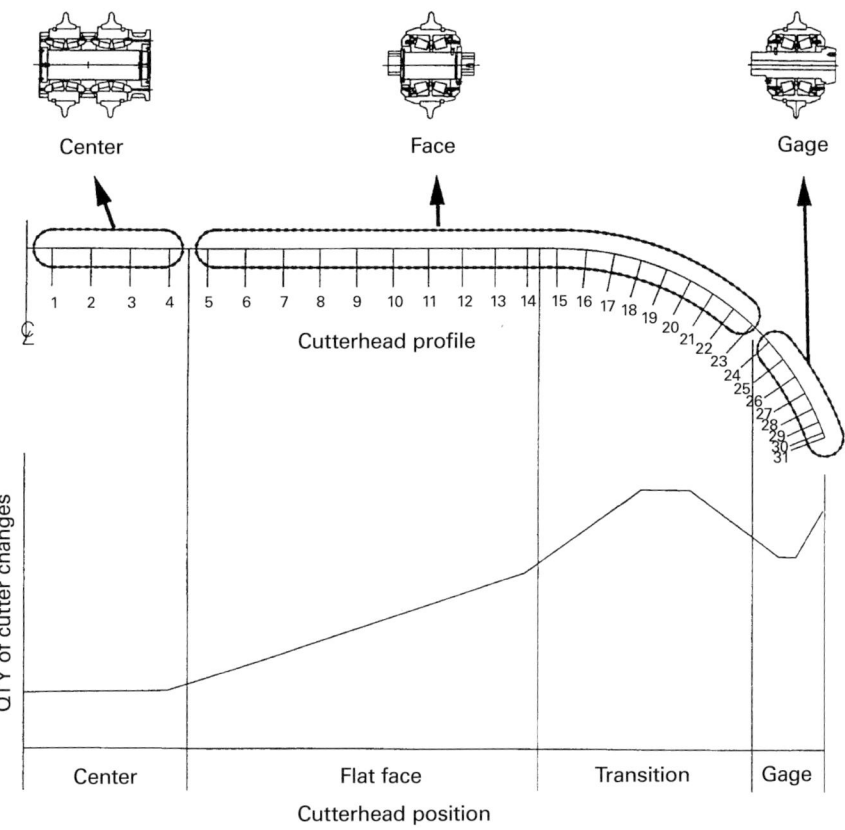

Figure 12.2 General disc cutter usage based on their positions on the cutterhead [8].

12.2 Determination of the abrasivity

Determination of abrasivity of the rock to be excavated in an excavation project is an important task at the feasibility and planning stages, since the project might suffer from the excessive costs of the cutters and/or other metal parts. The first developed and still the most widely used rock abrasivity determination method today for determination of the life of cutters excavating rocks is the Cerchar abrasivity test, which was originally developed in France in the early 1970s. Many other methods such as NTNU abrasion test (AVS), Mohs hardness, Schimazek abrasivity index, Tabor abrasivity index, modified Schmidt hammer abrasivity, LCPC test, Sievers J value test (bit wear index) and the Bohme test have been used for abrasivity determination.

The Cerchar abrasivity index (CAI) indicates the degree of rock abrasivity, for classifying cutter wear rate and costs for different cutter types such as drag cutters and roller cutters. Currently, there are two standard used for this test: ASTM (2010) and ISRM (2014) [9, 10]. The procedures suggested by West (1989) [11] are also used by some institutes and researchers.

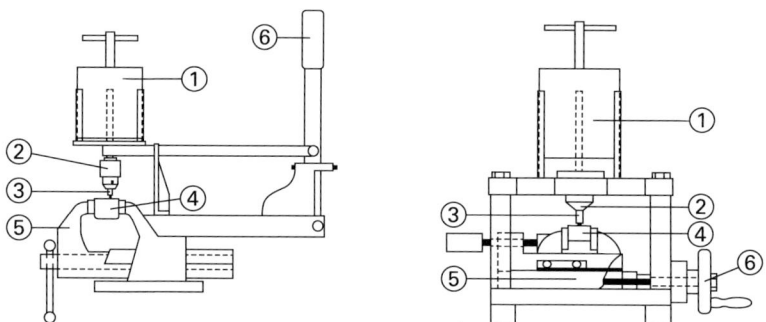

Figure 12.3 Drawings of the most widely used Cerchar abrasivity testing devices [10].

The Cerchar abrasivity tests are performed on freshly broken rock surfaces, free of weathering effects. The remnant pieces from indirect (Brazilian) tensile strength tests are usually used for this purpose. However, it is also possible to convert the test results performed on sawn cut rock surfaces by using relationships given in the literature. Both ASTM (2010) [9] and ISRM (2014) [10] suggest one of two types of Cerchar abrasivity testing devices, as shown in Figure 12.3. Rock samples are clamped in a vice with the surface facing upward. A sharp conically pointed (90° cone angle) hardened steel pin with Rockwell hardness (HRC) of 55±1 is fastened in a 70 N head and set carefully on the rock surface. Values for the Cerchar abrasivity index using a pin hardness other than HRC 55 can be converted using a method given in literature. Then the pin is drawn/dragged/scratched 10 mm across the sample surface in 1 s ± 0.5 s for the original Cerchar device and 10 s ± 2 s for the West type Cerchar device [9]. A number of measurements, five tests on each specimen each time with a new pin (in practice, three scratches perpendicular and two scratches parallel to bedding planes; if there is no obvious anisotropy, three scratches in one direction and two other scratches perpendicular to that direction) in [9] and all perpendicular to bedding planes in [10], should be made on a single specimen of rock to give a reasonable mean value for the rock abrasiveness. Each scratch should be separated from the others by at least ~5 mm, ensuring no interaction between the scratches. A sample photograph should be taken before and after testing.

The tested pin is taken off the apparatus and two perpendicular diameters of the abraded flat area of the pin are measured under a microscope. The microscope should have a minimum magnification of 30×, according to ASTM (2010). The length or diameter of the wear flat, d, should be based on optical and digital methods using a microscope having a minimum magnification of 25× according to ISRM (2014). Measurements may be executed by using a side or a top view setting in ISRM (2014). The readings for each sample are then averaged to obtain Cerchar abrasivity index (CAI) as in Equation (12.1):

$$CAI = \frac{10}{n} \sum_{i=1}^{n} d_i$$
(12.1)

Where i is the reading number (i = 1, 2, ..., n) and d is the respective reading of wear flat in (mm). A wear flat of 0.1 mm means CAI = 1.

Finally, the rocks are classified based on their Cerchar abrasivity indices as given in Table 12.4 by ASTM (2010) and Table 12.5 by ISRM (2014).

Table 12.4 Rock abrasivity classification based on Cerchar abrasivity index [9].

Average CAI	Abrasivity class
0.3–0.5	Very low abrasiveness
0.5–1.0	Low abrasiveness
1.0–2.0	Medium abrasiveness
2.0–4.0	High abrasiveness
4.0–6.0	Extremely high abrasiveness
6.0–7.0	Quarzitic

Table 12.5 Rock abrasivity classification based on Cerchar abrasivity index [10].

Average CAI	Abrasivity class
0.1–0.4	Extremely low
0.5–0.9	Very low
1.0–1.9	Low
2.0–2.9	Medium
3.0–3.9	Height
4.0–4.9	Very high
≥ 5.0	Extremely high

Determination of abrasivity of soils is a relatively new idea, even though some soils might actually be more abrasive than rocks. The one method of determining the abrasiveness of any mixture of grain sizes in the laboratory is the LCPC test, which was developed especially for 'Granulate' by the Laboratoire Central des Ponts et Chaussees in Paris [12]. It was also stated by [12] that LCPC abrasivity is correlated to both equivalent quartz content as described in [13] and to the Cerchar abrasivity index.

Gharahbagh et al. (2011) [14] and Rostami et al. (2012) [15] suggested another test method for determining abrasivity of soils. The testing system was configured to simulate the working condition of the cutting tools in an excavation chamber of pressurized face shields. This included the high contact stresses between the tool and the soil, maintaining the original soil size distribution, field moisture conditions and possibility of applying high ambient pressures, as well as soil conditioners.

Another soil abrasivity testing method worth mentioning is the NTNU soil abrasion test (SAT) [16–18], which is a modified version of the NTNU abrasion test used for rocks (AVS) and University of Torino Polytechnique [19].

Abrasivity determination methods for soils are quite new and require larger databases for a complete evaluation of soil abrasivity and prediction of cutter consumption. It should be stated that there is currently no generally accepted method or standard for determination of soil abrasivity.

12.3   Empirical prediction methods for disc cutter consumption

There are not many methods, or the available ones are limited to particular conditions or disc cutter types and diameters for predicting disc cutter consumption for excavating rocks. The most widely used empirical models based on large field databases are briefly introduced below.

12.3.1   Colorado School of Mines (CSM) model for CCS type 17-inch single-disc cutters

The CSM method is based primarily on experimental Cerchar abrasivity index (CAI) of the rock to be excavated to obtain a basic cutter life (CL) in rolling meters, which is then used for estimating cutter life and cost depending on some mechanical and operational parameters of TBMs [20]. The method is only valid for 17-inch single-disc cutters of constant cross-section (CCS) shape and predicts only abrasive wear, not other types of cutter consumptions such as blocked or chipped cutters. Estimations are as follows:

1) Measure CAI values of rocks by Cerchar abrasivity test.

2) Estimate cutter life in rolling meter (CL):

$$CL = \frac{2 \cdot 10^6}{CAI} \text{ meters} \tag{12.2}$$

3) Calculate the number of cutterhead revolutions before cutter needs replacement (NR):

$$NR = \frac{CL}{2\pi \cdot R \cdot SF} \text{ revolutions} \tag{12.3}$$

R = radius of cutter travel circle (meter).

SF = Severity factor defined by location of cutters. SF can be taken as between 10 and 20 for center cutters, 1.0 for the face cutters, and between 1.2 and 1.5 for the gauge cutters.

4) Estimate cutter life in hours (CLH):

$$CLH = \frac{NR}{RPM \cdot 60} \text{ hours} \tag{12.4}$$

RPM = cutterhead rotation per minute

5) Calculate production per cutter change (PPCC):

$$PPCC = \frac{IPR \cdot A \cdot CLH}{Nc}, \; m^3 \tag{12.5}$$

IPR = Instantaneous penetration rate (m/h)

A = tunnel cross-section area (m^2)

Nc = Total Number of Cutters

6) Calculate cost of each cutter change (CPC):

$$CPC = \left[\frac{HubCost + \left(\dfrac{Rings}{Hub} \cdot RingCost \right)}{\dfrac{Rings}{Hub}} \right] \cdot 1.15 \, USD \tag{12.6}$$

Rings/Hub parameter indicates the number of cutter rings replaced before each hub replacement. It may be assumed that 5–7 cutter rings are consumed for each hub. HubCost and RingCost indicate the cost of a hub and a cutter ring, respectively. The coefficient of 1.15 indicates an additional 15% of miscellaneous cutter part cost.

7) Estimate cutter cost per cubic meter (CCCM):

$$CCCM = \frac{CPC}{PPCC} \, USD/m^3 \tag{12.7}$$

12.3.2 Norwegian Institute of Technology (NTNU) model

NTNU method uses a cutter life index (CLI), which is based primarily on experimental Siever's J-value (SJ) and abrasion value steel (AVS) of the rock to be excavated [21]. CLI is then used for estimating the cutter life by normalizing with TBM- and rock-related parameters. The method is valid only for single-disc cutters of CCS type with different diameters and only predicts abrasive wear.

This model assumes that the TBM is being operated at a thrust level resulting mainly in abrasive wear of the cutter rings. It is also assumed that the amount of blocked cutters and cutter rings worn by ring chipping is less than 10–20% of the total number of replaced cutters. The average life of cutter rings can be estimated by [21] for a given rock type:

$$H_h = \frac{H_0 \cdot k_D \cdot k_Q \cdot k_{rpm} \cdot k_N}{N_{tbm}} \; \text{hours/cutter ring} \tag{12.8}$$

$$H_m = H_h \cdot I_n \; \text{meters/cutter ring} \tag{12.9}$$

$$H_f = \frac{H_h \cdot I_n \cdot \pi \cdot d_{tbm}^2}{4} \; \text{solid cubic meters/cutter ring} \tag{12.10}$$

where

H_0 = basic average cutter ring life in hours

H_h = average cutter ring life in hours

H_m = average cutter ring life in meters

H_f = average cutter ring life in solid cubic meters

I_n = Net penetration rate

N_{tbm} = Number of cutters on TBM

d_{tbm} = TBM diameter

k_D = correction for TBM diameter

k_{rpm} = correction for cutterhead RPM

k_N = correction for number of cutters

k_Q = correction for quartz content

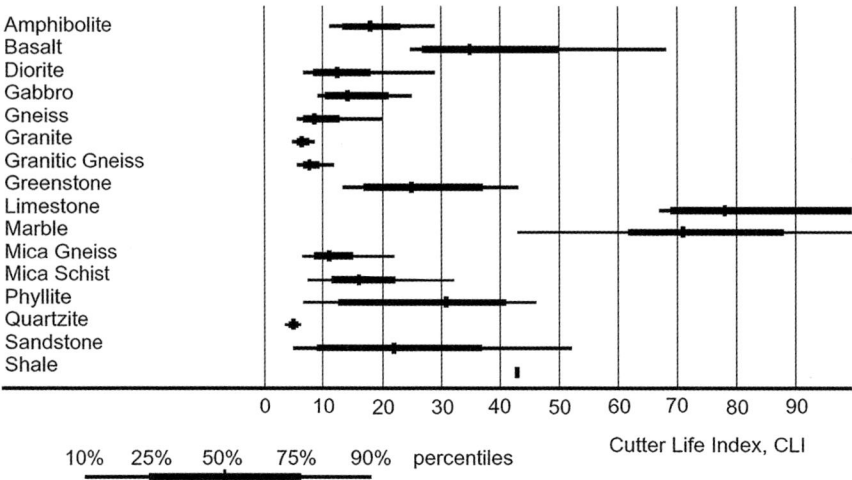

Figure 12.4 Cutter life index (CLI) values of different rocks [21].

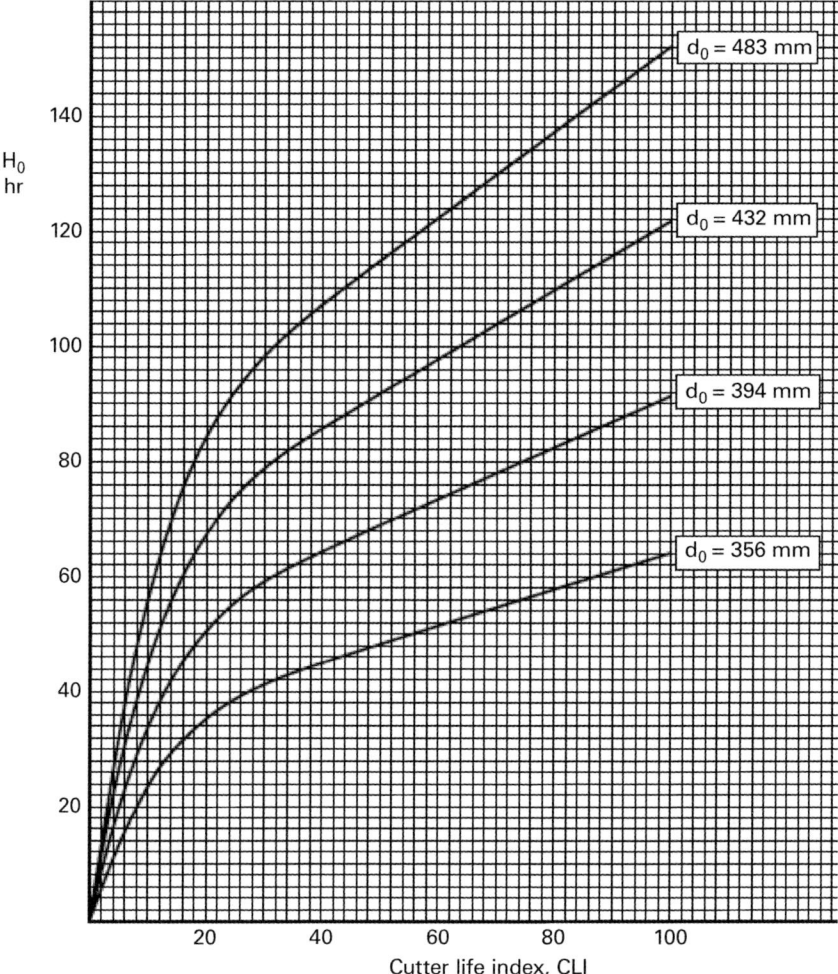

Figure 12.5 Relationship between cutter life index (CLI) and basic average cutter ring life (H0) depending on disc cutter diameter [21].

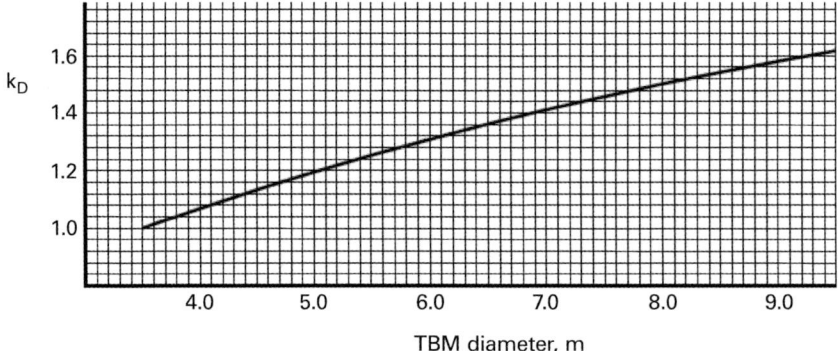

Figure 12.6 Correction for TBM diameter (kD) [21].

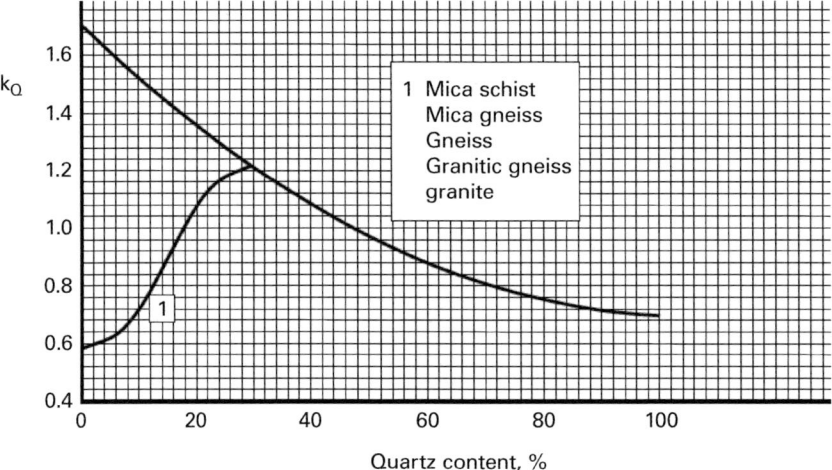

Figure 12.7 Correction for quartz content (kQ) [21].

Cutter life indices of different rocks are summarized in Figure 12.4, based on experimental results from NTNU. The relationship between cutter life index (CLI) and basic average cutter ring life (H_0), depending on disc cutter diameter, is presented in Figure 12.5. Corrections for TBM diameter (k_D) and quartz content (k_Q) are presented in Figures 12.6 and 12.7, respectively. When using Figure 12.7, it is suggested that the content of epidote and garnet may be included in the quartz content. The rock types defined as Group 1 in Figure 12.7 indicates a different characteristic. Corrections for cutterhead RPM (k_{rpm}) and number of cutters (k_N) are presented in Equations (12.11) and (12.12), [21].

$$k_{rpm} = \frac{\left(50 / d_{tbm}\right)}{RPM} \tag{12.11}$$

$$k_N = \frac{N_{tbm}}{N_0}$$ (12.12)

where

RPM = cutterhead rotational speed in rpm as seen in Figure 12.8.

N_0 = Standard number of cutters as seen in Figure 12.9.

H_0 expresses the life of one individual cutter ring in the average cutter position ($\approx 0.59\ r_{tbm}$). H_h, H_m and H_f express average cutter life for the cutterhead or the tunnel. For example, if H_h is equal to 10 meters/cutter ring, it means that on average, someone should change one cutter for every 10 m of tunnel bored.

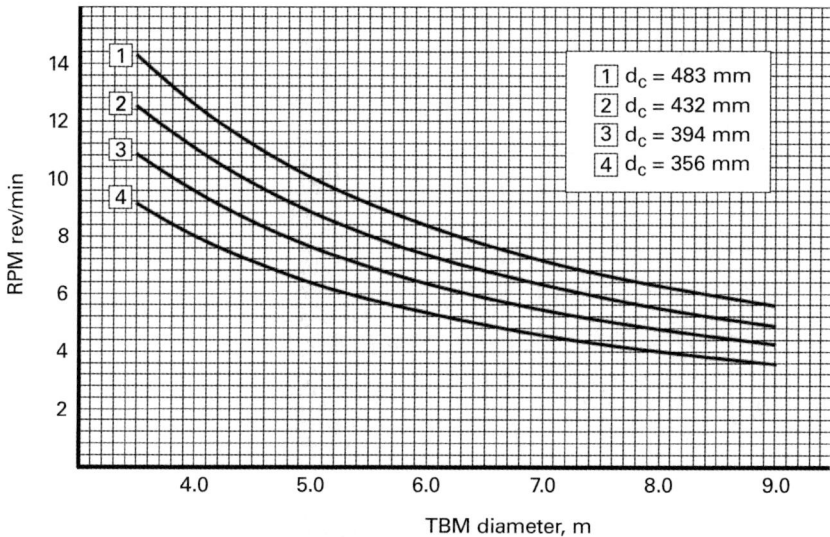

Figure 12.8 Cutterhead rotational speed vs TBM diameter as a function of cutter diameter [21].

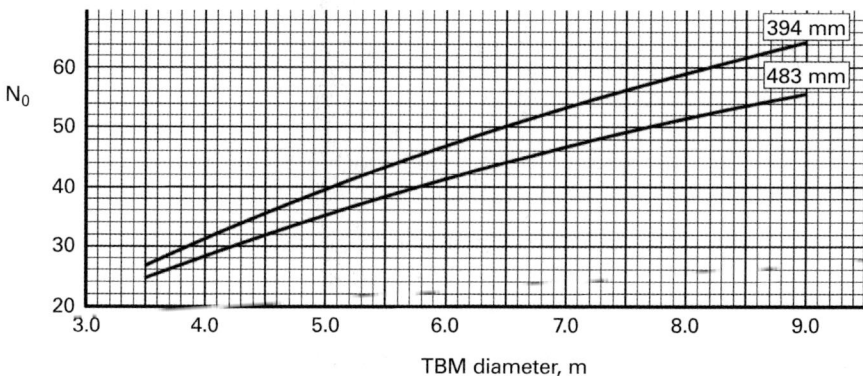

Figure 12.9 Standard cutter number vs TBM diameter as a function of cutter diameter [21].

12.3.3   Maidl et al. (2008) model for CCS type 17-inch single-disc cutters

Maidl et al. (2008) [22] presented a chart for predicting the life of CCS type 17-inch single-disc cutters based on Cerchar abrasivity index and uniaxial compressive strength of the rocks, as seen in Figure 12.10.

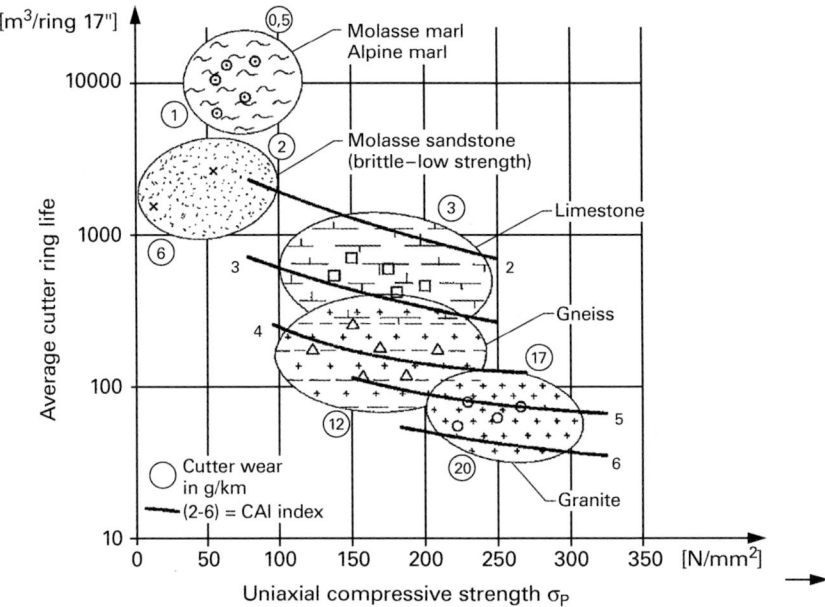

Figure 12.10 Graph for predicting CCS type 17″ single-disc cutter consumption as a function of uniaxial compressive strength and Cerchar abrasivity index [22].

12.3.4   Frenzel (2011) model for CCS type 17-inch single-disc cutters

Another valuable empirical method was developed by Frenzel (2011) [5], based on a large database of field applications using 17-inch single-disc cutters of CCS type, which predicts cutter consumption rate caused by both primary (wear) and secondary (blockage, chipping, split ring breakage, etc.). The model of Frenzel (2011) is based primarily on a wear coefficient, which can be related to CAI of the rocks, and uses TBM related and operational parameters.

12.3.5   Gumus et al. (2016) model for CCS type 12-inch monoblock double-disc cutters of an EPB-TBM

This model was developed by Gumus et al. (2016) [23] for CCS type 12-inch monoblock double-disc cutters used in an EPB-TBM excavating different rocks which can be classified as difficult ground. The detailed information related to this model will be given in the following section as a case study.

12.4 Examples of cutter consumptions on TBMs in Turkey

Parameters or terminology for analyzing and identifying the cutter consumption rates are summarized below after some revisions on Akgul et al. (2015) [24]:

The average linear cutter ring life ($LCRL_i$) per linear tunnel length can be estimated by:

$$LCRL_i = \frac{L_{tunnel}}{NCRR_i + 1} \tag{12.13}$$

where i is the index showing the respective cutter ring position ($i = 1, 2, \ldots, k$), k is the total number of the cutter rings mounted (positioned) on the cutterhead, $LCRL_i$ is the linear cutter ring life for the ith cutter ring position, L_{tunnel} is the total tunnel length and $NCRR_i$ is the total number of cutter ring replacements for the ith cutter ring position. The additional number of (+1) found in the denominator of Equation (12.13) takes into account the first cutter mounted on the cutterhead at the ith position at the beginning of the excavation or the remaining tunnel length up to the final breakthrough (or assuming the final cutter ring replaced is disposed of after breakthrough).

The life of a cutter ring can also be estimated by its rolling life (rolling distance) for a certain replacement and/or as average. The average rolling life for the ith cutter ring position (RL_i, in meters of linear rotational distance of the cutter ring) can be estimated by:

$$RL_i = \frac{1}{n+1} \cdot \sum_{j=1}^{n+1} RL_{i,j} \tag{12.14}$$

where j is the index showing the number of replacement ($j = 1, 2, \ldots, n, n+1$), n is the last replacement number for the respective (i^{th}) cutter ring position ($n = NCRR_i$), (+1) accounts for the remaining tunnel length up to the final breakthrough, and $RL_{i,j}$ is the rolling life of the (i^{th}) cutter ring position for the respective (j^{th}) replacement (in rolling meters/cutter ring).

The rolling life of the (i^{th}) cutter ring position for the respective (j^{th}) replacement ($RL_{i,j}$) can be estimated by:

$$RL_{i,j} = 2 \cdot \pi \cdot r_i \cdot L_{i,j} \cdot \frac{RPM_{i,j}}{IPR_{i,j}} \tag{12.15}$$

where r_i is the distance of the ith cutter ring position to the center of cutterhead (meters), $L_{i,j}$ is the linear excavated tunnel length between two successive replacements of the $(j-1)^{th}$ and the jth (meters), $RPM_{i,j}$ is the average rotational speed of the cutterhead between the tunnel chainages of two successive replacements (revolution per minute) and $IPR_{i,j}$ is the average instantaneous (net) penetration rate of the TBM between the tunnel chainages of two successive replacements (meter per minute). It should be noted that '$j-1=0$' indicates the time before excavation starts; therefore, $RL_{i,0} = 0$. $IPR_{i,j}$ can be estimated by:

$$IPR_{i,j} = L_{i,j} / t_{i,j} \tag{12.16}$$

where $t_{i,j}$ is the operational time between two successive replacements of the cutter ring at the ith position (minutes). Then, Equation (12.15) can be rewritten as:

$$RL_{i,j} = 2 \cdot \pi \cdot r_i \cdot RPM_{i,j} \cdot t_{i,j} \tag{12.17}$$

The volumetric cutter ring life for the ith cutter ring position and the jth replacement ($VCRL_{i,j}$, in solid bank m³/cutter ring) can be estimated by:

$$VCRL_{i,j} = L_{i,j} \cdot \pi \cdot (r_i^2 - r_{i-1}^2) \tag{12.18}$$

It should be noted that in Equation (12.18), when $i = 1$, $r_{i-1} = r_0 = 0$ m corresponds to the center of the cutterhead. Then, the average volumetric cutter ring life for the ith cutter ring position ($VCRL_i$, in solid bank m³/cutter ring) can be estimated by:

$$VCRL_i = \frac{1}{n+1} \cdot \sum_{J=1}^{n+1} \left(VCRL_{i,j} \right) \tag{12.19}$$

Alternatively, Equation (12.19) can be written as:

$$VCRL_i = LCRL_i \cdot \pi \cdot \left(r_i^2 - r_{i-1}^2 \right) \tag{12.20}$$

12.4.1 Tuzla-Akfirat wastewater project in Istanbul

This section is a summary based on Gumus et al. (2016) [23]:

Tuzla-Akfirat wastewater tunnel, with a length of 8178 m and finished diameter of 2.20 m has recently been commissioned to Nas-Akad Construction JV by Istanbul Water and Sewage Authority (ISKI). The basic aim of the Tuzla-Akfirat wastewater tunnel project is to collect the wastewater and transfer it to the Tuzla Wastewater treatment plant. The tunnel site is located in the Anatolian part of Istanbul. The excavation job started on 7 July 2014 and the first shaft (S9-237) was reached on 14 July 2015 after excavating 2189 m through basically hard and abrasive rocks by an EPB-TBM.

A ring of the segmental lining used in the construction includes six segments. Inner and outer diameters of the segment rings are 2.6 and 2.9 m, respectively. The length of each segment ring is 1.0 m. Two locomotives are used for muck and material transportation. One locomotive convoy includes four muck cars each having 4 m³ capacity, one grout mixer and one car for material and personnel. The working pattern during the construction is seven days per week, three shifts per day and eight hours per shift.

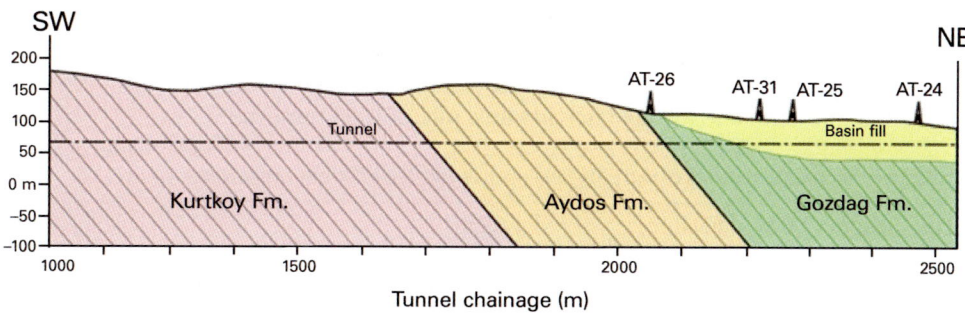

Figure 12.11 Geological cross-section of the Tuzla wastewater tunnel between chainages of 1+000 and 2+500 km [25].

The geological cross-section of the tunnel alignment is presented in Figure 12.11 just for chainages 1+000 to 2+500 km [25]. The tunnel alignment includes from bottom to upper levels, sandstone, conglomerate and shale of Kurtkoy formation (from 0+000 to 1+700 km); quartz arenite and quartz conglomerate of Aydos formation (from 1+700 to 2+100 km); laminated shale with quartz lances of Gozdag formation covered by Cenozoic aged fill material including clay, sand and gravel. These units are cut frequently by andesite and diabase dykes with thicknesses of up to 10 meters and significantly fractured contact zones. The overburden is in the range 10–90 m.

An EPB-TBM is used for excavation. It is a new machine manufactured by Herrenknecht and designed to work in closed and open modes. Its basic characteristics are summarized in Table 12.6. The cutterhead is shown in Figure 12.12.

Figure 12.12 General view of cutterhead of the EPB-TBM used In the Tuzla-Akfirat wastewater tunnel [23].

The TBM started excavation on 7 July 2014 at shaft S9-237. The overall excavation performance from 11 November 2014 (0+564 km) to 11 June 2015 (1+982 km) is summarized in Table 12.7 for 211 calendar days. It should be noted that there were a total of 88 non-working days due to different stoppages during this period, giving 123 working days. Average daily advance rate was 6.7 m/day, average monthly advance rate 200.2 m/month and average net advance rate of 9.63 mm/min.

Table 12.6 Technical features of the EPB-TBM used in the Tuzla-Akfirat wastewater tunnel [23].

Total length (TBM + backup)	62 m
Excavation diameter	3,151 mm
Shield diameter	3,095 mm
Number of push cylinders	8
Stroke of push cylinders	1,700 mm
Main bearing thrust capacity	3,500 kN
Maximum torque (continuous)	620 kNm
Maximum torque (intermittent)	780 kNm
Variable rotational speed	0–9.2 rpm
Installed power	800 kW
Cutterhead power	$2 \times 225 = 450$ kW
Number of disc cutters	4 single removable (center) + 9 double (face) (mostly monoblock) + 2 single removable (gauge)
Diameter of disc cutters	12 inch (305 mm)
Number of scrapers + buckets	16 scrapers + 6 buckets
Nominal diameter of screw conveyor	500 mm
Rotational speed of screw conveyor	0–27 rpm

Table 12.7 Summary of excavation performance in the Tuzla-Akfirat wastewater tunnel [23].

Chainage	0+564 to 1+982 km
Analysis period	11 November 2014 to 11 June 2015
Total calendar days	211 calendar days
Segment rings installed	1413 segment rings
Excavated tunnel volume	11,058 m^3
Average daily advance	6.7 rings (6.7 m/day)
Best daily advance	18 rings (29 February 2015)
Average monthly advance	200.2 m
Best monthly advance	253 m (December 2014)
Machine utilization time	45%

Segment installation time	15%
Other stoppages	30%
Stoppages due to cutter replacement	10%
Average cutterhead RPM	4.4 rpm
Average net advance per minute	9.63 mm/min
Average thrust force	2,132 kN (gross)
Average cutterhead torque	182 kNm (gross)

Each monoblock double-disc cutter and single-disc cutter is replaced with a new one if the wear of a ring reaches at 15 mm (10 mm for gauge cutters). The tip width of the original cutter rings is 13 mm. Anti-wear foaming agent was used throughout the excavation, which is known to reduce wear rates by about 15% [26].

A total of 263 cutter replacements (145 monoblock double discs, 44 removable double cutters and 74 removable single-disc cutters on positions 1, 2, 3, 4, 14, 15) were made between 0+564 and 1+982 km (1418 m, ~11,000 m³ excavation). At positions 8, 10 and 13, only monoblock double discs were used as a precaution, as they have more durability than replaceable ones, since these positions indicated higher consumption rates. Although each ring of the removable double discs could be replaced separately, it was seen that both rings were usually worn out and replaced together, except for a few of them. In this case, all removable double-disc cutters can be assumed to be monoblock disc cutters in terms of cutter life estimations.

Thirty of the removable double- and single-disc cutters (out of total 118 removable double- and single-cutter replacements) showed blockage type damage, and 54 of them indicated normal wear, of which 32 were single ring abrasion (basically on positions 1, 2, 3, 4, 14, 15) and 22 of the replacements were double ring abrasion (22 × 2 = 44 rings). The 80 of the monoblock double-disc cutter replacements (out of total 145 monoblock replacements) indicated normal wear (55%) and 38 of them indicated blockage type of damage (26%), and 27 of them indicated different types of damages (19%) such as spacer ring damage, split ring damage and ring breakage. Examples of unusual damage types (sharpening on a ring) are seen in Figure 12.13.

Figure 12.13 Unusual abrasion on the rings of a monoblock double-disc cutter [23].

The contractor company used three different cutter brands with different prices: one is high quality brand (high life) with a high price, one is medium quality brand (medium life) with a medium price and the other is low quality brand (low life) with a low price. Their general consumption ratios by the contractor company were 15~20% high quality, 25~30% medium quality, and 50~55% low quality brands. Using different qualities of cutters at the same time makes evaluations of cutter consumption rates more difficult (Namli et al., 2014).

In total, 170 replacements were made between the chainages 1+200 and 1+982 km. Of these, 64 were single removable disc cutters at positions 1, 2, 3, 4, 14, 15 and the remaining 106 were double monoblock disc cutters. General cutter life became 35.7 solid bank m^3/replacement (4.6 m of tunnel advance/replacement). Different cutter life realizations are summarized for the chainages between 1+200 and 1+982 km in Figures 12.14–12.18.

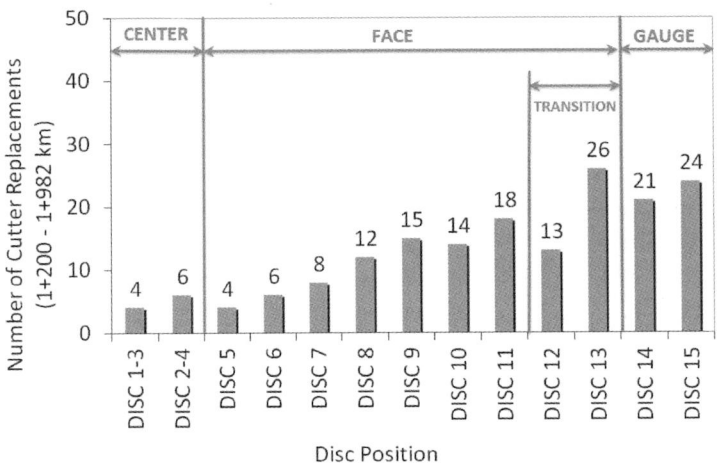

Figure 12.14 Disc cutter replacements based on cutter positions between 1+200 and 1+982 km [23].

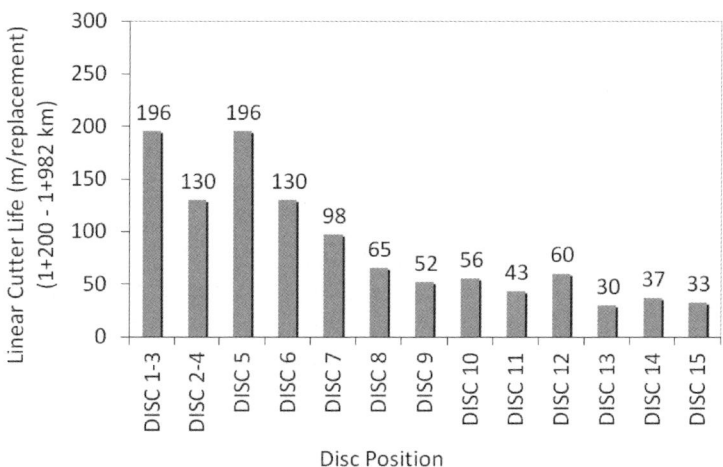

Figure 12.15 Linear cutter life based on cutter positions between 1+200 and 1+982 km [23].

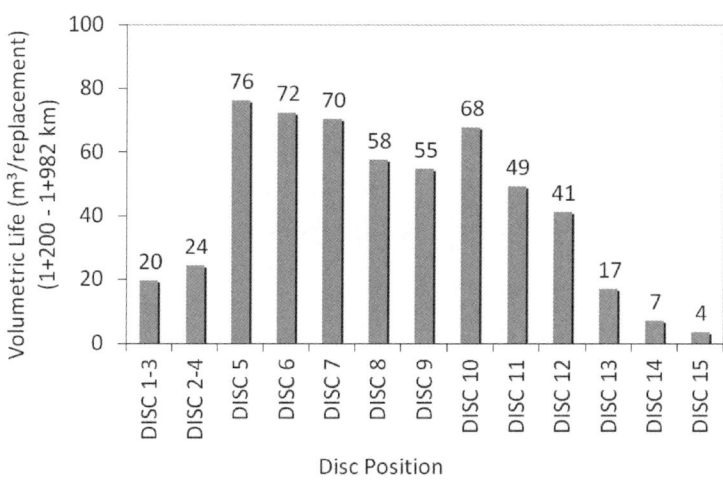

Figure 12.16 Volumetric cutter life based on cutter positions between 1+200 and 1+982 km [23].

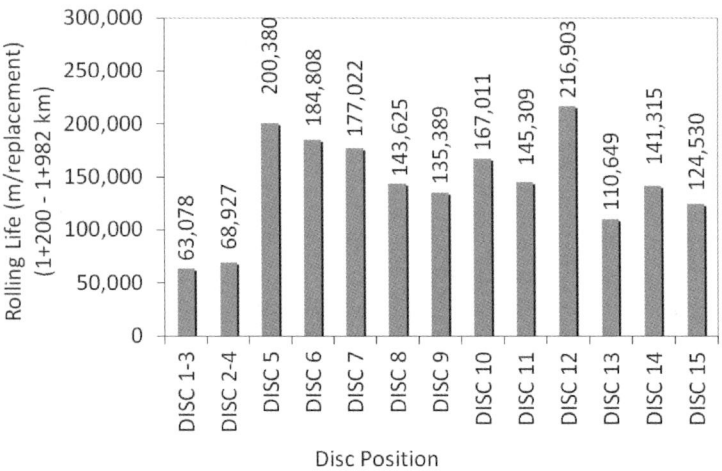

Figure 12.17 Rolling life based on cutter positions between 1+200 and 1+982 km [23].

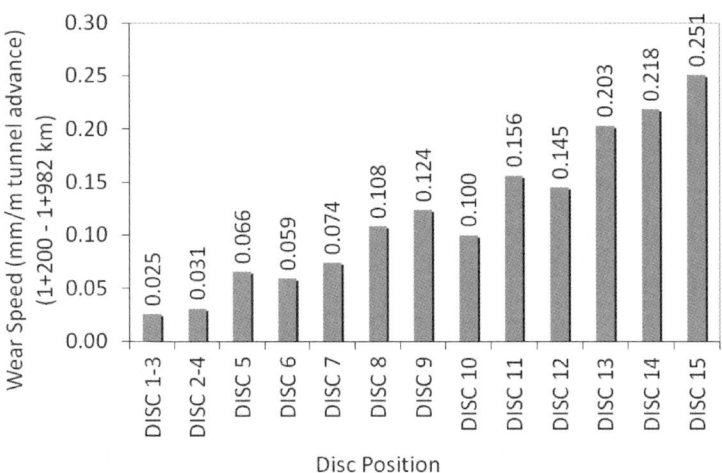

Figure 12.18 Wear speed based on cutter positions between 1+200 and 1+982 km [23].

Single variable regression analyses indicate that there are no meaningful relationships between rock properties (uniaxial compressive strength and Cerchar abrasivity index) and cutter life. Also, there are no meaningful relationships between rock properties and TBM operational parameters, although general expected trends are observed. This is mostly because the operator always tries to keep the thrust force at the same (safe) levels, being usually 2,000–2,500 kN. Studies also indicate that the TBM operational parameters of penetration per revolution and field specific energy are moderately correlated to cutter life and can be used for predicting the life of double-disc cutters as given below.

The relationship between uniaxial compressive strength and Cerchar abrasivity index values are presented in Figure 12.19 which shows the variation of rock properties along the chainage between 1+200 and 1+982 km. It is seen that uniaxial compressive strength varies between 30 MPa and 215 MPa, while Cerchar abrasivity index is in the range 1.0–5.5.

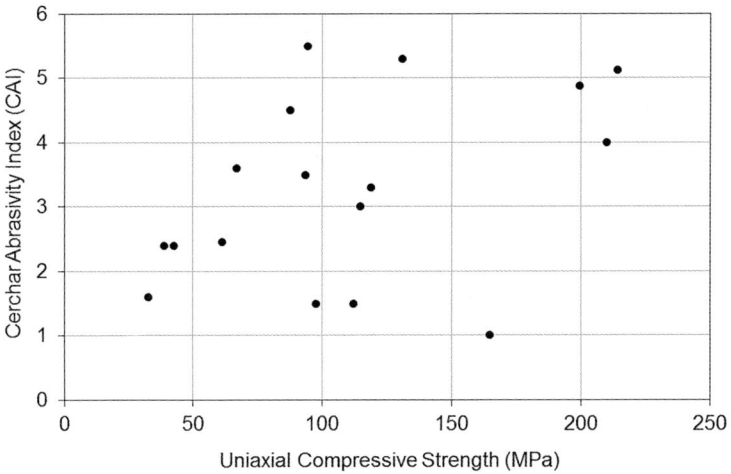

Figure 12.19 Relationship between uniaxial compressive strength and Cerchar abrasivity index between 1+200 and 1+982 km.

The relationships between penetration per revolution and wear speed and volumetric cutter life are seen in Figures 12.20 and 12.21, respectively. Cutter life increases with increasing penetration per revolution. Penetration per revolution does not indicate any meaningful relationship with rolling life. When line spacing between cutter rings is considered as being around 80–85 mm, the penetration values given in Figures 12.20 and 12.21 seem to be very low (usually with high line spacing to penetration ratios: over 20), since thrust capacity of the TBM main bearing (for a small diameter cutterhead) is usually not enough for the hard rocks excavated. This indicates that even not very hard and abrasive rocks can be considered as difficult ground for a small diameter TBM and small diameter double-disc cutters.

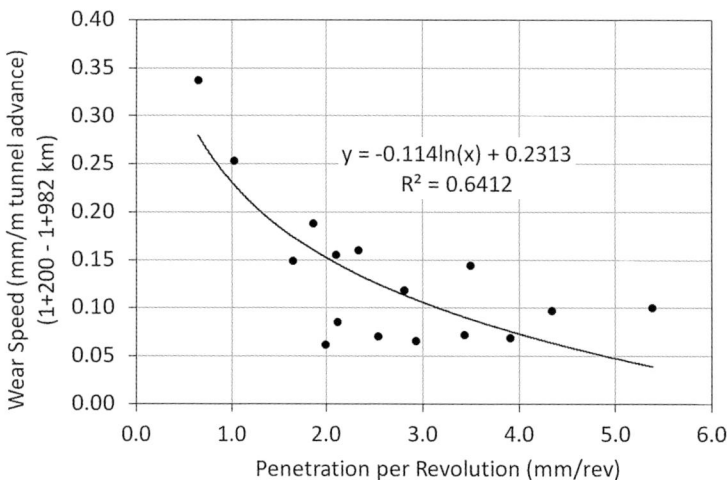

Figure 12.20 Relationship between penetration per revolution and wear speed between 1+200 and 1+982 km [23].

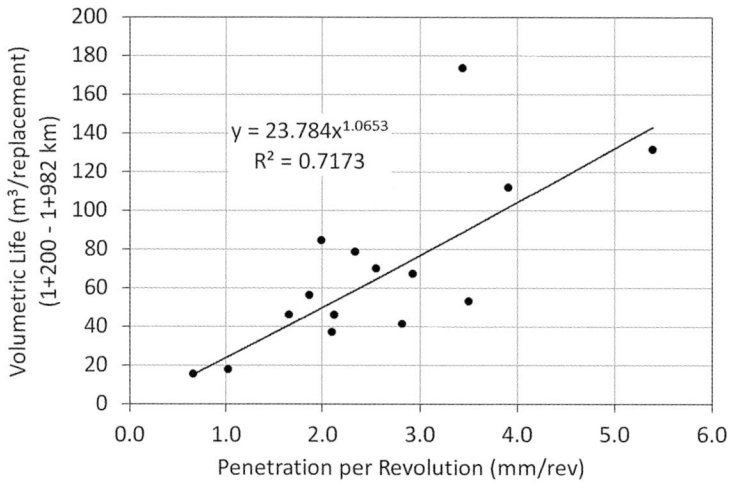

Figure 12.21 Relationship between penetration per revolution and volumetric cutter life between 1+200 and 1+982 km [23].

The relationships between field specific energy and cutter life are seen in Figures 12.22 and 12.23. Since the linear cutter life and volumetric cutter life are directly correlated to each other, only volumetric life is given in Figure 12.23. Cutter life decreases with increasing field specific energy. Field specific energy does not indicate any meaningful relationship with rolling life. It is also indicated that wear speed is highly correlated to the cutter position (distance to cutterhead center) as shown in Figure 12.24.

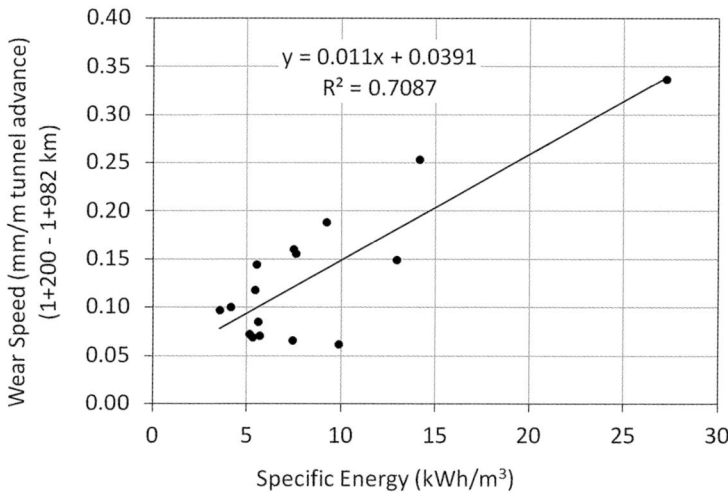

Figure 12.22 Relationship between field specific energy and wear speed between 1+200 and 1+982 km [23].

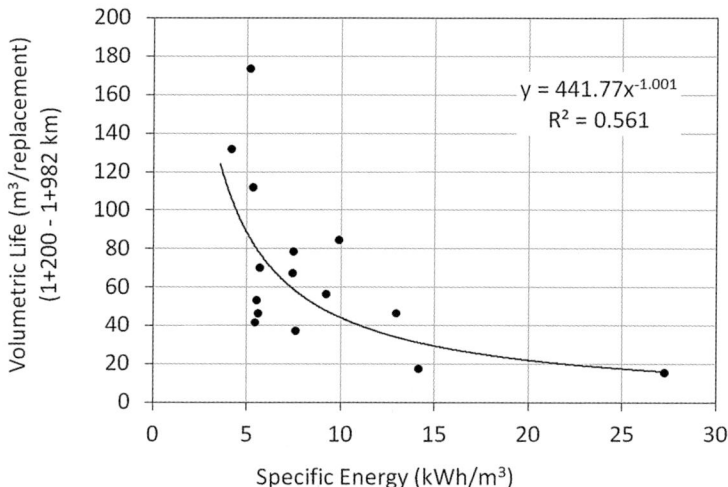

Figure 12.23 Relationship between field specific energy and volumetric cutter life between 1+200 and 1+982 km [23].

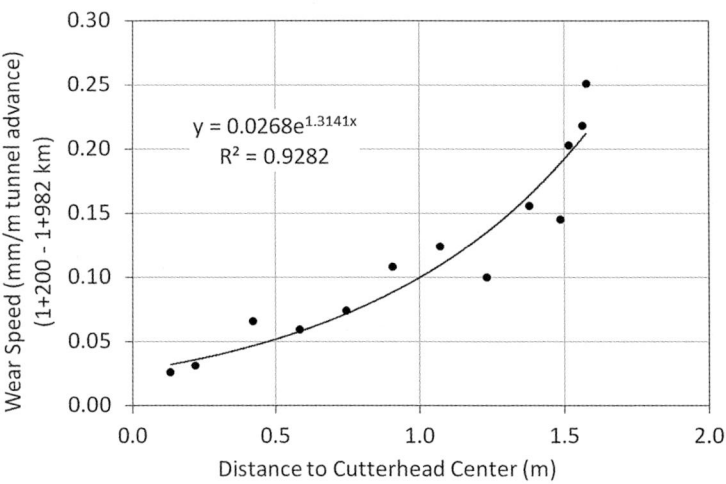

Figure 12.24 Relationship between wear speed and distance to cutterhead center (disc position) between 1+200 and 1+982 km [23].

Single parameter regression analyses indicate that penetration per revolution and field specific energy may be used for predicting life of constant cross-section double-disc cutters. However, additional data including different TBM diameters, disc cutter diameters and rock types is required to generalize the findings of this study.

As a summary, detailed analyses at chainages 1+200 to 1+982 km indicate that average cutter consumption rate is 35.7 m^3/replacement (4.6 m/replacement). Cerchar abrasivity index and uniaxial compressive strength does not indicate any meaningful relationship with cutter life. Penetration per revolution and field specific energy might be used for predicting lives of constant cross-section double-disc cutters as mentioned above.

The cutter lives observed in this project are considered as quite low. This is mostly because the ground conditions encountered can be considered as difficult ground conditions for the EPB-TBM used in the Tuzla-Akfirat wastewater tunnel, since it is considered that a larger diameter TBM equipped with larger diameter single-disc cutters would excavate these rocks without any significant problem. Since the EPB-TBM used in the Tuzla-Akfirat wastewater tunnel has a small diameter, it has a limited thrust and power for the rocks excavated. The disc diameter of 12″ limits the applied load and thrust on the cutters, and thus the penetration rate. It is also known that applied thrust on double discs is split into two halves, reducing its penetration capability into the rock face and increasing tool consumption rates. In addition, using a low quality brand of cutter (50–55%) contributes to the reduced cutter life.

12.4.2 Yamanli II HEPP project in Adana

Field observations on replacements of 17″ constant cross-section single-disc cutters and parameters affecting cutter consumption in construction of the Yamanli II HEPP tunnel excavated by a double-shield TBM with 4.305 m excavation diameter are summarized in this section, based on Akgul et al. [24].

The contractor company NTF completed excavation of the 6.62 km tunnel between 27 October 2012 and 8 December 2013 (between chainages 6+950 and 0+317 km). The geology consisted of mainly massive and laminated limestones with different lamination thicknesses having uniaxial compressive strength of 30–80 MPa. A total of 58 cutter replacements were realized. The total average cutter ring consumption rate was 1661 m^3/cutter ring (114 m/cutter ring). Although the cutter consumption rate can be considered as relatively low (high cutter life), the lamination thickness highly affected the type of cutter damage, being mainly cutter ring chipping, ring breakage and cutter blockage, and most of the cutter consumption was due to lamination effect.

The geological cross-section of the studied zone is presented in Figure 12.25. Geological and geotechnical properties of the three geological zones encountered along the tunnel alignment are summarized in Table 12.8, including some of the average operational parameters of TBM obtained during excavation.

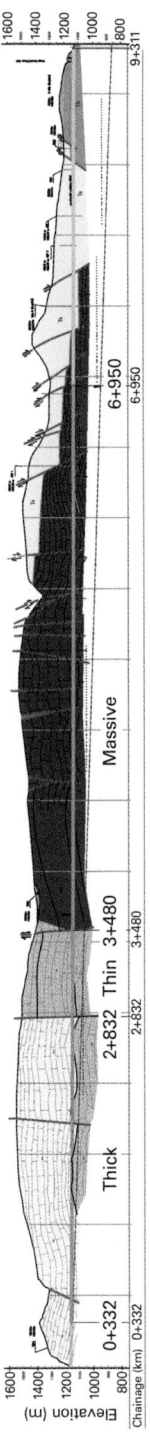

Figure 12.25 Geological cross-section of the Yamanli II HEPP tunnel [24].

Approximately 3,470 m of the alignment includes massive limestone (Zone 3). The rest includes thin to medium laminated limestone having thickness around 6~20 cm and medium to thick laminated limestone having thickness around 20~60 cm. The dip of the layers varies between around 5 and 75°. Average RQD is 75–80%. Average overburden height is around 400 m. Uniaxial compressive strength of the limestone is 30–80 MPa. Water ingress is in the levels of dripping. Average Q rock class is around 10.53 and RMR class is around 40–80. Five fault zones with crushed limestone and a few clay layers having thickness of 1–6 m were also encountered. Photographs of the outcrop of the excavated formation are presented in Figure 12.26.

Figure 12.26 Photographs of the outcrop of the excavated formation in the Yamanli II HEPP tunnel [24].

Table 12.8 Some of the geological and geotechnical properties of the excavated zones including some operational parameters of the double-shield TBM in the Yamanli II HEPP tunnel [24].

Rock mass zone	Zone chainage (km)	Zone length (m)	Zone comp. strength (MPa)	Zone RMR class	Average cutterhead rotational speed (RPM)	Average instantaneous penetration rate (mm/rev)
Zone 1	0+332–2+832	2500	30–50	40–50	8.75	6
Zone 2	2+832–3+480	648	45–70	50–60	9	7
Zone 3	3+480–6+950	3470	65–80	65–80	9	7

Zone 1: Medium to thick laminated (20–60 cm) limestone; Zone 2: Medium to thin laminated (6–20 cm) Limestone; Zone 3: Massive limestone

A double-shield TBM manufactured by Robbins was used for construction of the first stage energy tunnel of the Yamanli II HEPP project. The TBM is a new one (not used before and not refurbished). A picture and basic features of the machine are presented in Figure 12.27 and Table 12.9. The excavation job started on 27 October 2012 in Zone 1 (medium to thick laminated limestone) and completed on schedule on 8 December 2013 in Zone 3 (massive limestone) in a total of 14 months. The excavation performance is summarized in Table 12.10. Monthly advance rates are presented in Figure 12.28.

Figure 12.27 Photograph of the Robbins double-shield TBM used in the first stage energy tunnel of the Yamanli II HEPP [24].

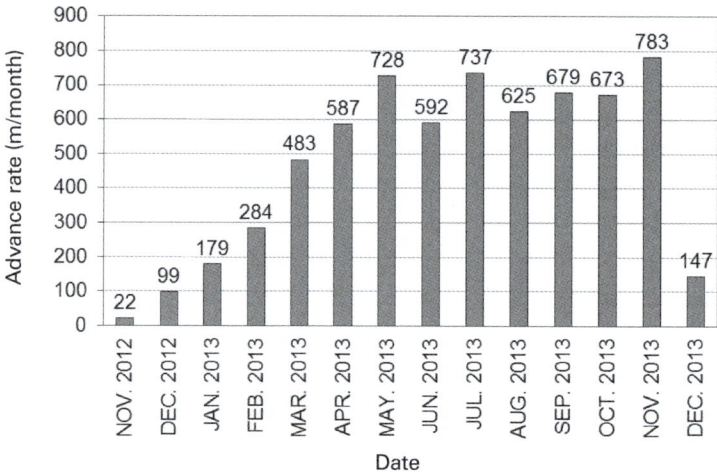

Figure 12.28 Monthly advance rates in the Yamanli II HEPP tunnel [24].

Total cutter ring replacement numbers and rates are summarized in Table 12.11 based on rock mass zones (lamination of limestone). A total of 58 cutter rings were replaced. Unit cutter ring consumption rate was around 114 linear meters of tunnel per cutter ring, which also gives 1,661 solid bank m^3/cutter ring. The lowest consumption rate was observed in the massive limestone zone, being 2,657 m^3 (or 183 m)/cutter ring. The highest rate of replacement is observed with the medium to thick laminated limestone, being 1,137 m^3 (or 78 m)/cutter ring.

Each ring of a double removable disc cutters is separately numbered in this study. Therefore, there are eight rings on the center, four single rings on the gauge, and remaining single discs on the face. The ring position 1 is the closest to and 28 is the furthest from the center of the TBM cutterhead. The transition cutters (cutter rings 20, 21, 22, 23, 24) are included within the face cutters.

The numbers of cutter ring replacements (including the rings of the complete cutter replacements) are summarized in Figure 12.29, based on cutter ring position. Cutter rings 1, 2, and 4 have never been replaced. The most replaced cutter rings are the rings 14, 15, 20 and 22 on the face (each was replaced four times).

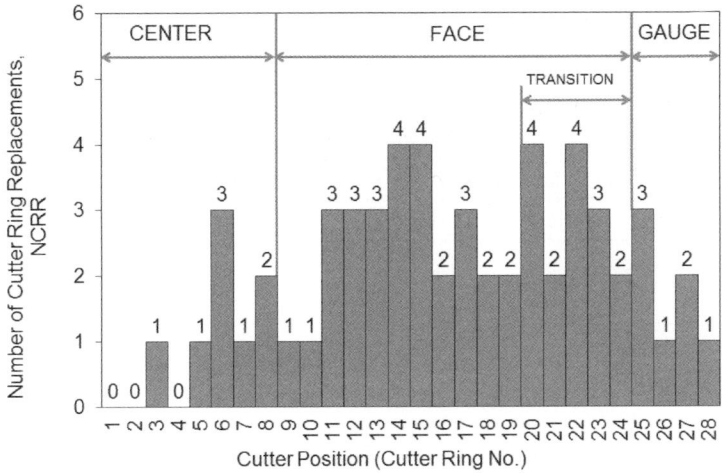

Figure 12.29 Number of cutter ring replacements based on cutter ring positions in the Yamanli II HEPP tunnel [24].

Table 12.9 Technical features of the Robbins double-shield TBM used in the Yamanli II HEPP tunnel [24].

Excavation diameter	4,305 mm
Maximum advance speed	130 mm/min
Number of disc cutters	4 double removable (center) + 16 single (face) + 4 single (gauge)
Diameter of disc cutters	17″ (432 mm) CCS
Tip width	19 mm
Ring hardness	55–57 HRC
Over-excavation capacity	50 / 80 mm
Number of main thrust cylinders	8 cylinders
Number of auxiliary thrust cylinders	11 cylinders
Maximum thrust of main cylinders	1,800 tons (17,660 kN)

Stroke of thrust cylinders	1,400 mm
Rotational speed of cutterhead	0–13.5 rpm
Maximum cutterhead torque	1,123 kNm
Maximum breaking torque	1685 kNm
Cutterhead power	$4 \times 330 = 1320$ kW
Installed total power	1,870 kW

Table 12.10 Summary of excavation performance in the Yamanli II HEPP tunnel [24].

Average daily advance rate	~ 16 m/day
Best daily advance rate	43.2 m/day
Average monthly advance rate	472.7 m/month
Best monthly advance rate	782.8 m/week
Machine utilization (excavation + support)	50% (35~60%)
Planned stoppages	18%
Stoppages due to cutter change	1.9%
Average RPM	8.75 ~ 9.25 rpm
Average instantaneous penetration rate	~ 3.4 m/h

Table 12.11 Total cutter ring replacement numbers and rates, based on excavated rock mass zones in the Yamanli II HEPP tunnel [24].

Rock mass zone	ETL (m)	NCRR (cutter rings)	ETL / NCRR (m/cutter ring)	ETV / NCRR (m³/cutter ring)
Zone 1	2,500	32	78	1137
Zone 2	648	7	93	1347
Zone 3	3,470	19	183	2657
TOTAL		58	Average 114	Average 1661

NCRR: Number of cutter ring replacements
ETL: Excavated tunnel length in related zone (m)
ETV: Excavated tunnel volume in related zone (m³)

Note: Tunnel length is 6,618 m; tunnel cross-section is 14.56 m²; total excavation volume is 96,358 m³.

The average linear cutter ring life is summarized in terms of linear tunnel excavation length in Figure 12.30, based on cutter ring positions. Since the cutter rings 1, 2 and

4 have never been replaced during the whole excavation, they have the highest life being 6,618 m of tunnel/cutter ring. Lives of the rest of the cutter rings on the respective positions vary between 1,324 and 3,309 m/cutter ring. The average rolling life and volumetric cutter life for each cutter position are presented in Figures 12.31 and 12.32, respectively.

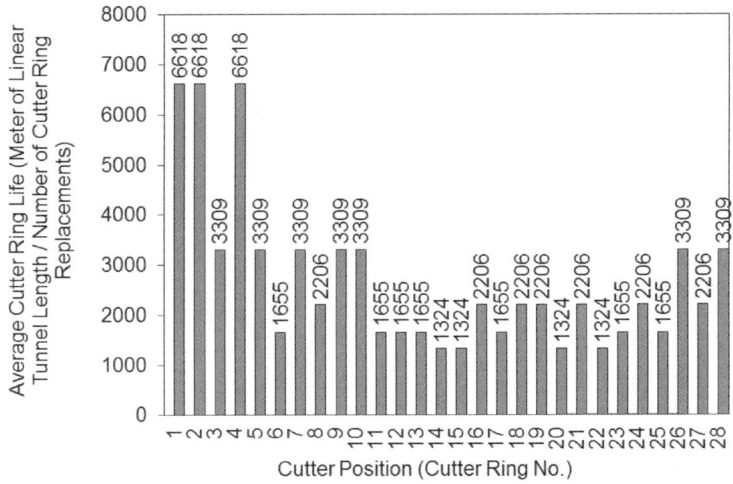

Figure 12.30 Average linear cutter ring life based on cutter ring positions in the Yamanli II HEPP tunnel [24].

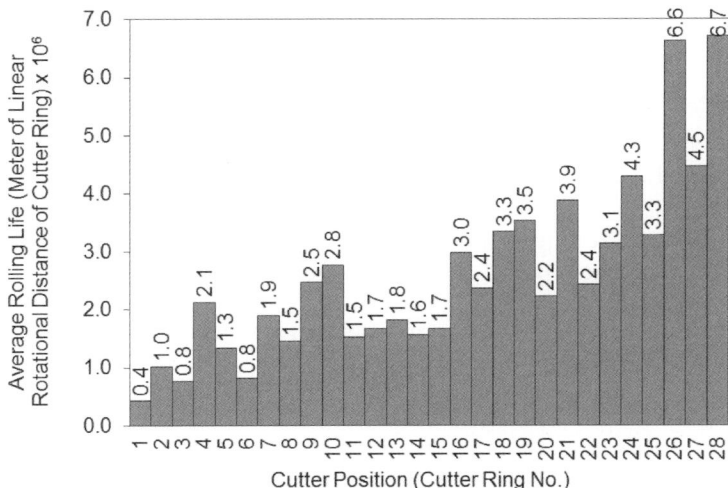

Figure 12.31 Average rolling ring life of cutters based on cutter ring positions in the Yamanli II HEPP tunnel [24].

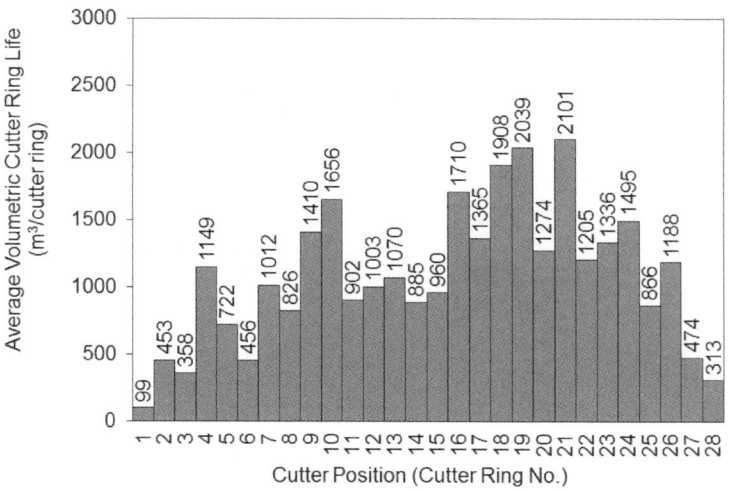

Figure 12.32 Average volumetric life of cutters based on cutter ring positions in the Yamanli II HEPP tunnel [24].

If the cutter positions are considered, the gauge cutters have the lowest average life (1,126 solid bank m^3/cutter ring) while the face cutters have the highest (1,819 solid bank m^3/cutter ring) in terms of volume excavated.

If the complete cutter (hubs + cutter rings) replacement numbers and rates due to blockages are considered, the highest complete cutter replacement rate was observed in the massive limestone, being 7,216 m^3 (or 496 m)/cutter. The lowest complete cutter replacement rate was observed on the medium to thin laminated limestone (Zone 2) being 9,428 m^3 (or 648 m)/cutter. The average complete cutter replacement rate was around 8,030 m^3 (or 552 m)/cutter. The rest (46 replacements) out of total 58 were only cutter ring replacements.

Distribution of the cutter damage types based on the rock mass zone is summarized in Table 12.12. As seen in Table 12.12, the thick laminated limestone (Zone 1) yields mostly the damage types of ring breakage and ring chipping, while the thin laminated limestone (Zone 2) yields mostly ring chipping type of damage. Massive limestone (Zone 3) mostly yields the damage types of cutter blockage and ring chipping.

Table 12.12 Number of cutter ring replacements based on cutter damage types and excavated rock mass zones in the Yamanli II HEPP tunnel [24].

Rock mass zone	Damage type according to Frenzel (2011)			Total
	ring chipping	ring breakage	blockage	NCRR
Zone 1	12	16	4	32
Zone 2	5	1	1	7
Zone 3	8	4	7	19
Total NCRR	25	21	12	58

Rock mass zone	Damage type according to Frenzel (2011)			Total
	ring chipping	ring breakage	blockage	NCRR
%	43	36	21	100

NCRR: Number of cutter ring replacement;

The encountered damage types are mainly ring chipping (43%), ring breakage (36%) and cutter blockage (21%). It is seen that the lamination of excavated formation generates cutter damage types of basically ring chipping, ring breakage and cutter blockage. Also, it can be deduced that although the cutter consumption rate can be generally considered as low, the lamination thickness has an important effect on cutter consumption rates. A picture of the excavated face showing the large cavities occurred due to lamination (discontinuity) is presented in Figure 12.33. The voids occurred due to lamination generating sudden impacts (shock loads) on the cutters and as a result damage on the cutters. Some pictures of the damaged cutters are presented in Figure 12.34. It was also observed in Yamanli that the protective metal pieces welded in front of each cutter improved the cutter life tremendously by preventing any hit of the excavated rock pieces (chips) to the cutters directly and/or any rock bridge occurring during the excavation, as seen in Figure 12.35.

Figure 12.33 Excavation face with voids (occurred due to lamination) affecting damage types in the Yamanli II HEPP tunnel [24].

Figure 12.34 Damaged cutters in the Yamanli II HEPP tunnel [24].

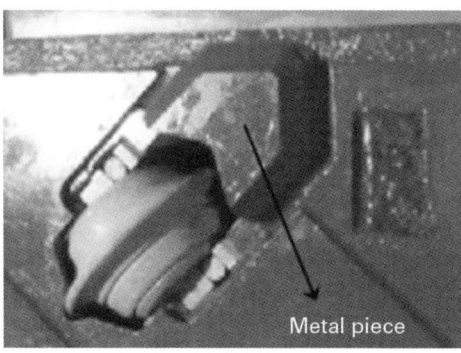
Metal piece

Figure 12.35 Protective metal piece increasing cutter life [24].

In summary, the data obtained from the Yamanli II HEPP Tunnel indicates that the lamination of the excavated formation affects the type of damage on the cutters, as well as consumption rates. The three main damage types were cutter ring chipping (43%), ring breakage (36%) and cutter blockage (21%) through all the excavated alignment. The thick laminated limestone yields mostly the damage types of ring breakage and ring chipping, while the thin laminated limestone yields mostly ring chipping type of damage. Massive limestone mostly yields the damage types of cutter blockage and ring chipping. The lowest average cutter consumption rate was observed in massive limestone as being 2,657 m³/cutter ring (183 m/cutter ring). The highest average cutter consumption rate was observed in thick laminated limestone as being 1,137 m³/cutter ring (78 m/cutter ring). The cutter consumption rate in the thin laminated limestone was close to that in thick laminated limestone, being 1,347 m³/cutter ring (93 m/cutter ring). Total cutter consumption rate for all alignment became 1,661 m³/cutter ring (114 m/cutter ring).

Since the excavated formation had a soft to medium strength and also did not consist of hard or abrasive minerals, the cutter cost due only to consumption and replacement (excluding replacement time losses due to stopping the excavation and additional cost of labor for replacement) was relatively low, being around 1.82 USD/m³. However, the low cutter cost cannot only be due to favorable ground conditions. Other contributing factors were the design of the cutterhead, the quality of the materials and workmanship, the optimized operational conditions and the fast advance rates.

12.4.3   Beykoz wastewater project in Istanbul

This section is a summary based on Bilgin et al. (2012) [27].

The sewerage tunnel between Kavacik and Beykoz in the northern part of the Istanbul Strait (Bosphorus) is part of an environmental protection project concerned with renewing the inadequate sewerage network around Beykoz, collecting the wastewater in a treatment plant and cleaning the polluted water for discharging into the Bosphorus.

The ground conditions change from soft ground to hard formations with water ingress in some sections (sometimes excessive). The total length of the tunnel was 7,253 m. Excavation of the tunnel started from the shaft AT2 on 9 January 2007, and 4,267 m of the tunnel alignment had been excavated by 26 January 2009. A single-shielded Robbins TBM having diameter of 3,175 m and housing 23 constant cross-section (CCS) single-disc cutters with a diameter of 356 mm was used (Figure 8.3). The cutting power of the machine was 400 kW having a torque of 254 kNm at 8.5 rpm and 525 kNm at 4.3 rpm.

The project area is situated in Beykoz in the northern part of the Istanbul Strait. Paleozoic aged Gozdag, Dolayoba and Kartal formations and alluvium are found in the area. The general geological cross-section of the tunnel between shafts AT2 and S6 is given in Figure 8.2.

Rock formations in the tunnel alignment consist of carbonated shale, sandstone, mudstone, clayey limestone, coral limestone, diabase dykes and frequent quartzite veins and some alluvial materials. Boulders were also encountered frequently during excavation of alluvium.

The machine passed through 1,200 m of very difficult ground conditions such as alluvium, artificial fills, mudstone and quite sludgy formations at chainages 900 to 2,100 m. A combination of chisel tools and discs was used on the cutterhead at chainages 800–1,765 m, and all the disc cutters were changed to chisel tools in shaft S9 (chainage 1,765 m), since quite uneven and excessive disc wear were encountered when passing the sludgy formation seen in Figure 12.36, and the rock was medium strength mudstone thereafter. Figure 12.37 also shows wear types encountered in sludgy formation: the disc cutters cannot roll freely in sludgy ground due to the low friction between the disc and the soft formation, and eventually this causes stalling and wearing out of the disc cutters [27–28]. The TBM was fully equipped with chisel tools for chainages 1,772–2,115 m from 1 December 2007 to 31 January 2008. In this zone, the rock formations had uniaxial compressive strengths of 10–76 MPa. All the cutters changed from chisel to discs after chainage 2,115 m, since limestone with high strength was expected to be excavated.

Figure 12.36 Sludgy formation encountered between the shafts AT2 and S6 in the Beykoz sewerage tunnel [27].

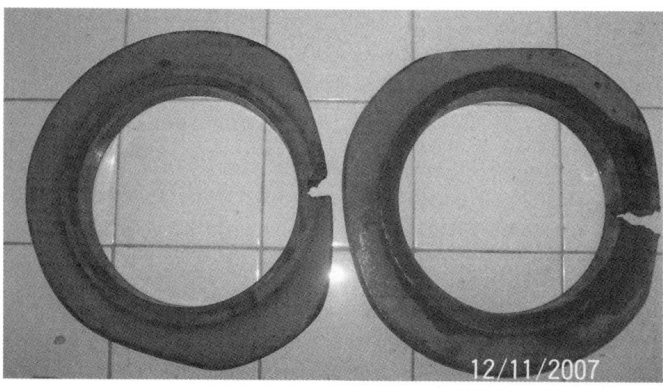

Figure 12.37 Typical disc cutter damage type in sludgy formation between the shafts AT2 and S6 in the Beykoz sewerage tunnel [27].

12.4.4 Buyukcekmece wastewater tunnel in Istanbul

This section is a summary based on Ozaydin et al. (2013) [28].

The Buyukcekmece wastewater tunnel is located in the European part of Istanbul, next to the Lake Buyukcekmece, and was constructed by Nas-Yse JV which was a subcontractor of Nas-Kaner JV. The length of the tunnel alignment is 3,702 m. The project also included excavation of another tunnel with 900 m length being 450 m away from the first tunnel. It was excavated by an EPB-TBM (Herrenknecht) with excavation diameter of 3.81 m (the finished diameter is 2.8 m), cutterhead power of 400 kW, cutterhead torque capacity of 1,014 kNm and thrust capacity of 13,600 kN (Figure 12.38). The TBM was equipped with 12 double-disc cutters and 42 ripper tools. Excavation of the tunnel started on 20 April 2013. The alignment passes through Oligocene aged Gurpinar formation (clay, sand and mudstone with RQD of 17–100% and uniaxial compressive strength of up to 1.4 MPa), Late Quaternary aged Kusdili formation (clay and sand), and Tersier-late Quaternary aged Sogucak formation (karstic limestone having 36 MPa average uniaxial compressive strength and RQD of up to 20% – highly fractured, at chainage 0+000 to 0+221 km). The Gurpinar formation consists of large rock blocks in the transition zone with the Sogucak formation.

Figure 12.38 EPB-TBM used in the Buyukcekmece wastewater tunnel [28].

While excavating the tunnel at 900 m in the Gurpinar formation, excavation speed of the TBM slowed right down. When the excavation chamber of the TBM was checked to determine the problem, it was seen that center cutter of the TBM was totally worn out (Figure 12.39) due to a 1–1.5 m of massive limestone block from the Sogucak formation. Experimental analysis indicated that the block had uniaxial compressive strength of 75 MPa and Cerchar abrasivity of 1.0. The center cutter was not replaceable in this project, which caused considerable difficulties. This case indicated the importance of replaceable center cutters.

Figure 12.39 New (left) and worn (right) center cutter in the Buyukcekmece wastewater tunnel [28].

12.4.5 Uskudar–Umraniye–Cekmekoy–Sancaktepe metro tunnel in Istanbul

This section is a summary based on Namli et al. (2014) [6].

Here, field observations on cutter wear and breakages between consecutive stations of Carsi and Umraniye of the Uskudar–Umraniye–Cekmekoy–Sancaktepe metro tunnel in Istanbul are detailed. The metro line of 20 km, having two tubes and 16 stations, was recently commissioned to Dogus Construction Company by Istanbul Metropolitan Municipality. Construction works started on 20 March 2012 and are still continuing (but nearing the end).

The geology between Carsi (S-08) and Umraniye (S-07) stations, having a total length of 860 m, includes siltstone-claystone, sometimes sandstone, and a diabase dyke of 50 m in length. They are lithological units of the Gozdag formation. The geological, physical and mechanical properties of the siltstone-claystone and diabase dyke are summarized in Table 12.13.

Table 12.13 Physical and mechanical properties of the siltstone-claystone and diabase between Carsi and Umraniye stations [6].

Lithology	Siltstone-claystone	Diabase dyke
Uniaxial compressive strength (MPa)	36.2 ± 12	
Static elasticity modulus (MPa)	3625–4120	
Cerchar abrasivity index	0.5–1.0	4.5–5.0
Rock quality designation, RQD (%)	0–100 (mean 36)	
Rock mass rating (RMR)	27–65	

Two EPB-TBMs (Table 12.14), with diameter 6.57 m, are used for the tunnel excavations. These TBMs were recently used in another metro project in Istanbul (Kartal–Kadikoy metro) and were refurbished before use in the Uskudar–Umraniye–Cekmekoy – Sancaktepe metro excavations. Technical properties of the TBMs are summarized in Table 12.14.

Table 12.14 Some of the technical features of the EPB-TBMs used in the Uskudar–Umraniye–Cekmekoy–Sancaktepe metro tunnel [6].

Excavation diameter	6.57 m
Total weight	355 tonnes
Total length	100 m
Thrust capacity	20,000 kN
Maximum rotation speed	5.5 rpm
Torque capacity	2912 kNm (5.5 rpm) 5200 kNm (3.1 rpm)
Cutterhead power	4×315 (1260) kW
Total power	2100 kW
Cutter numbers	26 single CCS discs with 17″ diameter (8 gauge cutters, 18 face cutters) 6 double disc cutters with 17″ diameter (center cutters)
	62 ripper tools + 16 buckets

There are two parallel lines excavated by the EPB-TBMs. Line 2 started excavation on 4 March 2013 and completed excavation on 11 June 2013 after installation of 573 rings. Line 1 started excavation on 21 June 2013 and completed excavation on 21 August 2013 after installation of 564 rings. Average values of the excavation performance and some of the operational parameters realized during excavation are summarized in Table 12.15 based on the TBMs' data loggers.

The cutter consumptions are summarized in Table 12.16. Overall, cutter consumption was slightly higher in Line 2, as seen in Table 12.16, which might be due to the learning period that was experienced in Line 2. Since the excavation works first started between Carsi and Umraniye stations in this construction project, the contractor company wanted to try different brands of cutters to make a decision for the rest of the tunnel. Different brands of cutters were used: three brands of double discs, four brands of single discs, six brands of ripper tools, three brands of buckets. However, after excavation of 140 rings, the lowest quality cutters were discarded. By using higher quality disc cutters, disc life increased two to three times.

Since the center cutters used on the cutterhead were double discs, when replacing them, usually they were replaced simultaneously. This increased the consumption rate per cutter ring or tool.

When using EPB-TBMs with cutterheads for mixed ground conditions, including both disc cutters and ripper cutters, for excavation of rock masses, ripper tool and bucket consumptions should also be considered at the planning stage.

Table 12.15 Excavation performance and some of the operational parameters realized during excavation of Lines 1 and 2 between Carsi and Umraniye stations [6].

Line 1 (21 June to 21 August 2013)	
Average daily advance rate	11.0 m/day
Best daily advance rate	25.5 m/day
Average weekly advance rate	103.5 m/week
Best weekly advance rate	137.5 m/week
Average Machine utilization	24%
Average Ring installation	21%
Stoppages	55%
Average RPM	2.63 ± 0.26
Average penetration rate	15 ± 2.53 mm/rev
Average penetration rate	2.38 ± 0.48 m/h
Average net cutting rate	80.52 ± 16.12 m^3/h
Average torque	2.7 ± 0.37 MNm
Average thrust	$11{,}124 \pm 1879$ kN
C_f, FER, FIR	3.1, 11.8, 56.7
Line 2 (4 March to 11 June 2013)	
Average daily advance rate	8.4 m/day
Best daily advance rate	27.0 m/day
Average weekly advance rate	61.1 m/week
Best weekly advance rate	121.5 m/week
Average Machine utilization	16%
Average Ring installation	17%
Stoppages	67%
Average RPM	2.67 ± 0.23
Average penetration rate	14.50 ± 3.36 mm/rev
Average penetration rate	2.31 ± 0.52 m/h
Average net cutting rate	78.26 ± 17.71 m^3/h
Average torque	2.42 ± 0.52 MNm
Average thrust	$11{,}438 \pm 2{,}313$ kN
C_f, FER, FIR	3.1, 13.9, 69.7

Table 12.16 Cutter consumptions during excavation of Lines 1 and 2 between Carsi and Umraniye stations [6].

Line 1: Excavation length: 846 m (564 segment rings), Volume: 28,681 m^3 (bank/in-situ)

Consumption quantity	m^3/C. ring	m/C. ring	C. ring/ 10,000 m^3	C. ring/ 100 m
Center: 12 double (24 rings)	1,195	35.3	8.37	2.84
Face + gauge: 34 rings (single)	844	24.9	11.85	4.02
Ripper tools: 133	216	6.4	46.37	15.72
Buckets: 31	925	27.3	10.81	3.66

Line 2: Excavation length: 859.5 m (573 segment rings), Volume: 29138 m^3 (bank/in-situ)

Consumption quantity	m^3/C. Ring	m/C. Ring	C. Ring/ 10,000 m^3	C. Ring/ 100 m
Center: 15 double (30 rings)	971	28.7	10.30	3.49
Face: 23 rings (single)	1,267	37.4	7.89	2.68
Gauge: 24 rings (single)	1,214	35.8	8.24	2.79
Ripper tools: 126	231	6.8	43.24	14.66
Buckets: 40	728	21.5	13.73	4.65

Overall (Lines 1 and 2) for disc cutters: 428 m^3/cutter ring, 7.9 cutter rings/100 m tunnel.

When the positions (segment rings or tunnel chainage) of the cutter replacements are considered for Line 2, it is observed that diabase dyke rapidly increases the cutter wear. Average FIR (foam injection ratio) was between 55% and 70%, which was enough for cutter and cutterhead lubrication and cooling purposes. Also, foam flow was very consistent during the excavation.

Some pictures showing the damage types for Line 2 are presented in Figure 12.40. As seen, abrasive wear is dominant followed by ring chipping. Wear and damage types of the disc cutters used between Carsi and Umraniye stations are classified based on Frenzel [5] as in Table 12.17 for Line 2.

Abrasive ring wear   Plastic deformation   Ring chipping

Secondary ring wear   Ring breakage   Cutter blockage

Hub breakage   Ring breakage   Cutter blockage

Figure 12.40 Classification of wear-damage types of the disc cutters for Lines 1 and 2 between Carsi and Umraniye stations [6].

This case study indicated that the quality of the cutting tools was very important in terms of tool consumption. By using higher quality disc cutters, disc life increased two to three times. Using double-disc cutters increases cutting tool consumption rates if the two of them have to be replaced simultaneously. When using EPB-TBMs with mix ground cutterheads for excavation of rock masses, ripper tool and bucket consumptions in addition to disc cutters should also be considered in the planning stage. It is observed that diabase dyke with high Cerchar abrasivity index rapidly increases the cutter wear rate. Abrasive wear is dominant followed by ring chipping for the studied zone.

Table 12.17 Cutter wear damage types of the disc cutters for Lines 1 and 2 between Carsi and Umraniye stations [6].

Wear type	Double disc	Single disc	Total quantity	Percentage (%)
Abrasive ring wear	2	29	31	36
Plastic deformation	6	2	8	9
Ring chipping	2	15	17	20
Secondary ring wear	0	15	15	17
Ring breakage	1	3	4	5
Cutter blockage	5	2	7	8
Hub breakage	0	4	4	5

12.5 Conclusions

The case studies mentioned in this chapter indicate that the abrasivity of the rocks excavated is not the only parameter to consider when projecting cutter consumption rate for a TBM tunneling project. In fact, other parameters affecting cutter consumption are more significant than abrasivity and strength parameters.

In the case of small diameter TBMs equipped with small diameter double-disc cutters (low thrust capacity) such as in the Tuzla-Akfirat wastewater tunnel, although the strength of the rocks excavated was not usually very high, this excavation case could be considered as a difficult ground condition. Small diameter TBMs require the use of small diameter double-disc cutters. Thrust capacity of the disc cutters is directly proportional to their diameters (small diameter disc means low thrust capacity). On the other hand, thrust force applied on double-disc cutters is divided in to two cutter rings, which is another thrust and penetration reducing factor.

It was also seen that lamination and discontinuities existing on the formations affect the cutter consumption, due to void occurrence and thus, sudden hitting of the cutters onto the rock, generating shock loads and increasing cutter consumption, as seen in the Yamanli II HEPP tunnel.

The case in Buyukcekmece wastewater tunnel indicated the importance of replaceable center cutters. The medium strength rock blocks encountered within a soft ground gave the TBM a very hard time by slowing down the excavation rate and losing time for replacement of the center cutter.

Another case, the Uskudar–Umraniye–Cekmekoy–Sancaktepe metro tunnel indicated the importance of the quality of cutter manufacturing (cutter brand). Using higher quality disc cutters increased the life of the 17″ single-disc cutters by two to three times. When using EPB-TBMs with mix ground cutterheads for excavation of rock masses, ripper tool and bucket consumption, in addition to disc cutters, should also be considered in the planning stage.

References

[1] Copur, H. (1999) Theoretical and experimental studies of rock cutting with drag bits toward the development of a performance prediction model for roadheaders. Colorado School of Mines, Dissertation.

[2] Bilgin, N., Copur, H., Balci, C. (2014) *Mechanical Excavation in Mining and Civil Industries*. CRC Press, London.

[3] Osborn, H.J. (1969) Wear of rock cutting tools. *Powder Metallurgy* **12** (24), 471–502.

[4] Basse, J.L. (1973) Wear of hard metals in rock drilling: a survey of the literature. *Powder Metallurgy* **16** (31), 1–32.

[5] Frenzel, C. (2011) Disc cutter wear phenomenology and their implications on disc cutter consumption for TBM. *ARMA American Rock Mechanics Association*, No: 11–211, 7 p.

[6] Namli, M., Savk, S., Bostanci, E., Copur, H., Balci, C., Bilgin, N. (2014) *Field analysis of disc cutter consumption in Uskudar–Umraniye–Cekmekoy–Sancaktepe metro tunnel in Istanbul*. World Tunnel Congress, 9–15 May, Foz do Iguaçu, Brazil.

[7] Deketh, H.J.R. (1995) *Wear of Rock Cutting Tools*. CRC Press.

[8] Askilsrud, O. (1997) *Tunnel Boring Machines – Atlas Copco Robbins Inc.* Mechanical Mining Short Course Notes, L. Ozdemir: Course Director, Colorado School of Mines, 19–21 March, Golden.

[9] ASTM (2010) Standard test method for laboratory determination of abrasiveness of rock using the CERCHAR method. Designation: D7625–10.

[10] ISRM (2014) Alber. M., Yarali, O., Dahl, F., Bruland, A., Kasling, H., Michalakopoulos, T.N., Cardu, M., Hagan, P., Aydın, H., Ozaraslan, A. ISRM Suggested method for determining the abrasivity of rock by the Cerchar abrasivity test. The ISRM Suggested Methods for Rock Characterization, Testing and Monitoring. 2007–2014, Part I. 101–107.

[11] West, G. (1989) Rock abrasiveness testing for tunnelling. *Int J Rock mech Min Sci & Geomech Abstr,* **26** (2), 151–160.

[12] Thuro, K., Kasling, H. (2009) Classification of the abrasiveness of soil and rock. *Geomechanics and Tunnelling* **2** (2), 179–188.

[13] Schimazek, J., Knatz, H. (1970) Der Einfluss des Gesteinsaufbaus auf die Schnittgeschwindigkeit und den Meißelverschleiß von Streckenvortriebsmaschinen. *Glückauf* 106, pp. 274–278.

[14] Gharaghbagh, E.A., Rostami, J., Palomino, A.M. (2011) New soil abrasion testing method for soft ground tunneling applications. *Tunn Undergr Sp Technol,* **26**, 604–613.

[15] Rostami, J., Alavi Gharahbagh, E., Palomino, A.M. and Mosleh, M. (2012) Development of soil abrasivity testing for soft ground tunneling using shield machines. *Tunn Undergr Sp Technol*, **28**, 245–256.

[16] Nilsen, B., Dahl, F., Holzhäuser, J., Raleigh, P. (2006): Abrasivity testing for rock and soils. *Tunnels and Tunneling International*.

[17] Nilsen, B., Dahl, F., Holzhäuser, J., Raleigh, P. (2007) *The new test methodology for estimating the abrasiveness of soils for TBM tunnelling*. In: Rapid Excavation and Tunneling Conference (RETC), USA.

[18] Jacobsen, P.D., Bruland, A., Dahl, F. (2013) Review and assessment of the NTNU/ SINTEF Soil Abrasion Test (SAT™) for determination of abrasiveness of soil and soft ground. *Tunn Undergr Sp Technol, 37*; 107–114.

[19] Peila, D., Picchio, A., Chieregato, A., Barbero, M., Dal, Negro. E. and Boscaro, A. (2012). *Test procedure for assessing the influence of soil conditioning for EPB tunneling on the tool wear*. World Tunnel Congress, Bangkok, Thailand.

[20] Ozdemir, L. (1995) Mechanical Mining Technology. Short Course, Colorado School of Mines June 11, Golden.

[21] Bruland, A. (1998) Hard rock tunnel boring – Advance rate and cutter wear. Project Report 1 B-98, Trondheim: the Norwegian Institute of Technology, 159 p.

[22] Maidl, B., Schmidt, L., Ritz, W., Herrenknecht, M. (2008) *Hard Rock Tunnel Boring Machines*. Ernst & Sohn, Berlin.

[23] Gumus, U., Altay, U., Bilgin, A.R., Bostanci, E., Copur, H. (2016) *Double-disc cutter consumption of an EPB-TBM in Tuzla wastewater tunnel*. World Tunnel Congress, 22–28 April, San Francisco, USA. (Accepted paper).

[24] Akgul, M., Akgul, E., Bostanci, E., Copur, H. (2015) *Analysis of disc cutter consumption of a Double Shield TBM*. World Tunnel Congress, May 22–28, Dubrovnik, Croatia.

[25] Akyuz, S., Copur, H. (2015) Investigations into rock and soil related problems encountered during construction of Tuzla Wastewater Basin (Omerli Dam) collectors and networks. *Unpublished Report submitted to Nas-Akad JV*, (in Turkish).

[26] Valenzuela, R. (2007) Anti-abrasion and anti-dust technology. *BASF TBM Conference*, 7–9 February, Istanbul.

[27] Bilgin, N., Copur, H., Balci, C. (2012) Effect of replacing disc cutters with chisel tools on performance of a TBM in difficult ground conditions. *Tunn Undergr Sp Technol*, **27**, 1:41–51.

[28] Ozaydin, Y.T., Avunduk, E., Copur, H. (2013) *EPB-TBM Performance in Excavation of Buyukcekmece Waste Water Tunnel*. The 3rd International Underground Excavations for Transportation Symposium and Exhibition, 29–30 November, Istanbul, Turkey (In Turkish).

13   Effect of methane and other gases on TBM performance

Grounds containing different gases such as methane, sulfur dioxide, carbon dioxide, radon and gas vapor (leaking from underground storage tanks and pipelines) are considered as difficult ground conditions in tunneling. Different precautions depending on the type and features of the gas on the tunnel environment should be taken based on local regulations for tunneling in these types of grounds. These underground gases can be noxious (toxic), flammable (combustible), explosive and asphyxiant (suffocating); they can create harm to the personnel working underground and the equipment used.

Two methane explosion cases that occurred in tunnels in Turkey are summarized here. Also, some more cases are summarized as examples of tunneling in gassy ground. However, properties of methane are briefly introduced first, since the two cases to be mentioned are related to methane explosion.

13.1   Properties of methane

Methane (CH_4) is a colorless, odorless and non-noxious gas. Since the density of methane (0.716 kg/m^3 at 0°C and 101.3 kPa) is lower than the density of air (1.293 kg/m^3 at 0°C and 101.3 kPa), it accumulates at the crown of a tunnel or gallery. Methane gas can be detected in a tunnel or gallery by using mines safety lamps, chemical analysis and methane detectors. When it mixes with air referred to as firedamp in mining, it becomes flammable and explosive, creating a danger that might result in the deaths of many people in mines and tunnels. When the concentration of methane within the air is between 5% (Lower Explosive Limit, LEL) and 15% (Upper Explosive Limit, UEL), it becomes explosive in closed areas. Methane in air at 9.8% by volume is the most explosive (stochastic) mixture. A source of temperature about 650–750°C is required to create an ignition of methane. Methane cannot explode out of LEL and UEL range but can cause asphyxiation due to lack of oxygen at over 15% concentrations. Methane burns in air with a pale blue flame and can be observed as flame of a mining safety lamp. Methane can be easily emitted through geological discontinuities, joints and pores of the ground, as shown in Figure 13.1 [1]. Methane can also migrate and solve in water (33.1 litres/cubic meter at 20°C) and noticed by a spurting or bubbles in water.

TBM Excavation in Difficult Ground Conditions. Case Studies from Turkey. First Edition. Nuh Bilgin, Hanifi Copur, Cemal Balci.
© 2016 Ernst & Sohn GmbH & Co. KG. Published 2016 by Ernst & Sohn GmbH & Co. KG.

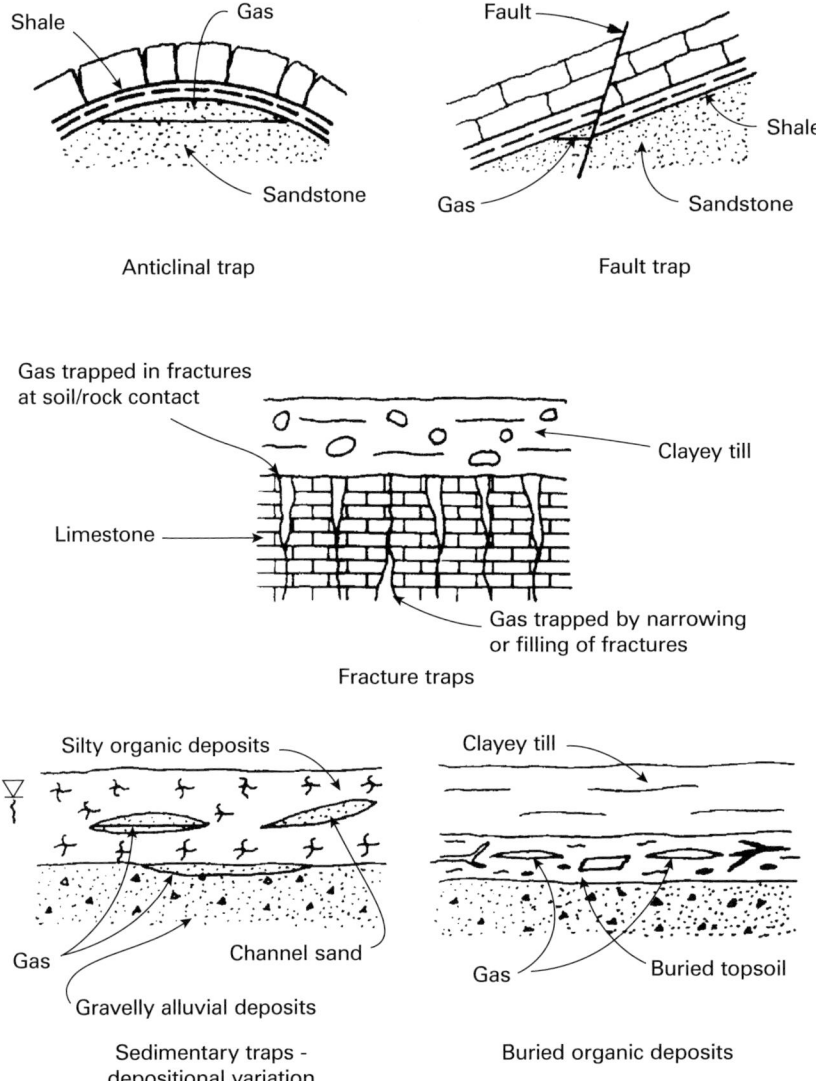

Figure 13.1 Structural and sedimentary gas traps [1].

Safety and health standards of the Occupational Safety and Health Administration (OSHA) of the USA require continuous monitoring of gases for underground construction when TBMs are used. OSHA states that when an air sample indicates for 5% LEL or more, air supply should be increased with continuous gas control; at 10% LEL or more, hot works such as welding or cutting should be suspended; at 20% LEL or more, power should be turned off and crew should be withdrawn [2].

13.2   Selimpasa wastewater tunnel, methane explosion in the pressure chamber of an EPB-TBM

13.2.1   Introduction to the Selimpasa wastewater project

The basic aim of the Selimpasa utility project is to construct tunnels along Silivri-Gu-zelce in the European part of Istanbul and transport the wastewater to the Selim-pasa treatment plant. The Selimpasa wastewater tunnel alignment, with a length of 11,687 m, runs almost parallel to the shore of the Sea of Marmara. The project owner is Istanbul Water and Sewage Authorization (ISKI) and the contractor is Ozka-Kalyon Construction JV. The geological map and location of the tunnel alignment are given in Figure 13.2 [3].

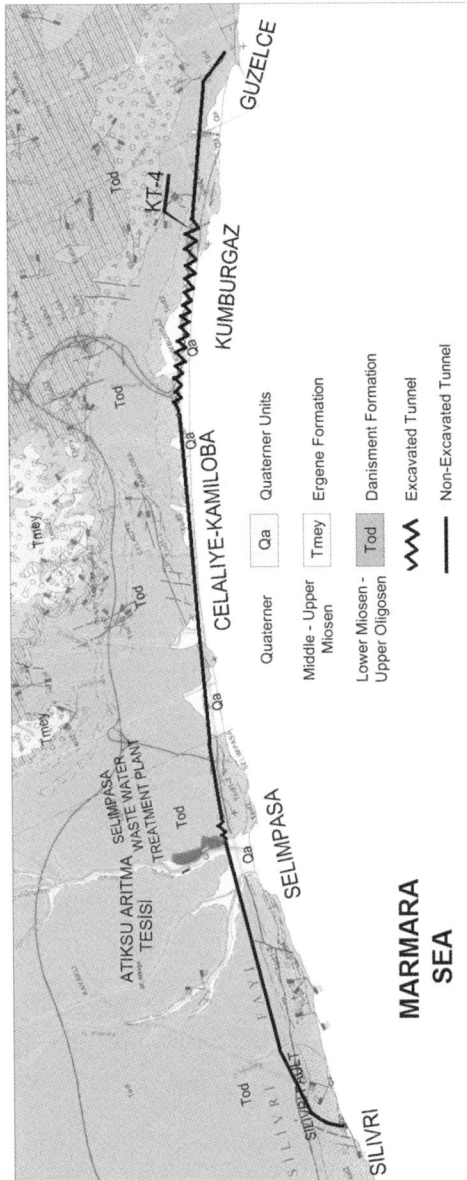

Figure 13.2 Structural geological map of the Selimpasa Tunnel alignment [3].

The tunnel is constructed through the Oligocene aged deltaic Danisment (Gurpinar) formation consisting mostly of sandstones and shales. This formation also includes thin coal bands and tuff intercalations, gray carbonated clays and claystones. The Danisment formation is known as a natural gas production basin in the Trakya area of the Marmara region of Turkey. Kusdili formation is unconformable on the Danisment formation, consisting of the late Quaternary aged sand, gravel and fossiliferous sand.

The units described as beach dunes, talus and alluviums are the other young units encountered in the region.

An EPB-TBM manufactured by Herrenknecht having an excavation diameter of 2,966 mm is used for excavation, and was transported to the site on 13 July in 2009. A picture of the TBM and its technical specifications are given in Figure 13.3 and Table 13.1, respectively. The TBM is designed for mixed ground conditions. Rings consisting of six pieces of precast segments each having lengths of 950 mm are used for tunnel lining having an inner diameter of 2,300 mm and outer diameter of 2,700 mm.

Figure 13.3 EPB-TBM used for excavation of the Selimpasa wastewater tunnel.

Table 13.1 Technical specifications of the Herrenknecht EPB-TBM.

Cutting (excavation) diameter	2,966	mm
Cutterhead power	2 × 200	kW
EPB-TBM weight	103.3	tonne
Cutterhead torque capacity (left / right)	1,020 / 1,020	kNm
Cutterhead rpm	0–6	rpm
Thrust capacity	12 x 4,580	kN
Nominal dia. of screw conveyor	500	mm
Torque capacity of screw conveyor	35	kNm
Screw conveyor rpm	0–29	rpm
Theoretical muck transportation capacity of screw conveyor	84	m^3/h

The tunnel is ventilated by two axial type fans each having power of 11 kW connected serially. They can theoretically provide 240 and 360 m^3/min of air, and generate pressure differences of 550 and 3,550 Pa separately. Flexible ducting with diameter of 600 mm hanging on the crown is used for delivering clean air to the face. Oxygen, methane, carbon dioxide, carbon monoxide and nitrogen oxides are measured and recorded in the tunnel air at the beginning of each shift by using a mobile detector. Excavation of the 10,537 m tunnel was completed in January 2012.

13.2.2 Occurrence and causes of methane explosion in the Selimpasa wastewater tunnel

A methane explosion occurred inside the excavation (working pressure) chamber of the EPB-TBM during excavation of the tunnel at chainage of 3+095 km on 20 May 2010 during the night shift [4–8]. The EPB-TBM was around 780 m away from the logistic shaft KT-6 and 40 m away from the shaft KT-4, as seen in Figure 13.4.

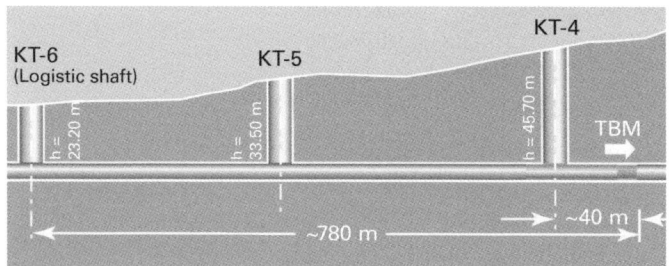

Figure 13.4 Position of the EPB-TBM when explosion occurred.

It is considered that the explosion was initiated by sparks created by friction between the screw conveyor and its casing. The explosion forced the muck in the chamber and some of the non-excavated ground to be blown through the screw conveyor and out of the discharge door. A fireball then engulfed the inside of the machine, reaching the rear end of the tail shield. The discharge door then closed automatically after the explosion.

The shift records indicated that while excavation of a ring was taking around 1 hour in the final few days before the accident, some very soft ground was suddenly encountered and the excavation speed increased up to 10–15 minutes/ring, just five or six rings before the explosion site. This indicated that the TBM was working through an unforeseen fault zone, which was not detected during the site investigations. The visual analysis at the excavation face performed after the accident indicated that the methane was being emitted through the fractures, joints and pores of the ground within this zone (Figure 13.5).

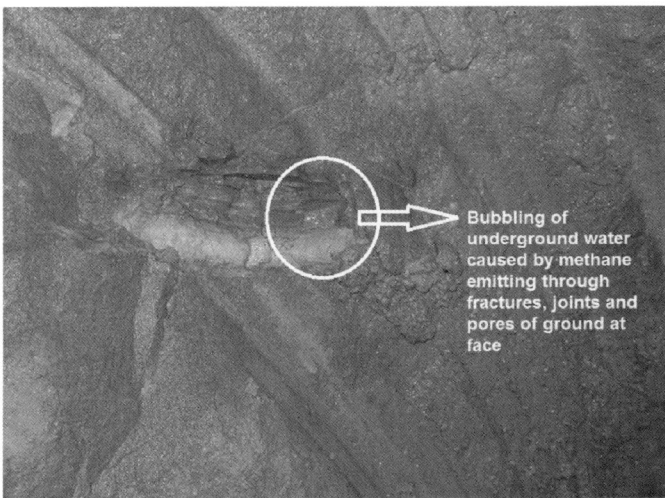

Bubbling of underground water caused by methane emitting through fractures, joints and pores of ground at face

Figure 13.5 Methane emission through the fractures and pores at the face.

The shift records also indicated that the gantry located at the backup of TBM was broken down while excavating the 13th ring in the final 15–20 minutes of the day shift of 20 May. The night shift started at around 20:00, just after the shift engineer measured the gases in the tunnel air with a mobile detector. The gantry was first maintained by welding, and excavation of the 13th ring was restarted. After excavation of around 10–15 cm in a few minutes, the methane explosion occurred at around 22:00. It is considered that the methane had accumulated in the working chamber during the stoppage of 2–2.5 hours for maintaining the gantry. The air required for explosion probably came from the foam injected into the chamber, which was most likely half-full of muck during the stoppage. Since there was excavated material waiting inside the chamber and blocking the entrance of the screw conveyor, the methane could not leak into the tunnel air, but accumulated inside the chamber. Since the TBM did not have any gas monitoring sensor inside the excavation chamber, it was not possible to detect methane accumulation with the mobile detector used in the tunnel at the beginning of each shift.

13.2.3 Consequences of methane explosion in the Selimpasa wastewater tunnel

The explosion caused extrusion of 3–4 m^3 of muck through the screw conveyor into the tunnel (Figure 13.6) in which 10 personnel were working and all of them were injured due to the fire that followed the muck blow-out. The TBM operator, who was closest to the discharging door of the screw conveyor, was severely burnt. Another four personnel working around the segment erector were moderately burnt, while the other five personnel working outside the rear shield and close to the shaft KT-4 were slightly injured after being knocked over by the pressure of the explosion. The eight personnel were hospitalized. Electricity was shut down as a result of the accident, which made the rescue efforts quite difficult. Severely injured personnel were carried away from the accident area by less badly injured personnel.

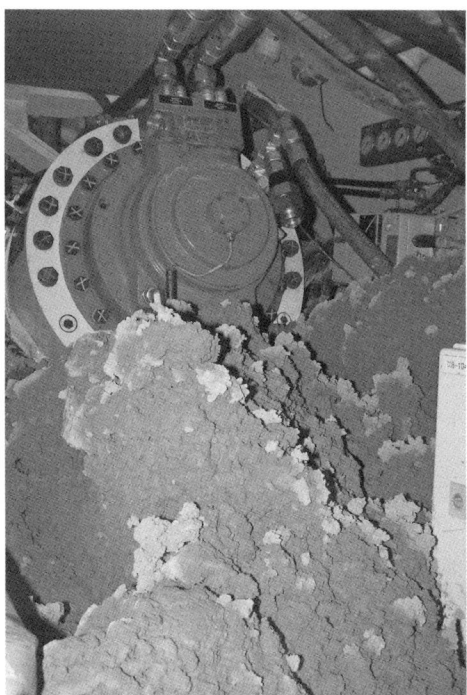

Figure 13.6 Muck blow out under
discharge gate of the screw conveyor.

The accident caused a 44-day delay to the tunnel construction which restarted on 3 July
2010. The speed of the construction was reduced as a precaution against another methane explosion, since the methane was still being emitted into the working chamber.

The personnel was quite psychologically affected by the accident. The representatives
of the contractor arranged educational programs for improving their motivation. Since
half of the personnel was in hospital, the activities toward restarting construction were
performed by one shift per day. All injured personnel came back to work in the second
half of July 2010, except the TBM operator who was in hospital until the end of August
2010 and came back to work at that time.

Some objects inside the shield such as safety warning charts and clothes of the personnel hanging on the wall burnt during the explosion. However, the explosion did not
cause any open fires. Some parts of the TBM such as steering system, data logger and
touch-tone LCD monitor of the control panel were damaged. There was no damage
to or collapse of the segment rings. It was also seen that the pressure of the explosion
caused the TBM to move backward about 60–70 cm, which was understood from the
positions of thrust cylinders.

13.2.4 Precautions against methane and excavation performance in the Selimpasa wastewater tunnel

Representatives of the contractor decided first to continuously measure the amount
of methane released into the working chamber and tunnel. To measure the methane

inside the chamber, one end of a piece of hose with 2.5 cm diameter was connected to a secondary water discharge valve located on the pressure bulkhead of the TBM; its other end was taken to the surface through the shaft KT-4. It was seen that the methane was emitting continuously at a rate of 4.5 m^3/h. The methane emission continued at the same rate in the chamber until the construction activities restarted on 3 July 2010. Since the hose was draining the methane from the working chamber and the door of the working chamber was kept closed, the tunnel air was free of methane. After restarting excavation, the door of the working chamber was kept open for effective dilution of the methane. The methane emissions gradually reduced and reached trace levels after the excavation of around 500 m beyond the explosion site.

The tunnel segments in the shaft KT-4 were broken out to create a hole of around 1×1 m^2 to improve the tunnel ventilation. Also, the ducting was refurbished to provide airtightness. Two boreholes of around 50 m depth (around 2 m deeper than the invert level) and 10 cm diameter were drilled 1 m and 6 m ahead of the TBM face on the 26 and 27 May, respectively, to help drain the methane. Emission of the methane through these boreholes continued until they were flooded with water on 28 May.

Cleaning of the muck heap that had accumulated inside the shield was started on 29 May and was completed on 1 June, with great care and at a slow pace. The damage inspections of the TBM and tunnel were started on 1 June and completed on 3 June. The damaged equipment was ordered through the machine manufacturer on 4 June. The damaged parts of the TBM were replaced and/or maintained by a local expert engineer of the machine manufacturer on 9 June.

The expert witnesses from the university analyzed the explosion and suggested further precautions for continuation of construction. The project owner and the contractor decided to inject a pressurized bentonite mixture into the working chamber to stop or at least reduce the methane emission before the excavation of each ring.

A special bentonite mixture consisting of water, sodium bentonite, sodium hydroxide, polyanionic cellulose and cement was injected into the chamber by around a few bar pressure on 26 June. While injecting the mixture, the methane squeezing inside the chamber was drained through the hose connected to the secondary water discharge valve. In order to increase the pressure and fill as many fractures, joints and pores inside the chamber as possible, the TBM was also pushed towards the face (it had moved backwards due to the explosion). The pressure inside the chamber increased to 5–6 bar and there was a delay for intrusion of the bentonite mixture through the fractures, joints and pores. But after a few hours, it was seen that the pressure inside the chamber reduced to 1.2 bar. Then, the TBM was again pushed forward to increase the pressure. Once more, the pressure dropped to 1–1.2 bar after a short while. The methane was measured on 28 June, and it was finally seen that there was no methane inside the excavation chamber. On the same day, trail excavations started, which was the first after the accident, and completed quite slowly and with great care. Measurements indicated that the methane emission was continuing after boring recommenced. Again, the pressurized bentonite mixture was injected into the chamber before excavation of the second ring. After excavation of the second ring, the methane emissions continued and the contractor decided to stop boring.

Measurements of the methane coming through the hose connected to the water valve indicated that the emission had decreased to 3.5 m^3/h between 29 June and 3 July. The excavation was restarted on 3 July and eight rings were completed by injection of pressurized bentonite mixture into the chamber after excavation of each ring.

An automated gas measurement, warning and shut-off system, which was designed by a local company and ordered a couple of weeks earlier, was installed on 4 July. The system measures and records the methane at three positions every minute (Figure 13.7): inside the excavation chamber, over the screw conveyor discharge door and inside the rear shield close to the crown. The system also measures carbon dioxide (CO_2) and hydrogen sulfide (H_2S) and automatically stops excavation when the methane concentration reaches 20% of the LEL (lower explosion limit).

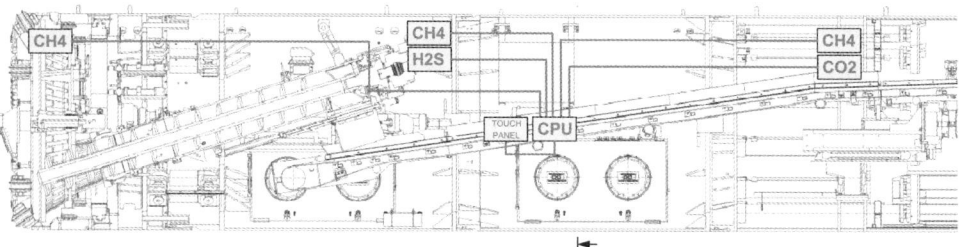

Figure 13.7 Automatic gas measurement and warning system.

After 4 July, the methane emission was reduced by keeping the door of the excavation chamber open, which is only possible for open mode operations on stable faces, and by ventilating the tunnel through the ducting tubes, whose front end was set very close to the pressure bulkhead. Also, the tunnel annulus was always grouted as soon as each ring was installed.

Injection of the pressurized bentonite mixture into the chamber was stopped on 7 July, since the ground had become harder and the mix could not penetrate it. Since the scraper cutting tools could not cut the ground, they were replaced with disc cutters on the same day. Although the ground was harder, the methane emission still continued. After excavation of around 500 m, the methane emission diminished gradually reaching trace levels. The daily advance rates of the TBM during this difficult gassy ground conditions are presented in Figure 13.8.

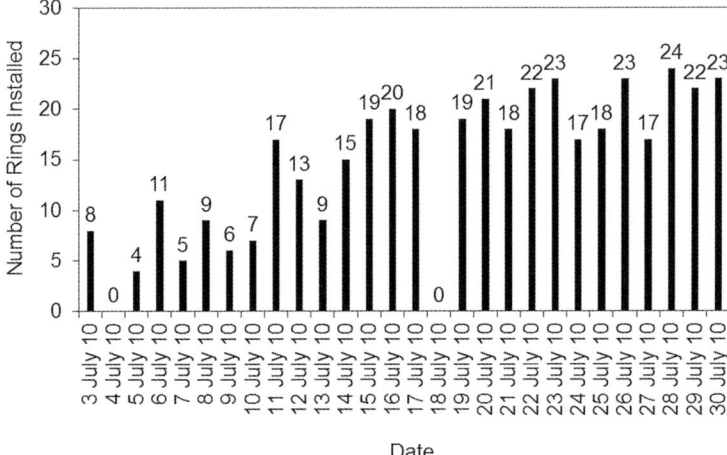

Figure 13.8 Daily ring installations in July 2010 after the accident in gassy ground conditions.

By 27 January 2012, the crew had completed excavation of the 10,537 m tunnel without any problem, even breaking the Turkish record for monthly advance rate set in April 2011 (Figure 13.9, Table 13.2). The final section of around 1,150 m was completed after a few months. Monthly progress rates during construction are presented in Figure 13.10.

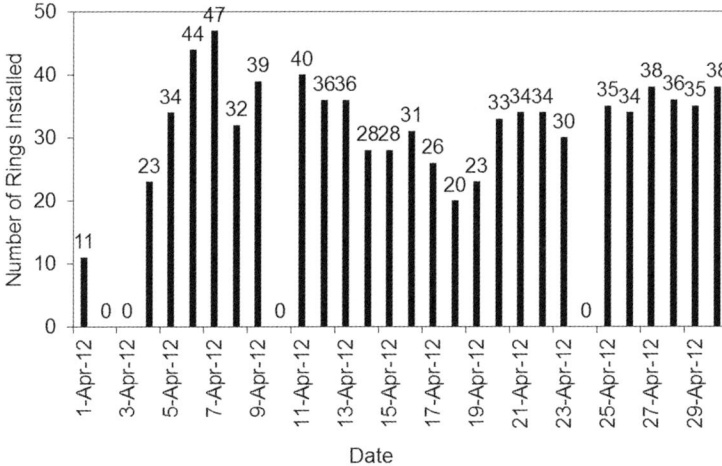

Figure 13.9 Daily ring installations in record-breaking month April 2011.

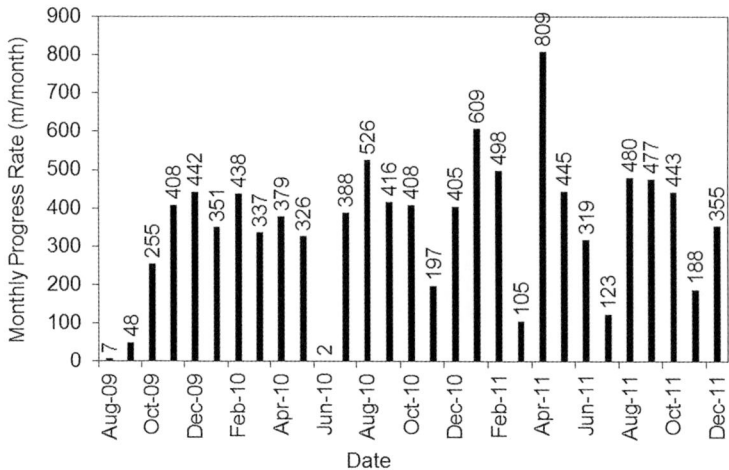

Figure 13.10 Monthly advance rates in the Selimpasa wastewater tunnel.

Table 13.2 Record excavation performance in April 2011.

Total monthly progress rate	809.40 m (852 rings)
Best daily progress rate	44.65 m (47 rings)
Best shift progress rate	22.80 (24 rings)
Average daily progress rate	27.0 m
Machine utilization rate	24%
Segment installation	20%
Muck transportation	23%
Rail installation	20%
Other stoppages	13%

This performance values show also that the crew regained their motivation after short but effective educational meetings. The contractor company was recently granted two short tunnels having a total of 6,015 m with the same diameter in the same region and having difficult gassy ground conditions.

13.3   Gas flaming in the Silvan irrigation tunnel

The basic aim of the Silvan irrigation project is to construct tunnels and channels between the Silvan Dam and agricultural lands in Diyarbakir, in the south-east of Turkey, to irrigate 245,372 hectare of agricultural land (Figure 13.11). The Silvan tunnel alignment, which is double tube, has a length of $2 \times 10,210$ m. The project owner is the General Directorate of State Hydraulic Works and the contractor is Sistem-Kayao-glu-Yertas-Intekar JV. The geological map of the tunnel alignment obtained from the site investigations is given in Figure 13.12 [9].

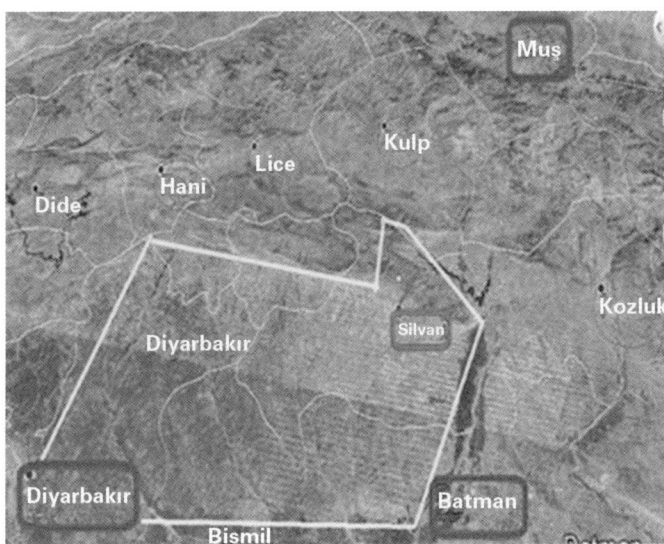

Figure 13.11 Location of the Silvan irrigation tunnel.

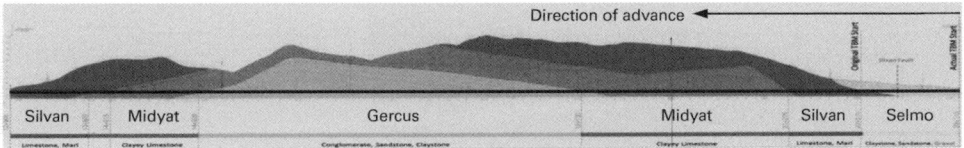

Figure 13.12 Structural geological map of the Silvan irrigation tunnel alignment [9].

The tunnel is constructed in four different formations [10]: Selmo, Silvan, Midyat and Gercus formations. Uniaxial compressive strength values of these formations vary between 45 and 93 MPa. RMR values vary between 22 and 62. Selmo formation includes claystones, sandstones and conglomerates; Silvan formation includes limestone and marl; Midyat formation includes clayey limestone; and Gercus formation includes claystones, sandstones and conglomerates.

After a gas flaming incident in the tunnel, it was discovered that the Turkish Petroleum Corporation (TP), a public institution, had already made some site investigations along the Silvan irrigation tunnel and found a natural gas and oil reservoir (anticline petroleum trap) with a capacity of 24,000 m^3/day in Alt Sinan formation by drilling five boreholes at around 300 m away from, and 350–400 m underneath the tunnel alignment (Figure 13.13). But neither the project owner nor the contractor had known about this dangerous situation. The findings of TP indicate that the natural gas reservoir is of anticline type, and the cap rock of the anticline has some tensile fractures due to folding of the formation especially under Gercus formation, which is the closest part of the tunnel to the natural gas reservoir cap rock. These tensile fractures reach up to the Gercus formation.

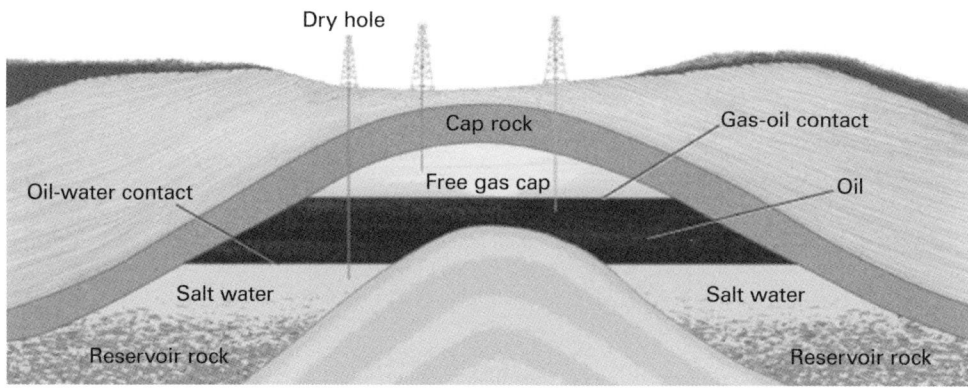

Figure 13.13 Anticline type petroleum and natural gas trap.

Figure 13.14 Double-shield TBM used for excavation of the Silvan irrigation tunnel.

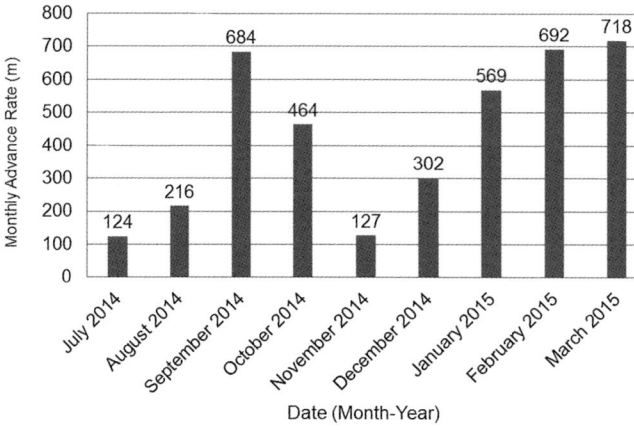

Figure 13.15 Monthly advance rates in the Silvan irrigation tunnel [10].

A double-shield hard rock TBM with an excavation diameter of 7.8 m was used for excavation (Figure 13.14, Table 13.3). Rings consisting of 5+1 pieces of precast segments each having lengths of 1,500 mm are used for tunnel lining having an inner diameter of 7000 mm and outer diameter of 7,600 mm. Excavation of the 4,668 m tunnel was completed on 21 April 2015 when the gas flaming accident occurred at the chainage 18+447 km. Monthly progress rate of the TBM, which started in July 2014 and ran to March 2015 is given in Figure 13.15 [10]. Extreme water ingress of around 1663 l/s reduced the progress rate in November 2014.

Table 13.3 Technical specifications of the Herrenknecht double-shield TBM used in the Silvan irrigation tunnel.

Cutting (excavation) diameter	7,850	mm
Cutterhead power	8×350	kW
Cutterhead torque capacity (nominal / limit)	5,384 / 8,346	kNm
Cutterhead rpm	0–8	rpm
Thrust capacity	35,600	kN

Totally 36 personnel were working in the tunnel during the gas flaming in 21st April 2015 at around 9:40 a.m. Thirteen of the personnel were burnt due to the gas flare, of whom five were seriously burnt. After evacuation of the personnel, entrance to the tunnel was restricted and as of December 2015, it is still restricted (no work). The fire/flame/flare engulfed the tunnel starting from the face area through the crown of the tunnel. An experienced mine rescue team from Turkish Hard Coal Enterprise (TTK) was invited to perform a survey within the tunnel. The team has measured hydrocarbon-based gas (basically methane) of around 28.6% at the tunnel face and 2.5% at the rear end of the TBM shield. The Turkish regulations for gassy tunnels and mines requires working with ex-proof (explosion-proof) equipment, the construction was stopped, since all of the equipment used in the tunnel are not ex-proof. The gas flaming extensively damaged the TBM, which cannot be used anymore without extensive maintenance and revisions.

13.4   More gas-related accident examples for mechanized tunneling

Some examples of gas-related accidents that occurred during (mostly) mechanized tunneling operations are summarized in Table 13.4.

Table 13.4 Examples of gas-related accidents that occurred during tunneling.

Project	Tunnel length (km)	Tunnel diameter and excavation method	Accident date	Accident type	Result of accident	Notes on causes of accident	Reference
Los Angeles water tunnel, USA	8.85 km	5.5 m hard rock TBM	1971	Methane explosion	17 died	Oilfields in the region, gas detector was malfunctioning	[11]
A tunnel in Higashimurayama, Japan	—	—	1978	Methane explosion	9 died, 2 injured	Power supply did not shut off automatically, natural gas deposits in the region	[12]
HEPP tunnel, Georgia	—	Shaft excavation	1979	Methane explosion	Damage to locomotive and cars	Power equipment was not ex-proof	[13]
Lancashire tunnel, Abbeystead valve house, England	—	—	1984	Methane explosion	16 died, some others injured	Methane dissolved in groundwater	[14–15]
Jay-Arnett wastewater tunnel, New York, USA	34 km	Hard rock TBM	1985	Methane in tunnel air, no accident	No fatality or injury	Natural gas deposits in the region, halted borehole drilling operations	[16]
Aqueduct tunnel in Carsington, England	8.5 km	2.4 m final dia., Hard rock TBM	1987	Methane in tunnel air, a few minor incidents	No fatality or injury	Methane dissolved in groundwater	[14]
Tunnel, Huyiki Town, Tokyo, Japan	—	EPB-TBM	1993	Methane explosion	4 died, 1 injured	Gas sensors were issuing alarms at the office outside the tunnel, but staff members were absent	[17]
Mill Creek tunnel, USA	4.65 km	7.8 m shielded TBM	2004	Methane in tunnel air, no accident	No fatality or injury; tunneling suspended for 8 months	Natural gas production by shallow wells in the region	[18]

Project	Length	TBM/method	Year	Gas/accident	Casualties	Description	Ref.
Electric cable tunnel, Hung Hom, Hong Kong	—	4.5 m TBM	2004	Methane in tunnel air, no accident	No fatality or injury	Borehole program was applied to measure methane	[19]
Metro line extension, Istanbul	—	(36 m^2) New Austrian tunneling method excavated by impact hammer	2007	Gasoline vapor incident in tunnel air	No fatality or injury	Excavation under a set of fuel-oil storage tanks, tunneling operations stopped for a while and then continued at slow pace	[20]
Zagros (Nousoud water trans.) tunnel, Iran	26 km excavated by TBM (total 48 km)	6.73 m, double shield TBM	2009	Hydrogen sulfide and methane in tunnel air, no accident	No fatality or injury; four month shutdown of tunneling in high H_2S, corrosion of TBM components and segments, TBM utilization reduced up to 7%	Formations in the region were the major oil and gas bearing basins; both hydrogen sulfide and methane were present as dissolved in groundwater	[21–24]
Variante de Pajeres tunnels, Spain	—	Single shield TBM	2010	Methane in tunnel air, no accident	No fatality or injury	TBM advance rate slowed down	[25]
Selimpasa wastewater tunnel, Istanbul, Turkey	10.63	3 m, EPB-TBM	2010	Methane explosion	8 injured; 45 day shutdown	Formation in the region was the major natural gas bearing basin	[4–8]
Silvan irrigation tunnel, Diyarbakir, Turkey	10.31	7.85 m, double shield TBM	2015	Methane combustion	13 injured, still not in operation as of April 2015	Natural gas reservoir was formerly discovered in 2009 under the tunnel alignment	[26]

13.5   Conclusions

These accidents indicate that the basic negligence starts during site investigations. Especially Carboniferous age formations are the basic formation types having potential gas hazard during underground operations. Also, the geotechnicians, authorities and contractors are not careful enough about the existence of natural gas and oil-bearing formations in the vicinity of the tunnel alignment, as well as natural gas and oil production, which should be always considered as a gas danger to underground operations. Sometimes authorities do use site investigations at different times, but they do not know each other's activities, as there is no communication among the earth-related institutions. In some cases, the authorities and contractors do not take enough precautions against known gassy ground conditions such as ex-proof equipment, very effective ventilation and automatic monitoring-warning-shut-off system. It should also be kept in mind that waterproofing is also important since tunnel gases of many types can be found dissolved in underground water [27]. Another important issue underlined in some of those accidents is that the tunnelers can utilize the experience of coal miners, who are involved every day in gassy ground conditions [28]. Also, tunneling in urban areas is always susceptible to leakages (gasoline vapor) from fuel oil storage tanks of fuel stations.

A recent successful application of EPB-TBMs in the excavation of two inclined drifts in a coal mine in Australia indicated that closed face TBM technology can be successfully used for excavation in gassy underground operations [29].

References

[1] Doyle, B.R., Gronbeck, M.P., Rose, J.P. (1991) *Construction of tunnels in methane environments.* In: Proceedings of the Rapid Excavation and Tunneling Conference, USA.

[2] Kissel, F.N. (2006) *Preventing methane gas explosions during tunnel construction, Handbook for Methane Control in Mining.* IC 9486. Chapter 14, pp. 169–184.

[3] Yurtsever, A., Caglayan, M.A. (2002) *1/100,000 Scaled geological maps of Turkey, Istanbul F-21 G 21 maps.* General Directorate of Mineral Research and Exploration (MTA).

[4] Copur, H., Cinar, M., Okten, G., Bilgin, N. (2012) A case study on the methane explosion in the excavation chamber of an EPB-TBM and lessons learnt including some recent accidents. *Tunn Undergr Sp Technol*, **27**, 159–167.

[5] Bilgin, N., Okten, G., Copur, H. (2010) Analysis of methane explosion on 20th May 2010 during excavation of wastewater tunnel. *Report submitted to Ozka-Kalyon Construction JV.* ITU Mining Faculty Project (in Turkish).

[6] Copur, H., Cinar, M., Okten, G., Bilgin, N. (2011) *A case of methane explosion in the excavation chamber of an EPB-TBM.* World Tunnel Congress, 21–26 May, Helsinki, Finland.

[7] Copur, H., Okten, G., Bilgin, N., Cinar, M. (2012) A methane explosion during EPBM tunneling. *Tunneling Journal of Turkish Tunneling Society*, March-April, 40–43.

[8] Okten, G., Cinar, M., Copur, H., Bilgin, N. (2014) A methane explosion during EPBM tunneling. *Magazine of Turkish Tunneling Society*, Issue: Jan-Feb, pp.14–19. (In Turkish)

[9] Bar, S. (2011) *Engineering Geology Report for Silvan Project Silvan 1st Part Irrigation Tunnels and Channel Construction.* (In Turkish)

[10] Inal, M.E., Inal, N. (2015) Investigation on TBM Performance in Silvan Tunnel depending on Ground Conditions. *Magazine of Turkish Tunneling Society*, Issue: March-April, pp.54–62. (In Turkish).

[11] Proctor, R.J. (2002) The San Fernando tunnel explosion. *Eng Geol*, **67**, 1–3.

[12] Kitajima, M. (2015) *Methane gas explosion hazard during construction of headrace tunnel for agriculture.* URL: http://www.sozogaku.com/fkd/en/cfen/CD1000099.html (15 November 2015).

[13] Vlasov, S.N., Makovsky, L.V., Merkin, V.E. (2001) *Accidents in transportation and subway tunnels.* Elex-KM Publications.

[14] Pearson, C.F.C., Edwards, J.S., Durucan, S. (1989) *Methane occurrences in the Carsington Aqueduct tunnel project-a case study.* In: Proceedings of the Rapid Excavation and Tunneling Conference, USA.

[15] Lockyer, J.W., Howcroft, A. (1997) *The Abbeystead explosion disaster.* Annals of Burns and Fire Disasters, 10, Sept; 1–4.

[16] Peters, J.F., Breu, F.A., Neyer, R.F., Gorman, R.F., Critchfield, J.W. (1985) *Sewer tunnel in gassy rock*. In: Proceedings of the Rapid Excavation and Tunneling Conference, USA.

[17] Kitajima, M. (2010) *Methane gas explosion hazard of an earth pressure type shield tunnel*. URL: http://shippai.jst.go.jp/en/ Detail?fn=0&id=CD1000098& (18th November 2010).

[18] Schafer, M., Pintabona, R., Lukajik, B., Kritzer, M., Janoska, S., Switalski, R. (2007) *Gas mitigation in the Mill Creek Tunnel*. In: Proceedings of the Rapid Excavation and Tunneling Conference, USA.

[19] Wightman, N.R., Mackay, A. (2008) *Gas ground investigation for tunneling works at Hung Hom freight depot, Hong Kong*. In: Proceedings of the World Tunnel Congress, Agra-India.

[20] Fisne, A., Okten, G. (2007) *The Effects of Leakage Petroleum Products from BP Maslak Station on Tunnel Air Quality*. In: Proceedings of Underground Excavation for Transportation, Istanbul, Turkey (in Turkish).

[21] Shahriar, K., Rostami, J., Hamidi, J.K. (2009) *TBM tunneling and analysis of high gas emission accident in Zagros long tunnel*. In: Proceedings of the World Tunnel Congress, Budapest-Hungary.

[22] Zarei H.R., Sharifzadeh, M., Uromyehic, A. (2011) *Gas ground risks and geological investigations for TBM tunneling in Iran*. In: Proceedings 1st Asian and 9th Iranian Tunneling Symposium, Tehran, Iran.

[23] Mirmehrabi, H., Ghafoori, M., Lashkaripour, G., Azali, S.T., Hassanpour, J. (2012) Hazards of mechanized tunnel excavation in H_2S bearing ground in Aspar tunnel, Iran. *Environ Earth Sci* **66**, 529–535.

[24] Taherian, A.R. (2015) Experiences of TBM operation in gas bearing water condition – A case study in Iran. *Tunn Undergr Sp Technol*, **47**, 1–9.

[25] Rodriguez, R., Lombardia, C.R. (2010) Analysis of methane emissions in a tunnel excavated through carboniferous strata based on underground coal mining experience. *Tunn Undergr Sp Tech* **25**, 456–468.

[26] General Directorate of State Hydraulic Works (2015) *Unpublished Reports*. (In Turkish).

[27] Doyle, B.R. (2001) *Hazardous Gases Underground: Applications to Tunnel Engineering*. Marcel Dekker Inc., New York, USA.

[28] Bilgin, N. (1985) *The most disastrous explosion in Turkish coalfields*. In: Proceedings of the 2nd U.S. Mine Ventilation Symposium, Nevada-Reno, USA.

[29] Scialpi, M., Ofiara, D. (2015) *Unique Hybrid EPB Design for use in Coal Mine Drifts*. World Tunnel Congress, Dubrovnik, Croatia.

14 Probe drilling ahead of TBMs in difficult ground conditions

14.1 Introduction

This chapter summarizes the results of probe drilling carried out ahead of TBMs in Melen and Kargi projects in a very complex geology. Melen water tunnel was excavated under the Istanbul Bosphorus within sedimentary rocks that are cut frequently by andesitic dykes, fracturing the surrounding rocks and creating a potential risk for water ingress into the tunnel. Probe drillings with petrographic analysis and strength tests were used with samples collected from the TBM muck. This analysis made it possible to identify some critical normalized probe drilling rate values for predicting potential weak zones created by andesitic dykes. These studies gave a sound basis for further interpretation of TBM and geological data for the tunnel. The second set of probe drilling analyses was from the Kargi tunnel. The North Anatolian Fault highly affected the tunnel excavation by fracturing the rock formations. Although the change in normalized probe drilling data was a good indicator of fractured zones, the diversity of rock formations made it difficult to interpret the data. On the basis of the probe drilling results at Kargi, an umbrella arch was used to strengthen the rock mass for easing tunneling operations and to stop the jamming of the TBM cutterhead. A summary of the probe drilling in Melen and Kargi has already been published, Bilgin and Ates [1]. More detailed information about this will be given here.

14.2 General information on probe drilling and previous experiences in different countries

Probe drilling is a technique used ahead of TBMs in difficult geological conditions as an indicator of water ingress into a tunnel, gas potential ahead of a tunnel, fault and transition zones. However, probe drilling rate is also related to the type of drill rig and drilling parameters such as feed, percussion and rotation pressures, in other words to the thrust and the torque of the drill rig. Drilling rate measurements should be carried out under constant, percussive and rotary pressures in order to make a meaningful comparison of the changing situations ahead of a tunnel. Before evaluating the measurements, drilling rate values should be normalized using thrust or rotary pressures in order to have comparable values. However, it should be emphasized also that in difficult ground conditions and in large diameter tunnels more than one probe drill hole will be necessary. In such cases, the interpretation of the drilling results with TBM operational parameters such as thrust and torque values may be a useful guide for predicting potential hazards ahead of the tunnel.

Tunnel face collapses and sudden and unexpected water ingress in a tunnel are always associated with fractures or other geological singularities such as faults. Schunnesson [2] developed a model for predicting RQD of rock mass ahead of the tunnel using the percussive drilling parameters. He concluded that the ability to predict RQD, based on drill parameters, offers a unique opportunity to utilize RQD not only for characterization of the entire tunnel section, but also to provide detailed knowledge of the structural geometry of the rock mass within the tunnel section. Schunnesson [3] also concluded that a major problem in such attempts is the analysis of the data. The monitored

TBM Excavation in Difficult Ground Conditions. Case Studies from Turkey. First Edition. Nuh Bilgin, Hanifi Copur, Cemal Balci.
© 2016 Ernst & Sohn GmbH & Co. KG. Published 2016 by Ernst & Sohn GmbH & Co. KG.

'raw' data is significantly affected by the operator, who often adjusts the drill settings in order to achieve the best drilling result. Furthermore, the advanced control systems on modern drill rigs adjust drill parameters (thrust and torque pressures) independently to avoid drilling problems and damage to the drill string, and as a result he concluded that drilling data should be normalized according to thrust and torque pressure values. Another problem in interpreting the drilling data is the wear of drill bits, which decreases drilling rates to a considerable extent in abrasive rocks. The drilling rate is also affected by the energy absorption of the drill string within long holes.

One of the most interesting probe drilling operations in TBM applications was carried out in the Lesotho Highlands water project, in delivery tunnel north. Probe drilling was carried out within a double-shield TBM with probe holes being 115 m long and 65 mm in diameter. They were angled at 5.5° to the roof of the tunnel. Probe holes, in addition to providing an indication of groundwater ahead of the face, also yielded information on the geology. For example, in the Ash River tunnel, they indicated the location of dykes, some of which were associated with groundwater. These required grouting: in the case of the Elim Dyke (20 m in width) the initial water ingress was as much as 400 l/min at 9 bar pressure. It took approximately 15 days to reduce these water inflows to 6 l/min. Around 120 m^3 of cement grout, with anti-washout and non-shrink additives, was injected in three concentric arrays of grout-holes at 15 m centers, which started about 50 m in front of the dyke, De Graaf and Bell [4].

Experience gained from mechanical ground probing in the Gotthard base tunnel contributed also to the development of probe drilling methodology applied in front of a TBM. The critical and most interesting working phase was the approach to the Piora basin, formed of sugar-grained dolomite, mixed with water at a pressure of up to 150 bar. The drilling installation, mounted on the TBM, was equipped with a sophisticated preventer system, composed by a 'blow-out preventer', as used in oil prospecting technology. The equipment was set to connect to a water pressure of up to 150 bars. The bores were put above the TBM crown, with variable inclination of 2–5°. During the driving of the fifth borehole in advance, at station 5,553 m, the sugar-grained, porridge-like dolomite of the Piora basin was met and the dolomite–water mixture poured into the tunnel through the 42 m long borehole, with 98 mm diameter, at high pressure, initially of more than 90 bars. Some parts of drilling equipment were thrown up to 30 m and water ingress reached a maximum outflow of 400–500 l/sec. The granulometric composition of the material corresponded to a fine to middle grained sand, Henke [5]. Flurry and Priller [6] reported also that in most cases core drilling was inevitable in this project.

Probe drilling operations used in the Arrowhead tunnels in USA were also very effective. Arrowhead tunnels project consists of two 5.7 m excavated-diameter tunnels, 9.6 km east and 8 km west tunnels. The tunnels were bored using two TBMs. The Arrowhead west tunnel, the more difficult of the two, took four years to bore. At one point, a flash flood temporarily submerged the TBM. The TBM in the east tunnel encountered water-bearing strata of metamorphic and granitic rock. The presence of the water, coupled with the depth of the tunnel below the surface, up to 700 m, forced the tunneling team to deal with water pressures in the tunnel heading in excess of 14 bars.

Also, much like the west, the TBM had to traverse branches of the San Andreas Fault. The tunnel was completed in May 2008. The automated probe drill monitoring system was proven to be a useful tool for evaluating ground conditions ahead of the TBM. The drill pressure data was analyzed, usually within a couple of hours after the completion of probing, Duke and Arabshahi [7].

Two important comprehensive studies on TBM drives in complex geology, on probe drilling and treatment of rock in front of a tunnel face are published by Barla and Pelizza [8] and Peila and Pelizza [9]. A report on this subject, edited by a group of contributors and published by AFTES [10] is also worth mentioning.

Steele et al. [11] concluded that the rock mass strength was affected by the weathering grade and discontinuity intensity of the rocks. A classification made by using the probe drilling specific energy provided indicative correlation with the weathering grade and localized zones of weak rocks. However, they emphasized that in a given rock mass a reduction in energy occurs with increased length of drill rode. They reported that the energy loss, referred to as the 'Trod' coefficient is ideally needed to be established from site-specific records to determine the energy loss over drilling length. Drilling rate was found to be decreased with drilling distance by about 2.5 cm/min per drilled (m) length.

14.3   Melen water tunnel excavated under the Bosphorus in Istanbul

The Melen tunnel is a water tunnel in Istanbul, crossing the Istanbul Bosphorus strait at a depth varying from 0 to 146 m, situated between Beykoz and Sariyer on the Asian and European sides, respectively. Probe drilling for advance exploration every 40 m allowed proper determination of geological structure, boundaries of unstable areas and inflow of water. The construction of tunnels started in February 2008 and finished in April 2009. Injections of cement grout and polymer compositions in front of the heading and behind the lining were applied to stop the water ingress.

The Melen water tunnel passes through Kartal formation, which is an alteration of calcareous shale, clayey limestone, sandy limestone and sandstone. RQD values change from 0 to 90%. There are minor and major faults in the area with andesite dykes. These magmatic intrusions or dykes are high strength rocks with compressive strength going up to 140 MPa. However, these intrusions are sometimes fractured and weathered, with varying thickness, from a few meters up to 50 m, making the contact zones very fractured and filled with fine materials. The presence of dykes, and the small intrusions cutting the Paleozoic sedimentary rocks in Istanbul region, is known from previously published data on TBM tunneling [12]. These andesitic rocks, are generally considered to be of Cretaceous age. Figure 14.1 shows the geological profile along the alignment. Potential problems in fractured dykes and blocky ground include high water inflows, difficulties with mucking and the installation of segmental lining, the annulus grouting and instability in front of the TBM [13, 14]

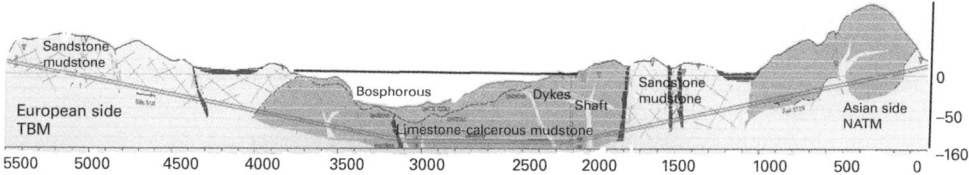

Figure 14.1 Geological cross-section of the Melen tunnel.

There is no systematic published data on the distribution and the orientation of these andesitic dykes, but Ozgorus and Okay [15] concluded in their recent studies that the distribution of the strikes is highly scattered with a few prominent directions with two possibilities. The first is that the Istanbul dykes constitute a local dyke swarm related to a yet unexposed pluton at depth. The second possibility is that the Istanbul dykes are from a regional dyke swarm. The contact between the main sedimentary rock and the dykes are weak, weathered and fractured. Problems expected are face collapses in front of TBM, jamming of the cutterhead, damage of the cutterhead and high water ingress [12]. A typical TBM cutterhead that jammed in the Beykoz–Istanbul tunnel due to a weak zone between dykes and the main rock is illustrated in Figure 14.2 [12]. These types of zones are very critical when excavating under sea, due to expected high water ingress. The description of dykes and their influences on probe drilling were the main objectives of the first pioneering studies, Anagnostou et al. [13].

Figure 14.2 TBM cutterhead jammed in the Beykoz–Istanbul sewerage tunnel within dyke zone.

14.4 Methodology of predicting weak zones ahead in the Melen water tunnel

In the tender document it was indicated that continuous probe drilling of diameter not less than 51 mm should be used ahead of the tunnel faces to locate gas or water-bearing areas and to detect changes in the geological conditions. Grouting was kept mandatory if water ingress from a single or multiple probe holes exceeded 5 l/min per drilled hole. Routine probe drilling was suggested to be carried out using rotary percussive drilling equipment. It was also demanded that additional core rotary drilling should be carried out if it was asked for by the engineer for extra geological characterization.

Drilling parameters and TBM operational parameters, including muck size, are affected directly by the rock mass properties. Figure 14.3 gives a typical example of the effect of geological discontinuities on muck size during TBM excavation in Kartal formation.

The methodology used in this study for predicting the weak zones ahead of the tunnel is summarized in Figure 14.4.

a) Muck size in competent rock

b) Muck size in fractured zone

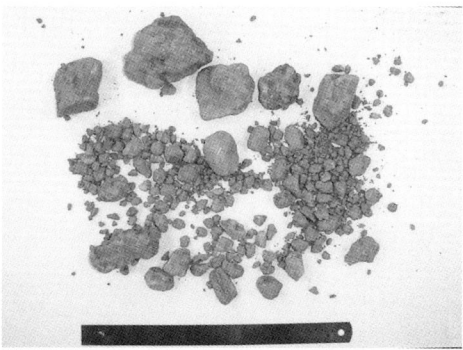

c) Muck size in week zone in kartal formation

Figure 14.3 Effect of discontinuities on muck size.

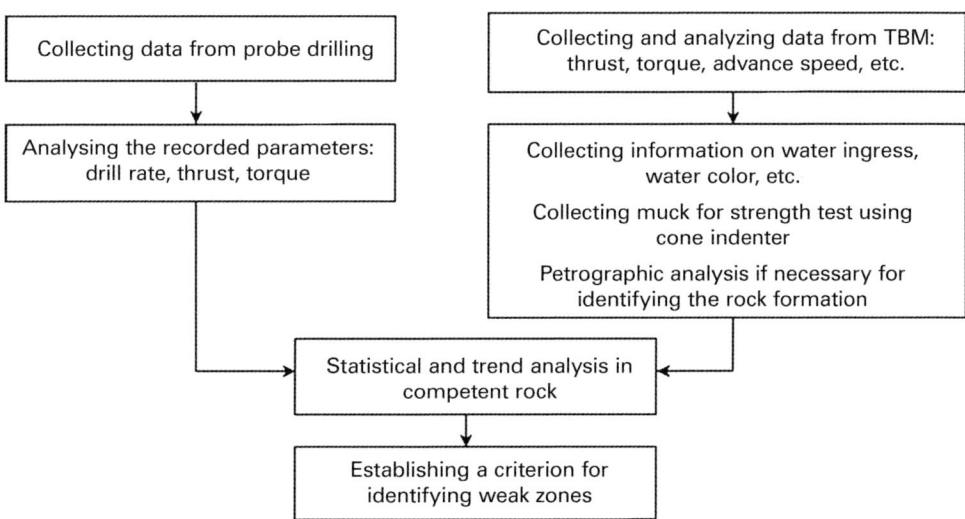

Figure 14.4 Methodology of using probe drilling in the Melen project.

At the beginning of the tunnel excavation extensive pioneering probe drilling operations were carried out between chainages 4+400 and 4+940 km to establish a criterion for further assessment of the probe drilling operations. The results presented in this study cover only the results of the site investigation within the chainages mentioned above. Later TBM crew took over the probe drilling operations, and first results of the data analysis were reported by Bakir et al. [16, 17]. Drilling parameters such as torque and thrust pressures and drilling rates were continuously recorded during probe drilling operations using an automated data acquisition system. Figure 14.5 shows the probe drilling equipment used in the Melen project.

Figure 14.5 Probe drilling equipment used in the Melen project.

To have consistent comparable results for data processing, drilling rate values were normalized in cm/min per drill thrust bar and in cm/min per drill torque (dividing drilling rate values by drilling thrust and drilling torque in bars), as Schunnesson suggested [2, 3].

14.4.1   Data analysis and results

Muck samples were collected systematically for petrographic analysis and strength tests. Several thin sections were prepared for petrographic identification of the samples under an optical microscope. This study also gave a unique opportunity to contribute to understanding the evolution of the geology under the Istanbul Bosphorus. Some of typical views of thin sections under the microscope are given in Figures 14.6–14.14, showing the diversity of petrographic characteristics of rock formations excavated under the Istanbul Bosphorus.

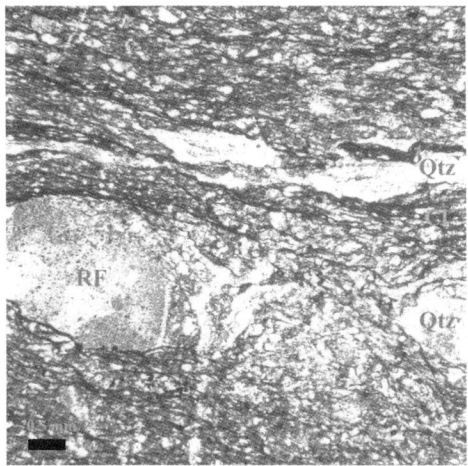

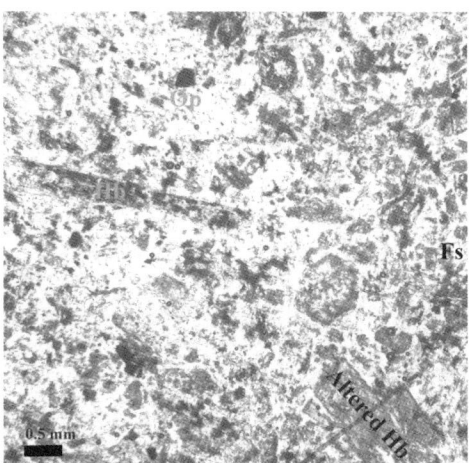

Figure 14.6 Meta-siltstone with typical cleavages (Ring 435, chainage 4+842 km). [Qtz: quartz, Cl: clay, RF: rock fragment].

Figure 14.7 Altered andesite (Ring 441, Chainage 4+836 km) [Op: opaque, Hb: hornblende, Fs: feldspar].

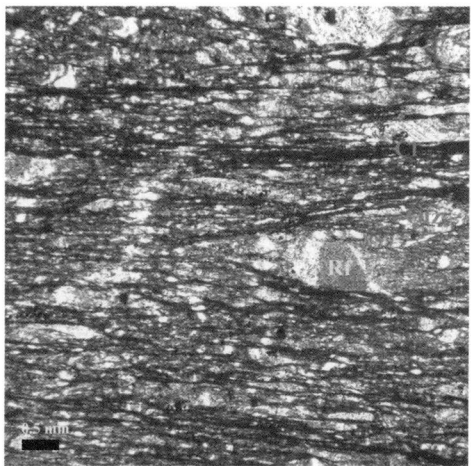

Figure 14.8 Meta-siltstone (Ring 464, Chainage 4+807 km) [Cl: clay, Qtz: quartz, Rf: rock fragment].

Figure 14.9 Siltstone with high carbonate content, 75% calcite (Ring 544, chainage 4+711 km) [Cal: calcite, Op: opaque, Qtz: quartz].

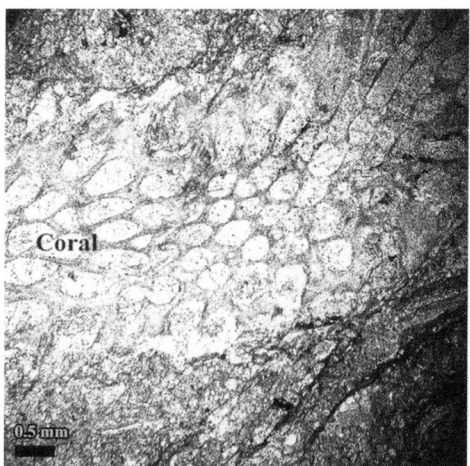

Figure 14.10 Carbonated siltstone, 50% carbonate, plenty Fossils, mainly coral (Ring 574, chainage 4+676 km).

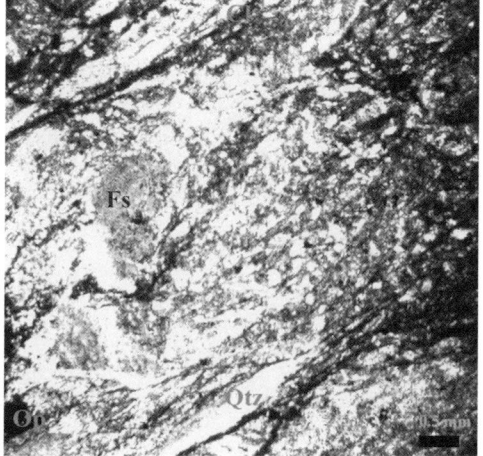

Figure 14.11 Siltstone, quartz content 30–40%, feldspat 20% (Ring 582, chainage 4+666 km) [Cl: clay, Fs: feldspar, Qtz: quartz, Op: opaque].

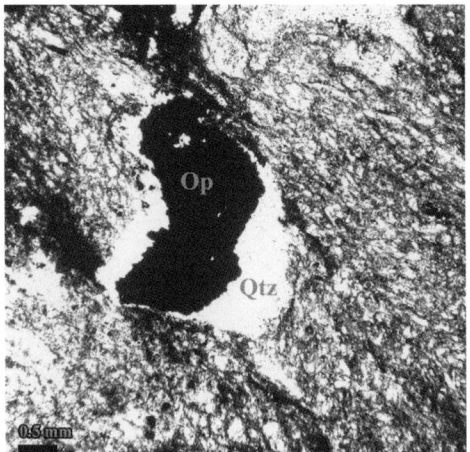

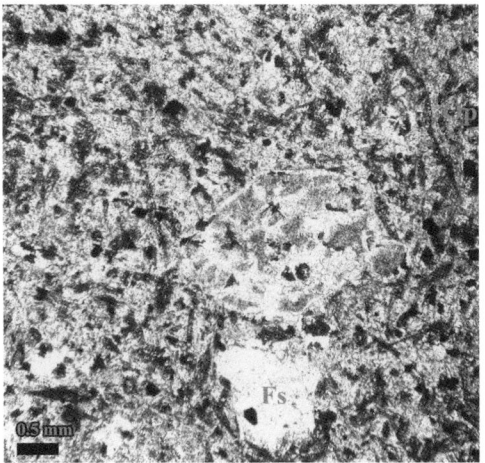

Figure 14.12 Siltstone, pyrite mineral with tension traces (Ring 605, Chainage 4+699 km) [Op: opaque, Qtz: quartz].

Figure 14.13 Alternated andesite (Ring 635 Chainage 4+603 km) [Op: opaque, Fs: feldspar].

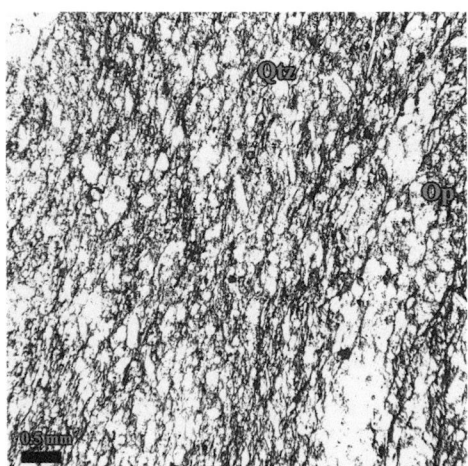

Figure 14.14 Fine-grained sandstone (Ring 660 Chainage 4+571 km) [Qtz: quartz, Op: opaque].

During sampling, it was observed that the rock material was broken into small pieces in the screw conveyor of the TBM, due to the brittle characteristics of the excavated material and existing fractures. Therefore, it was decided to carry out strength tests on small samples using an NCB cone indenter, as shown in Figure 14.15. This test developed by the National Coal Board is found to be very reliable for small rock samples [12]. The compressive strength values estimated from the cone indenter test are given Table 14.1.

Figure 14.15 NCB cone indenter used for strength tests.

Table 14.1 Compressive strength test results calculated after an NCB cone indenter test.

Ring No	Chainage (km)	σ_c (MPa)	Rock
435	4+842	72	Meta-siltstone, typical cleavages
436	4+841	92	Meta-siltstone
437	4+837	64	Siltstone
438	4+836	93	Siltstone
439	4+835	105	Dyke (D1), altered andesite
441	4+831	120	Dyke (D1), altered andesite
464	4+807	110	Meta-siltstone
479	4+789	78	Meta-siltstone
490	4+776	72	Siltstone
500	4+764	59	Siltstone
518	4+743	86	Siltstone
540	4+717	68	Siltstone
544	4+712	111	Carbonated siltstone
549	4+706	66	Siltstone
551	4+703	49	Siltstone
560	4+693	96	Siltstone
563	4+689	61	Siltstone

Ring No	Chainage (km)	σ_c (MPa)	Rock
565	4+685	55	Siltstone
580	4+669	84	Siltstone
582	4+666	74	Siltstone with quartz content of 35%
591	4+675	104	Dyke (D2), altered andesite
595	4+651	101	Siltstone with
605	4+639	48	Siltstone, pyrite minerals, tension traces
615	4+627	43	Siltstone
621	4+620	67	Siltstone
627	4+613	80	Siltstone with cleavages
635	4+603	68	Dyke (D3), altered andesite
640	4+596	64	Siltstone
643	4+594	92	Siltstone
646	4+590	101	Fine grained sandstone
685	4+543	66	Siltstone
690	4+537	76	Siltstone
710	4+513	117	Siltstone
725	4+495	54	Siltstone
735	4+483	58	Siltstone
745	4+471	113	Dyke (D4), diabase
755	4+459	116	Siltstone
760	4+454	132	Fine grained sandstone

Petrographic analysis and observations made during tunneling operations showed that the occurrence of andesite and diabase dykes is roughly one every 100 m of tunnel chainage. In one case, however, two andesite dykes within 100 m were encountered. Under the microscope it was observed that andesite dykes were altered, which reduced their strength (compressive strength changing from 67 to 120 MPa). However, it should be noted that dykes with compressive strength up to 200 MPa were also excavated. These types of dykes caused tremendous problems to the TBMs working in Istanbul as noted above.

During the application of rotary percussive drilling, it is expected that the drilling rate will decrease due to the percussive energy absorbed in the drill rods, as noticed by Steel at al. [11]. In the Melen and Kargi projects it was observed that this was not apparent due to the quick changes in the geological characteristics of the rock mass.

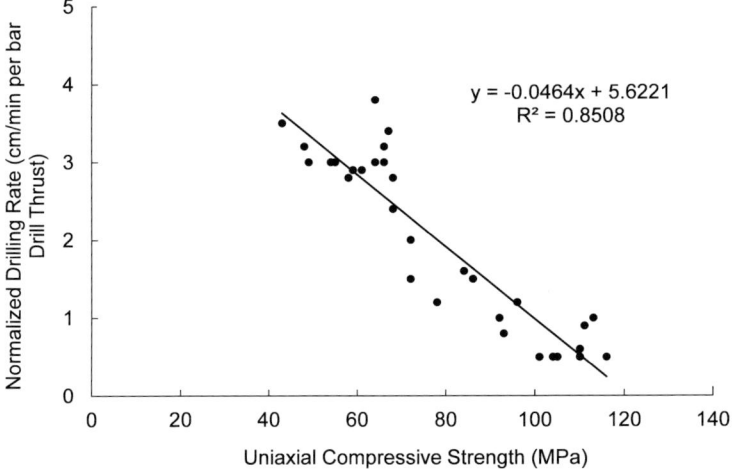

Figure 14.16 Variation of normalized drilling rate per bar thrust with uniaxial compressive strength of rock samples.

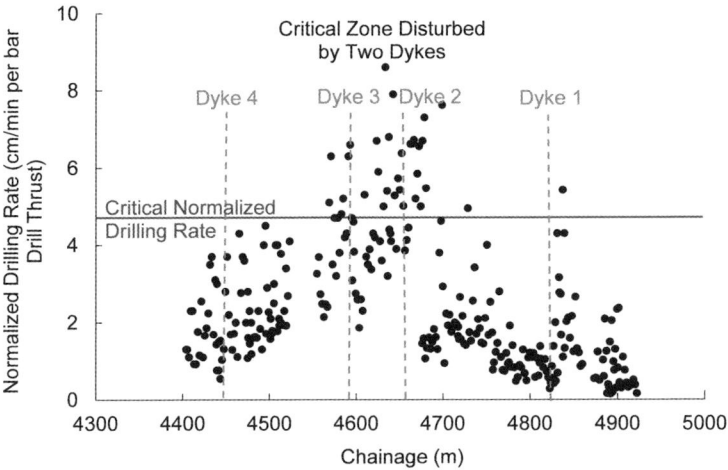

Figure 14.17 Variation of normalized probe drilling rate per bar thrust with tunnel chainage.

The variations of normalized drilling rate in cm/min per drill thrust bar with uniaxial compressive strength of rock samples tested and tunnel chainage are given in Figures 14.16 and 14.17. The variation of normalized drilling rate in cm/min per drill torque bar with compressive strength of rock samples and with tunnel chainage are given in Figures 14.18 and 14.19. It can be seen from Figures 14.17 and 14.19 that drilling rates were minimum in the areas where dykes were being excavated, due to the high strength characteristics of these magmatic inclusions. However, it is well-known that dykes generate disturbed and fractured contact zones. It is clearly seen from Figures 14.17 and 14.19 that the normalized drilling rates are higher between chainages 4+700 and

4+600 km in the influence zones of two dykes. This is very important since it shows that when excavating under the sea, these fractured and disturbed zones will result in high quantities of water inflow from the Sea of Marmara.

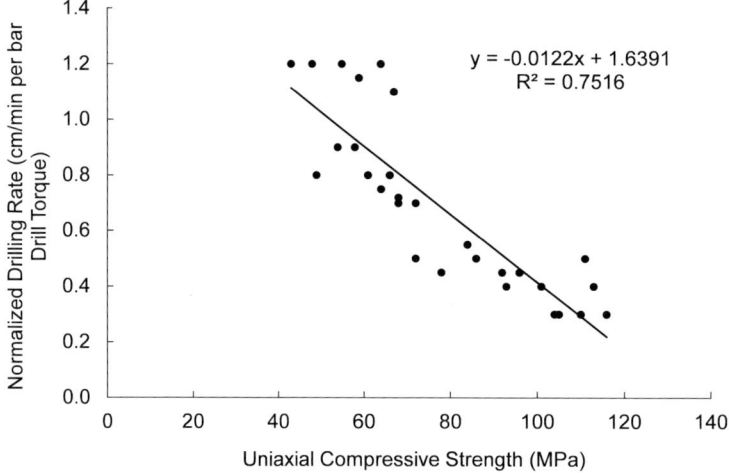

Figure 14.18 Variation of normalized drilling rate per bar torque with uniaxial compressive strength of rock samples.

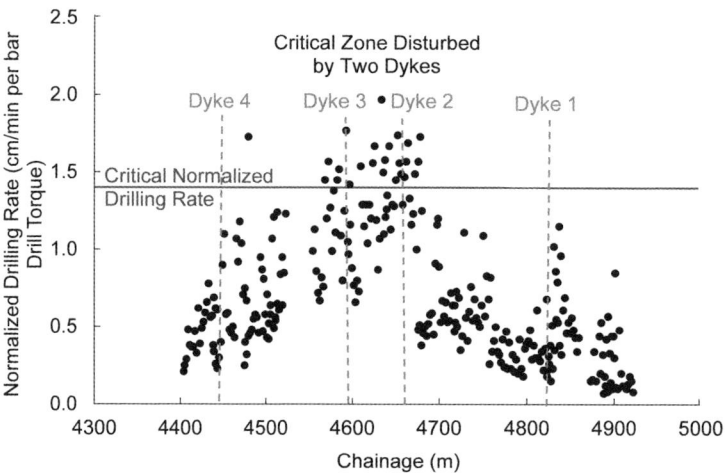

Figure 14.19 Variation of normalized probe drilling rate per bar torque with tunnel chainage.

Previous observations in the tunnel drivages in Istanbul showed clearly that the weathered contact zone between the andesitic dykes and the main rock formation is weak, and in most cases this zone has compressive strength of less than 20 MPa. Extrapolating the trend lines of Figures 14.16 and 14.18 to the left for compressive strength 20 MPa, the normalized drilling rate per thrust bar is greater than 4.71 cm/min per drill thrust bar and normalized drilling rate per drill torque bar is greater than 1.4 cm/min per torque bar. These values are considered as critical for drill thrust and torque. This emphasized that the zone between chainages 4+570 and 4+680 km was a critical zone between the two dykes. It is interesting to note that excessive water ingress was encountered in this zone, and chemical grouting was applied. After interpretation of these results it was concluded that these two critical normalized probe drilling rate values should be taken into consideration as predicting critical geological zones.

TBM thrust and torque values are also good indicators of rock mass properties. However, they should be also normalized, taking account of the penetration per revolution. The normalized probe drilling rate and normalized TBM thrust values are given in Table 14.2. The variation of these two variables normalized (drilling rate per thrust bar with TBM thrust index in kN/mm penetration per revolution values) is given in Figure 14.20. This figure shows the close relationship between these two variables, suggesting that TBM thrust index is also a good indicator of critical zones, since for TBM thrust index a critical value for weak zones is reached when the value reaches more than 550 kN/mm penetration. This value corresponds to the normalized drill rate per thrust bar values greater than 4.71 cm/min per drill thrust bar.

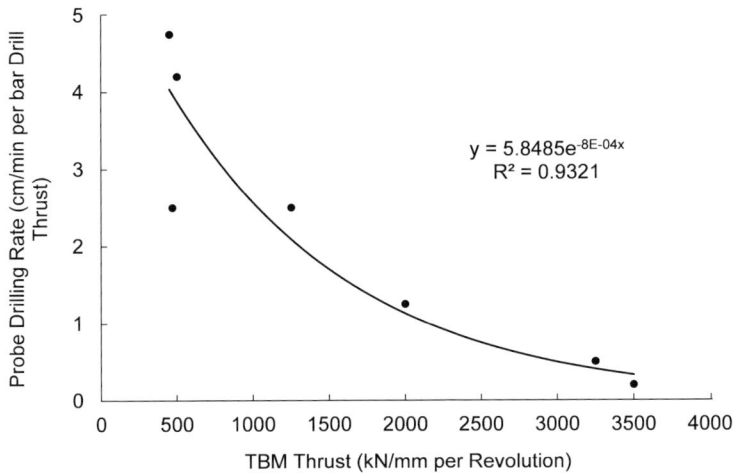

$$y = 5.8485e^{-8E\text{-}04x}$$
$$R^2 = 0.9321$$

Figure 14.20 Variation of normalized probe drilling rate per bar thrust with TBM thrust index.

Table 14.2 The variation of normalized probe drilling rate with normalized TBM thrust.

Chainage (km)	TBM thrust [kN/mm per revolution]	Probe drilling [cm/min] per bar thrust
4+905	470	2.50
4+874	3,500	0.20
4+847	1,250	2.5
4+825	500	4.2

Chainage (km)	TBM thrust [kN/mm per revolution]	Probe drilling [cm/min] per bar thrust
4+797	2,000	1.25
4+773	3,250	0.5
4+723	3,250	0.50
4+700	450	4.75

14.5 Kargi energy tunnel

14.5.1 General information on the Kargi project

The Kargi hydropower project was developed in order to utilize the energy potential of the Kizilirmak River between the towns of Osmancik and Boyabat in Turkey. The excavation of an 11.8 km tunnel has been recently finished, 7.8 km of which was excavated with a double-shield Robbins TBM of 9.84 m diameter, and 4 km opened with drill and blast, Home [18, 19]. The job owner was Stadkraft, and the contractor was Gulermak. The geology consists of Eocene Beynamaz volcanites, mainly of agglomerate, andesite, basalt, tuff and metamorphites and a 2500 m melange of Kiraztasi-Kargi ophiolites and graphitic schist. The geological cross-section of the Kargi tunnel is given in Figure 14.21.

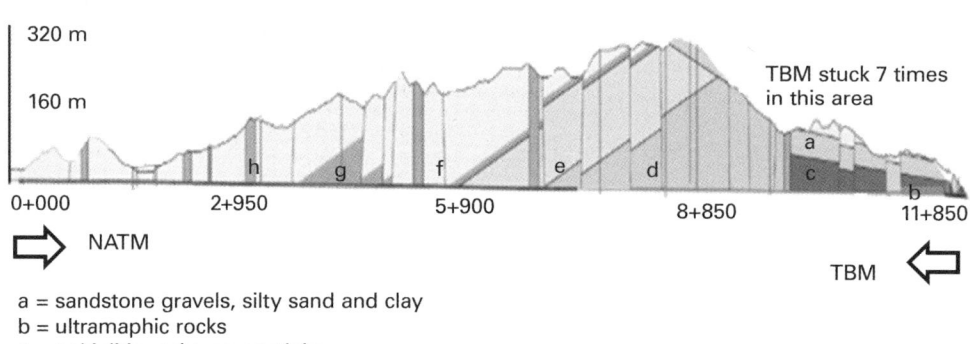

a = sandstone gravels, silty sand and clay
b = ultramaphic rocks
c = ophiolitic rocks, serpentinite
d = metapelites, micaschist, graphitic schist
f = andesitic basalt
g = agglomerate
h = dykes

Figure 14.21 Geological cross section of Kargi Tunnel.

The tunnel was intended to be finished in 30 months, but the TBM stuck several times due to geological difficulties related to the North Anatolian Fault. Bypass galleries were opened in seven different places to free the blocked TBM. In order to reach the scheduled job duration, time it was decided to open another tunnel from the other part of the tunnel line by conventional tunneling [18, 19, 20].

14.5.2   Probe drilling operations

Two Atlas Copco COP 1838 were used for probe drilling operations. Drilling rates and percussive and rotary pressures of the drill rigs were continuously recorded during the probe drilling operations, and there were 17 ports within the shield. Mostly, drill rods T38 with a length of 3 m and drill bits of 64 mm diameter were used. From the beginning of excavation, more than 5 km length of probe drilling was used.

In most cases, probe drilling from one of the upper ports was used, but if necessary extra probe drilling was carried out from the other ports. Probe drilling started 16 m behind the cutterhead and usually a length of 45 m probe drilling was used. Drilling was executed at an inclination of $7°$, which made a 4.4 m of upgrading for a drill length of 29 m and actual drilling of the face started from 0.7 m up from the tunnel periphery.

Probe drilling was applied systematically during the tunnel excavation. However, it was noticed that geological formations were disturbed tremendously by the North Anatolian Fault, from the entrance of the tunnel (chainage 11+873 km) up to a big dyke, which was encountered over an 80 m long section at chainage 9+700 km. The TBM was blocked in these areas, and side galleries were opened to rescue the TBM. It was also noticed that the behavior and the performance of the TBM in this area was completely different from in the area encountered thereafter. Due to the North Anatolian Fault activities, the geological formations were fractured in blocky form up to this big dyke. This blocky characteristic of the ground was making it extremely difficult to evaluate the probe drill data. Explaining the interrelations between the TBM data, the probe drill results and the strength of the rock samples was found to be extremely important for identifying critical weak zones. In some cases, an umbrella arch was used to stop the tunnel face collapsing and jamming the TBM cutterhead.

14.5.3   Analysis of probe drilling data in the Kargi project

Probe drilling rates were carefully recorded for each meter of the drill length. In most cases, feeding pressure of 80–85 bar and percussive pressure of 140 bar was maintained constant during measurements. In the Kargi project the interpretation of the probe drilling data was carried out by simultaneously observing probe drilling rate values with muck coming from the drill holes, including TBM thrust and torque values. Increasing probe drilling rate values and decreasing TBM thrust and torque values were always interpreted as crossing a weak zone, transition or fault zone. Typical recorded probe drilling and TBM data are plotted in Figures 14.22 and 14.23 for different chainages.

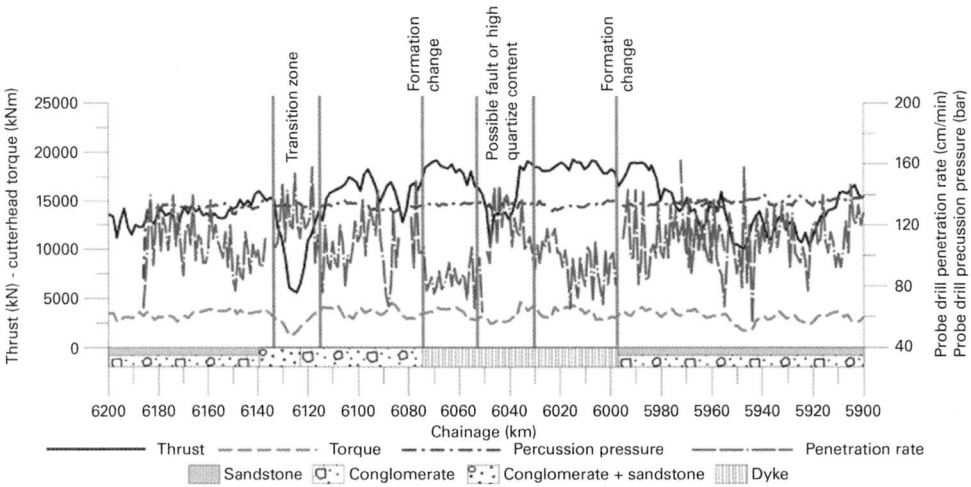

Figure 14.22 Variation of TBM and probe drill parameters with geological formation between chainages 6+200 and 5+900 km.

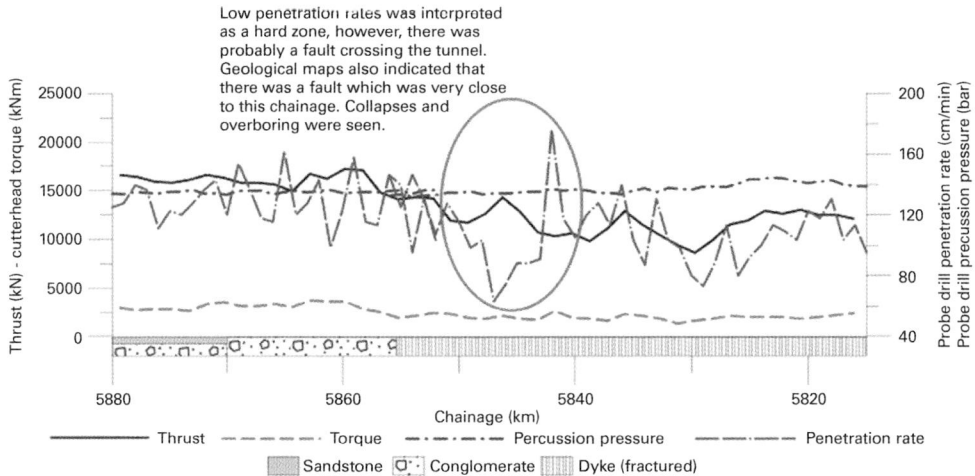

Figure 14.23 Variation of TBM and probe drill parameters with geological formation within chainages 5+880 and 5+820 km.

The interpretation of the data was made as follows.

1. As seen from Figure 14.22, probe drilling rate values increased and TBM thrust and torque values decreased between chainages 6+135 and 6+105 km. This was inter-preted as passing a weak zone, related to a transition zone from conglomerate plus sandstone to conglomerate.
2. Probe drilling rate values decreased and TBM thrust and torque values increased between chainages 6+075 and 6+000 km. This was interpreted as passing into a hard rock formation, sandstone.

3. Probe drilling rate values increased and TBM thrust and torque values decreased between chainages 6+050 and 6+035 km. This was interpreted as a fault zone within sandstone formation.
4. Probe drilling rate values decreased and TBM thrust and torque values increased between chainages 5+850 and 5+841 km. This was first interpreted as passing a hard zone. However, collapses and over-boring were observed in front of the cutterhead, indicating that the TBM was crossing a fault and crushed zone. The geological profile also indicated a fault zone in this area.

At the end of tunnel excavation, it was decided to make a final analysis of the probe drill data to make a sound basis for the contribution to future probe drilling operations. During the final analysis of probe drill data, some areas having different geological characteristics were chosen as the basis for the analysis. The variations of TBM and probe drill data (TBM thrust, TBM torque, probe drill percussion pressure, probe drill penetration drill) with geological formation were investigated in 17 different selected areas. Figures 14.22 and 14.23 are two typical examples from these selected areas.

Histograms of mean values of probe drilling rate in cm/min, TBM thrust in kN and TBM torque in kNm were drawn for 17 selected areas for different rock formations. Typical histograms of these parameters for geological formations found between chainages 7+420 and 7+230 km are given in Figures 14.24–26.

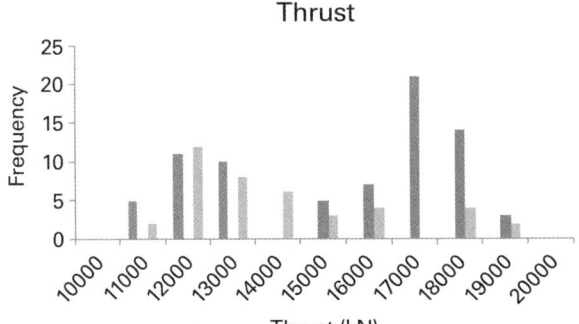

Figure 14.24 Frequency of TBM thrust for different rock formations within tunnel chainages of 7+420 and 7+320 km.

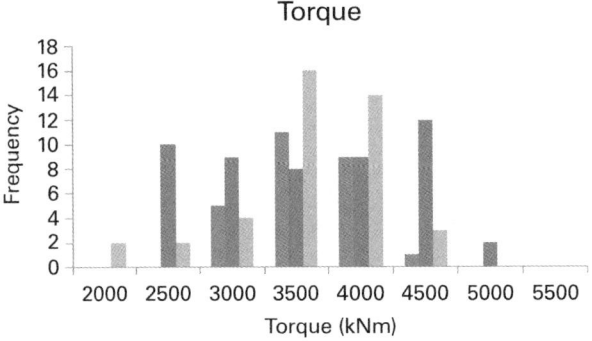

Figure 14.25 Frequency of TBM thrust for diff erent rock formations within tunnel chainages of 7+420 and 7+320 km.

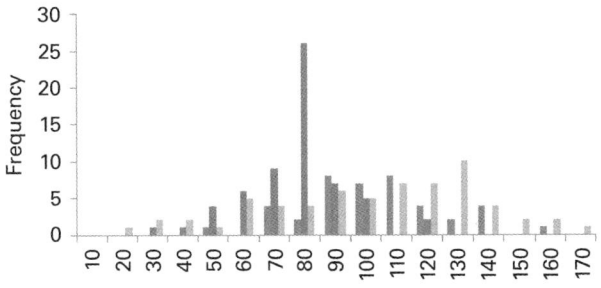

Figure 14.26 Frequency of probe drilling rate for different rock formations within tunnel chainages of 7+420 and 7+320 km.

Variations of TBM thrust and torque with probe drilling data derived from histograms for different geological units are given in Table 14.3. The values given are mean values derived from the histograms.

Table 14.3 The variation of TBM thrust and torque with probe drill data obtained from histogram within different tunnel chainages representing different geological formations.

Chainage (km)	Formation	Probe drill rate (cm/min)	Thrust (kN)	Torque (kNm)
9+000/8+875	Schist, calcite, clay	102.1	10,415	2,063
8+875/8+830	Schist, calcite, clay, Meta-siltstone	106	12,410	3,055
8+335/8+245	Micritic limestone	71.4	15,500	4,072
8+245/8+220	Schist, calcite, clay, meta-siltstone	120.3	10,294	2,681
7+420/7+405 7+365/7+305	Conglomerate,	96.6	13,534	3,438
7+405/7+365	Basaltic dyke, serpentinite	98.8	11,762	3,388
7+305/7+320	Limestone, marble	72.4	16,608	3,428
7+110/7+045	Limestone, marble	59.8	18,233	4,100
7+032/7+005	Andesitic dyke	100.4	10,290	3,448
7+035/7+990	Schist	119.4	10,290	3,448
6+910/6+865 6+785/6+725	Graphitic schist	120.7	10,454	2,857
6+865/6+785	Limestone-marble	74.7	16,755	3,124
6+200/6+075 5+990/5+900	Conglomerate, Sandstone	110.2	14,296	3,361
6+075/5+995	Dyke	97.6	17,357	3,496
5+880/5+857	Conglomerate, sandstone	128.3	16,094	3,050
5+857/5+795	Dyke fractured	111	12,115	2,165
5+795/5+750	Dyke fractured, sandstone	102.4	13,529	3,024

Data analysis was based on the new concepts of thrust of the TBM (kN)/probe drilling rate (cm/min) and TBM torque (kNm)/probe drilling rate (cm/min).

Figures 14.23 and 14.24 show the relationship between TBM thrust and probe drilling rate and between the TBM thrust/probe drill rate ratio and the probe drilling rate, respectively. Figures 14.27 and 14.28 show the relationships between probe drill rate, TBM torque and TBM torque/probe drill penetration rate. The relationships between TBM thrust with probe drilling rate values and TBM thrust/probe drill rate ratio with probe drilling rate values are given in Figures 14.27 and 14.28. The relationships between probe drill rates values, TBM torque and TBM torque/probe drill penetration rate values are given in Figures 14.29 and 14.30. As seen from these figures, TBM thrust and torque values divided by probe drilling rate values give better correlations with probe drilling rate values. As seen from these figures TBM thrust and torque values divided by probe drilling rate values give better correlations with probe drilling rate values.

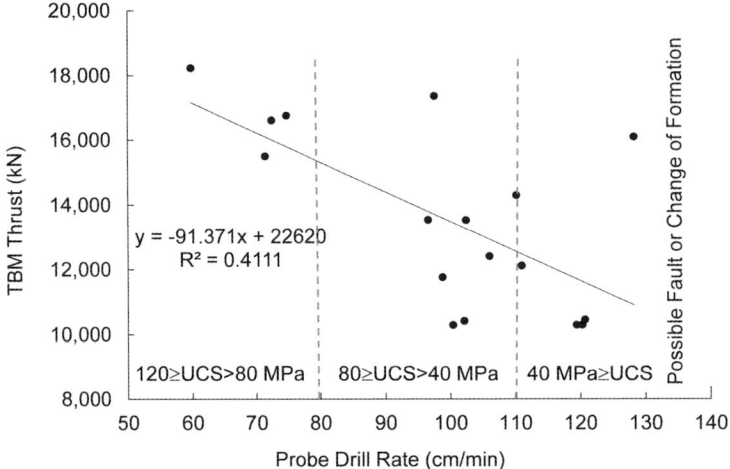

Figure 14.27 Variation of TBM thrust with probe drill rate.

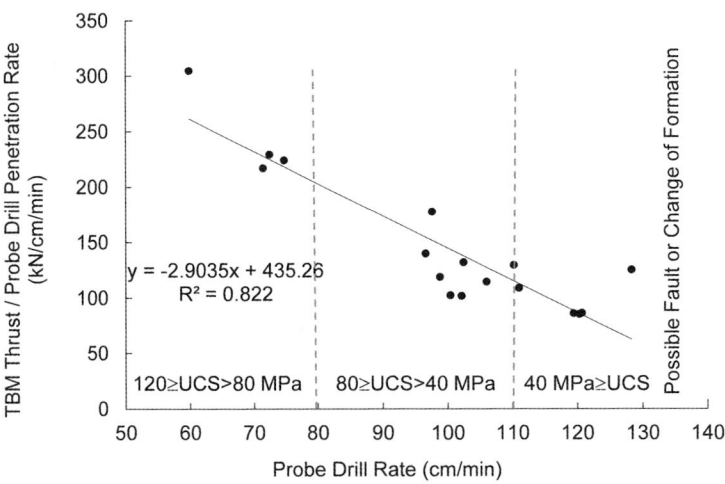

Figure 14.28 Variation of TBM thrust/probe drilling rate ratio with probe drilling rate.

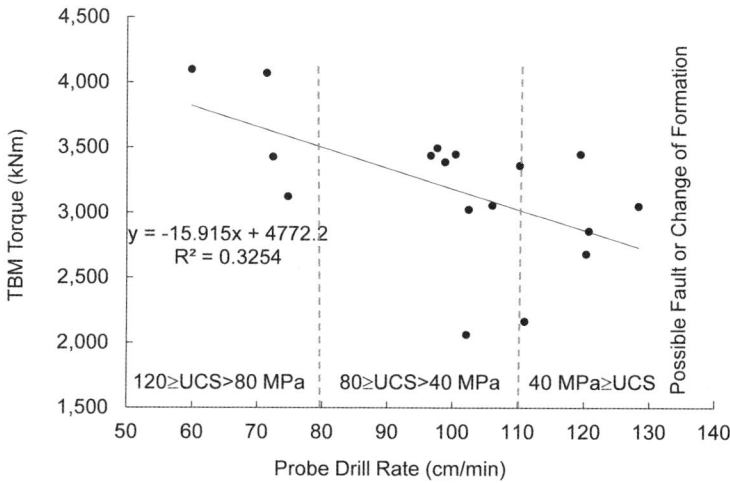

Figure 14.29 Variation with TBM torque with probe drilling rate.

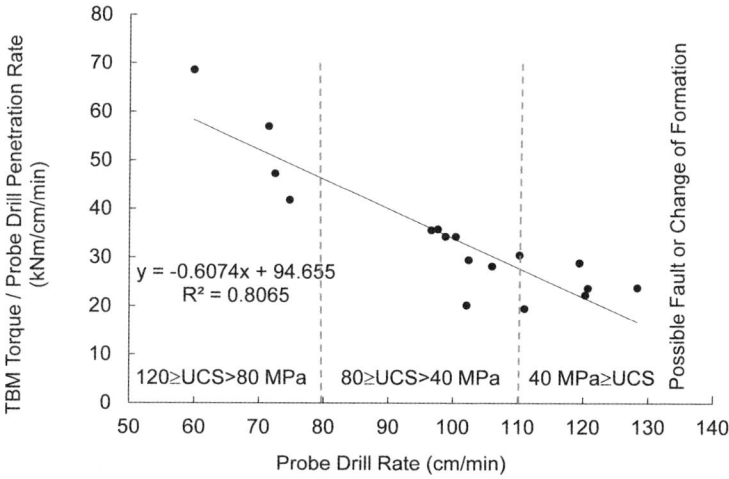

Figure 14.30 Variation of TBM torque/probe drill rate ratio with probe drilling rate.

As noted above, the PDR = f (FT/PDR) and PDR = f (Torque/PDR) show statistically more reliable relations. In these figures the effect of rock compressive strength on given relationships is also indicated. However, it is also important to note that compressive strength values are mean values of related areas.

Statistical analysis carried out on probe drilling and TBM data shows clearly that probe drill data, TBM thrust, TBM torque and the strength of the ground are interrelated, and Figures 14.26–28 may be used for this purpose. The above method of analysis and Figures 14.26 and 14.28 also made it possible to make a classification of difficulties in TBM tunnel boring as given in Table 14.4. This table may be used as a guide for further probe drilling applications.

Table 14.4 Classification for difficulty in TBM excavations, based on mean values taken from frequency of variables.

TBM thrust/probe drilling rate (kN/cm/min)	TBM thrust/probe drilling rate (kN/cm/min)	UCS (MPa)	Classification for difficulty in TBM excavations
Greater than 200	Greater than 45	120–80	Class 1 Competent rock
200–125	45–25	80–40	Class 2 Competent rock, few problems due to minor weak zones
Less than 125	Less than 25	Less than 40	Class 3 Weak zones, transition zones, fault zones

14.6   Conclusions

Probe drilling ahead of a TBM is a time-consuming and tedious operation. If it is not interpreted correctly, it may give misleading results in complex geology. This research study has revealed that for a correct interpretation of the drilling data, muck from the excavated area must be collected continuously for petrographic identification and strength tests. It should be always remembered that muck size, especially coarseness index, is always a good indicator of the TBM performance parameters, Tuncdemir et al. [21]. In most cases where a shielded TBM is used it is a problem to get big rock samples from the TBM face for strength characterization of the strata which will serve to correlate probe drilling and TBM performance parameters with rock strength. In this case, muck samples obtained from the TBM may be used, and the limited size of these samples will necessitate the use of point test, Shore Scleroscope hardness or NCB cone indenter tests. This chapter has also showed that TBM normalized thrust and torque data coupled with normalized probe drilling data may give more reliable results for identifying critical weak zones, fault zones and squeezing zones.

In summary it may be concluded that normalized penetration rate of a probe drill is a good indicator for detecting weak zones ahead of a tunnel. Probe drilling rate changed with changing geological formations and showed sharp increases and decreases in transition zones. Probe drilling rate is also very much affected in fault zones. In some cases, probe drilling rate increased within fault zones due to swelling and squeezing characteristics of the gauge materials within the fault zones.

The method of analysis of probe drilling rate data in the Melen project permitted the defining of a critical normalized probe drilling value. Values higher than the predetermined critical value point out critical geological zones susceptible to pressurized water ingress.

TBM thrust and torque values increased with decreasing probe drill penetration rates in competent hard rock formations in the Kargi energy tunnel. However, this study introduced a new concept of TBM thrust/probe drilling penetration rate and TBM torque/probe drilling penetration rate ratios explaining the variation of penetration rates in a statistically more reliable manner. This concept showed that by observing closely the variation of TBM thrust and torque values it is possible to predict the transient and fault zones, dictating the carrying out of umbrella arching, which will strengthen the weak zone in front of the tunnel.

References

[1] Bilgin, N., Ates, U. (2016) Probe drilling ahead of two TBMs in difficult ground conditions in Turkey. *Rock Mechanics and Rock Engineering*, **49** (7), 2763–2772.

[2] Schunnesson, H. (1996) RQD Predictions based on drill performance parameters. *Tunn Undergr Sp Technol*, **11** (3), 345–351.

[3] Schunnesson, H. (1998) Rock Characterization using percussive drilling. *Int J Rock Mech Min Sci*, **35** (6), 711–725.

[4] De Graaf, PJH, Bell, FG. (1997) The delivery tunnel North, Lesotho highlands water project. *Geotechnical and Geological Engineering*, **15**, 95–120.

[5] Henke, A. (2005) *Experience from the ground probing in the Gothard-Base Tunnel and their applicability in the Gibraltar Strait Crossing*. IV UN-ITA Workshop Gibraltar Strait Crossing, January 20, Madrid, pp. 20.

[6] Flury S, Priller, A. (2009) Successful penetration of the Piora Basic in the Faido Part-Section. *Tunnel*, **4**, 15–18.

[7] Duke, SK., Arabshani, J. (2007) *Evaluating ground conditions ahead of a TBM using probe drill data*. In: Rapid Excavation and Tunneling Conference, USA.

[8] Barla G, Pelizza, S (2000) *TBM tunnelling in difficult ground conditions*. GeoEng, Melbourne, Australia.

[9] Peila D, Pelizza S (2010) Ground probing and treatments in rock TBM tunnel to overcome limiting conditions. *Journal of Mining Sciences,* **45** (6), 602–619.

[10] AFTES (2014) Forward probing ahead of tunnel boring machines. Tunnels and espace souterrain, May/April 2014, 132–164.

[11] Steele, DJ., Wu, KK., Ishii, M., Mackay, AD., Kameyama, K. (2014) *The use of advanced percussive drilling techniques to predict ground conditions ahead of sub-surface excavations*. In: Underground Singapore, 25–26 September, organized by TUCSS, Tunnelling and Underground Society, Singapore 12 pages.

[12] Bilgin, N., Copur, H., Balci, C. (2014) Mechanical Excavation in Mining and Civil Industries. CRC Press, New York.

[13] Anagnostou, G,, Ozdemir, L., Bilgin, N. (2008) Greater Melen Project – Bosphorus Water Tunnel *Report on the site visit of 27th June 2008 and on the first results of drilling data analysis*.

[14] Anagnostou, G. (2010) Some rock mechanics aspects of subaqueous tunnels. Rock Engineering in Difficult Ground Conditions – Soft Rocks and Karst – Vrkljan (ed), Taylor & Francis Group, London.

[15] Ozgorus, Z., Okay, IA. (2005) Orientation of the andesitic dykes in the Istanbul Region An approach to the cretaceous stress distribution. *Mineral Res. Exp., Bull.,* **130**, 17–27.

[16] Bakir, A., Eris, I., Daglioglu, S. (2011) *Probe drilling in Bosphorus Tunnel (SP-7).* In; proceedings for Atilla Yalcin Symposium., İstanbul, ITU.

[17] Bakir, A., Eris, I., Daglıoglu, S. (2012) *Interrelationships between TBM performance and probe drill data in Melen Water Tunnel (SP-7) in Istanbul.* In. Proceedings, Geomechanic, symposium for Mahir Vardar, Istanbul, ITU.

[18] Home, L. (2014a) *Case studies on mechanized rock tunneling.* In: Proceeding of World Tunnel Congress, Tunnels for Energy, 9–10 May, Brazil, pp 1–10.

[19] Home, L. (2014b) *The Kargi Challenge, a study of TBM boring vs. conventional excavation.* In: third short course organized by Turkish Tunnelling Society, 28 August, Istanbul, Turkey.

[20] Yurt, M., Ozturk, A., Arslan, Z., Nuhoglu, C., Oystein, L., Erdogan, E., Atlar, B., Palakci, Y., Bilgin, N. (2014) *Factors affecting the performance of a double shield TBM in a very complex geology in Kargi Turkey.* In: proceedings World Tunnel Congress 2014 – Tunnels for a better Life. Foz do Iguacu, Brazil.

[21] Tuncdemir, H., Bilgin, N., Copur, H., Balci, C. (2008) Control of rock cutting efficiency by muck size. *Int J Rock Mech Min Sci,* **45**, 278-288.

[22] Grandori, R. (2006) Abdalajis east railway tunnel (Spain) -Double Shield Universal TBM cope with extremely poor and squeezing formations. *Tunn Undergr Sp Technol,* **21**, 268–280.

15 Application of umbrella arch in the Kargi project

15.1 Introduction

The aim of this chapter is to introduce one of the most comprehensive and successful umbrella arch (UA) operations used ahead of a TBM in the Kargi tunnel in blocky ground and complex geology in Turkey. The increase of TBM torque and the change in penetration rate of the probe drill were used as criteria for the application of UA. The project, the geology, the characteristics of the TBM and the probe drilling operations in this tunnel were summarized in the previous chapter and elsewhere [1–5]. Past experiences of UA will first be summarized, and the methodology and the results obtained will then be explained.

15.2 General concept of umbrella arch and worldwide application

The UA method is a pre-reinforcement technique which improves the stability of the tunnel crown and face. Pipe umbrella systems have effectively been used as an additional support system in the working area in weak ground conditions for the past few decades. Thus, a considerable number of applications have been applied in urban tunneling with shallow overburden [6]. However, this support system has also been successfully used when a tunnel alignment crosses fault zones and sediments, with overburdens from shallow to deep [7, 8, 9]. The pipe umbrella support system is termed differently around the world, synonyms being canopy tube umbrella method, umbrella arch method and forepoling method [10]. The arch supports the rock mass and the tunnel face predominantly by transferring loading longitudinally through the interaction of the support and surrounding ground condition. There are three categories of UA. Categorization depends on the type of support element used, including spiles, forepoles and grouts [11].

Using an umbrella arch ahead of a TBM cutterhead is a process usually used after probe drilling, and the change in probe drilling rate is sometimes used as a criterion for the application of UA. Two important comprehensive studies on TBM drives in complex geology, on probe drilling and treatment in rock in front of tunnel face have been published by Barla and Pelizza [12] and Peila and Pelizza [13]. A report on this subject, edited by a group of contributors, and published by AFTES, (2014) is also worth mentioning.

It is reported that the UA method was used to cope with extremely poor and squeezing formations in the Abdalajis tunnel. The 7 km long Abdalajis tunnel was one of the most complex construction projects on the 152 km section of the high-speed rail line between Cordoba and Malaga. The double-tube tunnel was driven using universal double-shield TBMs with an excavation diameter of 10 m. The passage of the predicted geological formation at the Abdalajis tunnel presented an extreme challenge, with variations from heavily disturbed zones to areas of very high formation pressure. A double-shield universal TBM had the facilities to treat the ground ahead of the face. In the first 2.5 km of tunnel, the encountered argillite formation was very unstable with rapid convergence of the tunnel walls and frequent face collapses. Large gas inflows with pressures up to 11 bar were measured in this section. In particular, for about 600 m at

TBM Excavation in Difficult Ground Conditions. Case Studies from Turkey. First Edition. Nuh Bilgin, Hanifi Copur, Cemal Balci.
© 2016 Ernst & Sohn GmbH & Co. KG. Published 2016 by Ernst & Sohn GmbH & Co. KG.

the contact with the adjacent limestone formation, the argillite rock was completely altered by tectonic action and transformed into a kind of flowing gravel with very low friction angle and little cohesion. To advance the TBM under these conditions it was necessary to make extensive use of expanding foams to fill large voids in front of and over the cutterhead. In addition, several patterns of chemically grouted micro-piles were executed from inside the TBM shield to improve the ground characteristics and stability ahead of the machine. Despite the extreme conditions and the restrictions to the tunneling operations as explained, the TBM was able to advance without requiring any bypass and without the shields getting trapped, this important event was reported in detail by Grandori [14]. The main characteristics of the UA technique used in this project are illustrated in Figure 15.1.

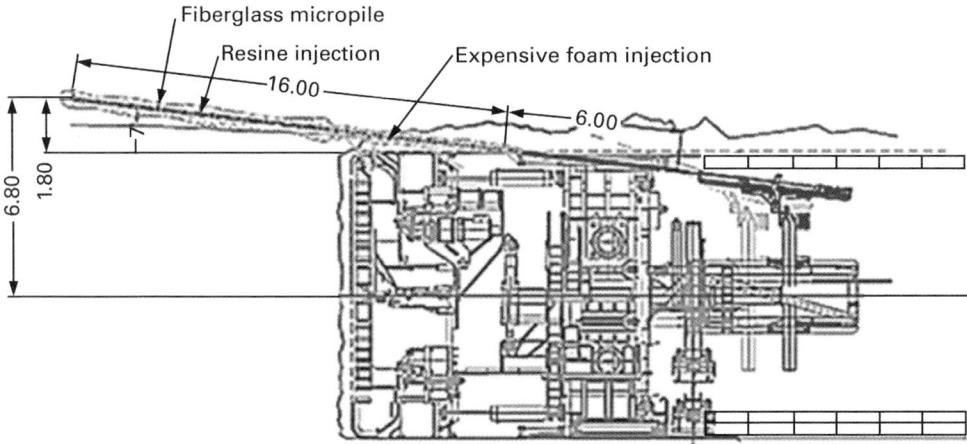

Figure 15.1 Main characteristics of umbrella arch technique used in Abdalajis tunnel Grandori [14].

Another interesting project for umbrella arch application was the Niagara tunnel project of 10.4 km (from the Sir Adam Beck generating complex to above Niagara Falls). A 14.4 m diameter Robbins main beam TBM was selected to bore the tunnel. After about 793 m of excavation the TBM entered the Queenston shale formation, where large rock blocks started to fall from the crown before rock support could be placed. In some cases, significant overbreak up to 3 m above the cutterhead support was reported. A ground support system to cope with the geology, which consisted of 9 m long pipe spiles in an umbrella pattern at the crown of the tunnel was used. Using the new spiling method, overbreak was limited to about 0.9 m above the normal tunnel diameter. Nearly 500 m of very difficult ground was excavated using this method, at average rates of about 3 m per day [15].

15.3   Methodology of using umbrella arch in the Kargi project

A double-shield Robbins TBM of diameter 9.8 m was used in the tunnel. As seen in Figure 15.2, there were eleven ports in the front shield for probe drilling and umbrella arching. Drillings for UA started approximately 3 m behind the cutterhead. At the beginning of the operation, one drill was used for the UA for all ports, and after the third application a second drill was also used to reduce drilling time. For the first seven umbrella arches, all ports on the shield were used, but on the last two umbrella arches, only five ports were used. Arrangements of drill rods and drilling operations for the UA in the Kargi tunnel are illustrated in Figures 15.3 and 15.4 [2].

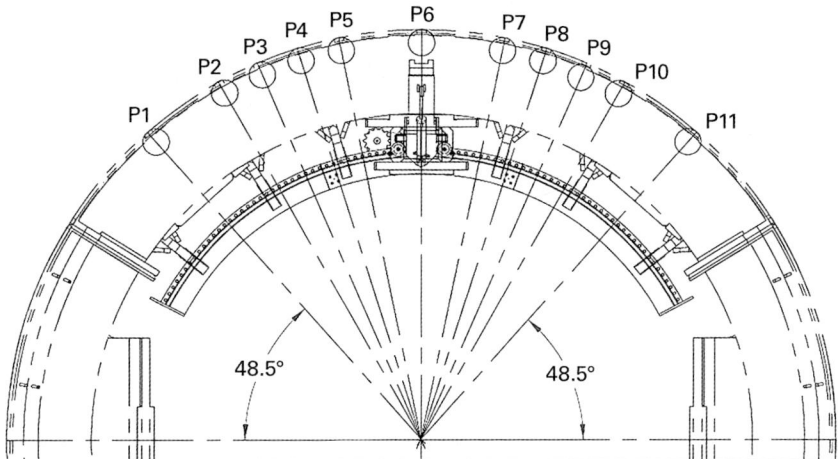

Figure 15.2 Ports for drilling around shield.

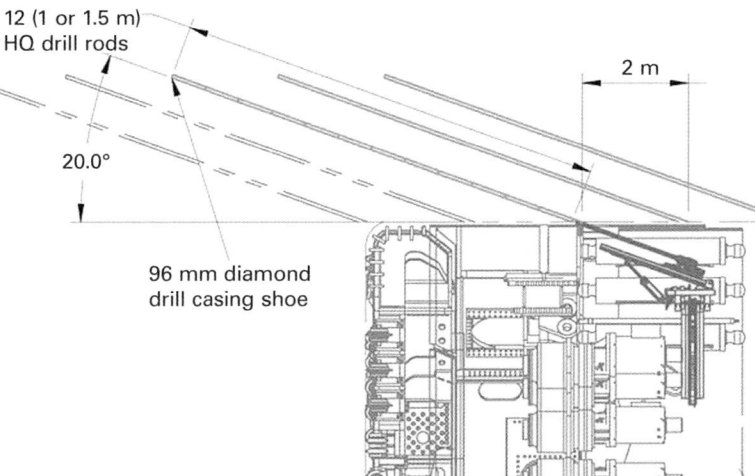

Figure 15.3 Arrangements of drill rods for umbrella arch in the Kargi tunnel, Clark [2].

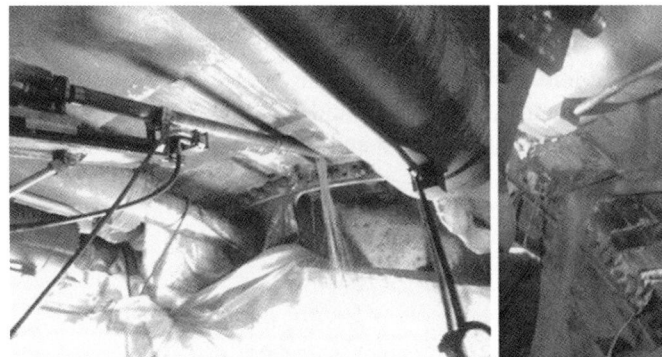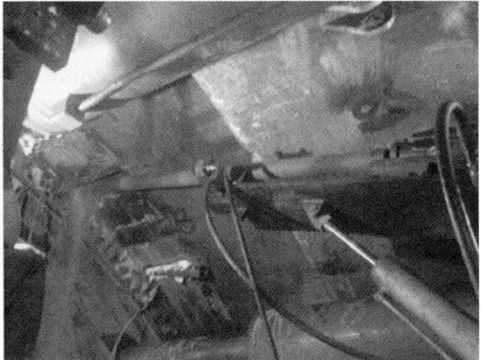

Figure 15.4 Drilling for the umbrella arch.

Drill pipes of lengths 1 and 1.5 m and diameter of 3″ were used for UA. At the beginning of the operation the holes were drilled with a drill bit attached to drill pipes, then the string was removed to recover the drill bit. After removing the drill bit, a string of perforated pipes for injection was placed into the drill hole. Perforated pipes help the penetration of the injection into the ground. Injection was done as up to 120 bar, depending on the injection material properties. For injection, fast-setting micro-fine Portland cement (Rheocem 650) was used. As well as microcement, two-component polyurethane injection resin (MasterRoc MP 355 and Geofoam) were also used. For filling very large voids, two-component urea-silicate (MasterRoc MP 367) injection was used. Generally, 15 m of drill holes at approximately 20° were used with 4 m overlaps between drills. It was seen that one umbrella arch drill permits the support of approximately four ring excavations, which is 6 m in length. As the distance between the drill pipes increases, the effect of the umbrella arch reduces and loose material starts flowing to the cutterhead between the pipes. Figure 15.5 shows the umbrella arch overlaps and how the pipes holds the loose rock in place. Injected material (polymer foam or microcement) fills the voids between the blocky grounds and creates bond between blocks, Figure 15.6. Also, the pipes installed on the UA, prevent loose rocks from flowing through the cutterhead.

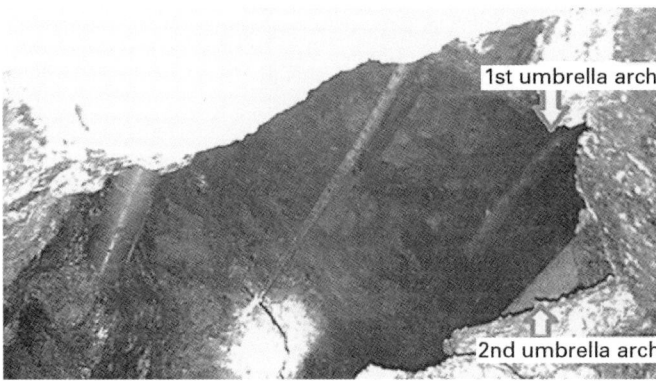

Figure 15.5 Overlapping of pipes for the umbrella arch.

Figure 15.6 Polymer binding loose rocks.

15.4 Criteria used for umbrella arch in the Kargi project and the results

The cutterhead of TBM was jammed seven times in the Kargi tunnel. Bypass tunnels were excavated to rescue the trapped TBM, taking 15–28 days each, with a total delay time of 142 days. A drawing of a typical bypass tunnel is seen in Figure 15.7.

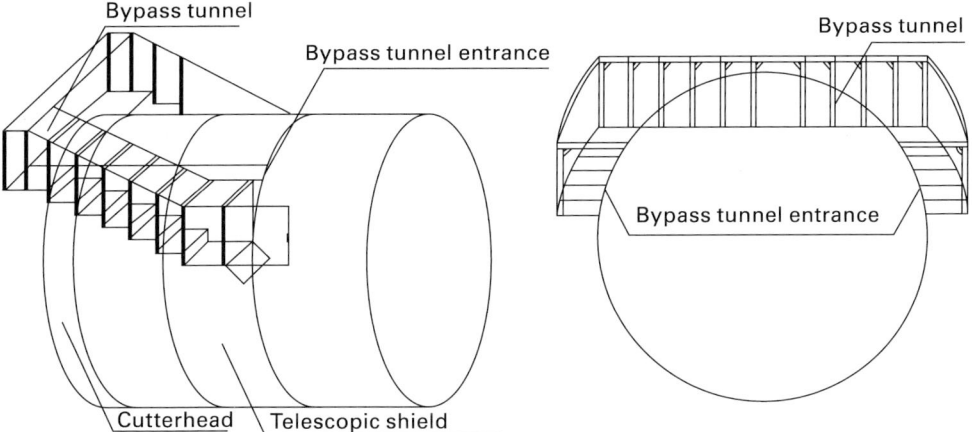

Figure 15.7 Typical drawing of a bypass tunnel in the Kargi tunnel, Clark [2].

One of the criteria for UA was the interpretation of the probe drilling results. Two Atlas Copco COP 1838 were used for probe drilling operations. Drilling rates, percussive and rotary pressures of the drill rigs were continuously recorded during the probe drilling operations. From the beginning of the excavation, a total of more than 5 km of probe drilling was used. For each a set of probe drilling operations, covering 11 ports, a map showing contour lines of equivalent probe drilling rate values was drawn, as seen in Figure 15.8. Low probe drilling rate values represent fractured rock mass and high drilling rate values represent big blocks within this fractured rock mass. The difference between higher and lower drilling rate values represents the degree of rock mass fracturing in front of the TBM. In Figure 15.8, it is apparent that between ports 11 and 10, within a very fractured rock mass, two big blocks, 1×2 m and 1.5×2.5 m in size, exist covering a distance 3.5 and 9 m in front of the cutterhead. The 2.5 m^3 of expanding polyurethane foam having an expansion factor of 1:10 was used in port 11. It can be easily seen in Figure 15.8 that there are also some big blocks within ports 1–6. A volume of 0.5–1.5 m^3 Portland cement with water of 1:1 ratio was used through ports 1–10.

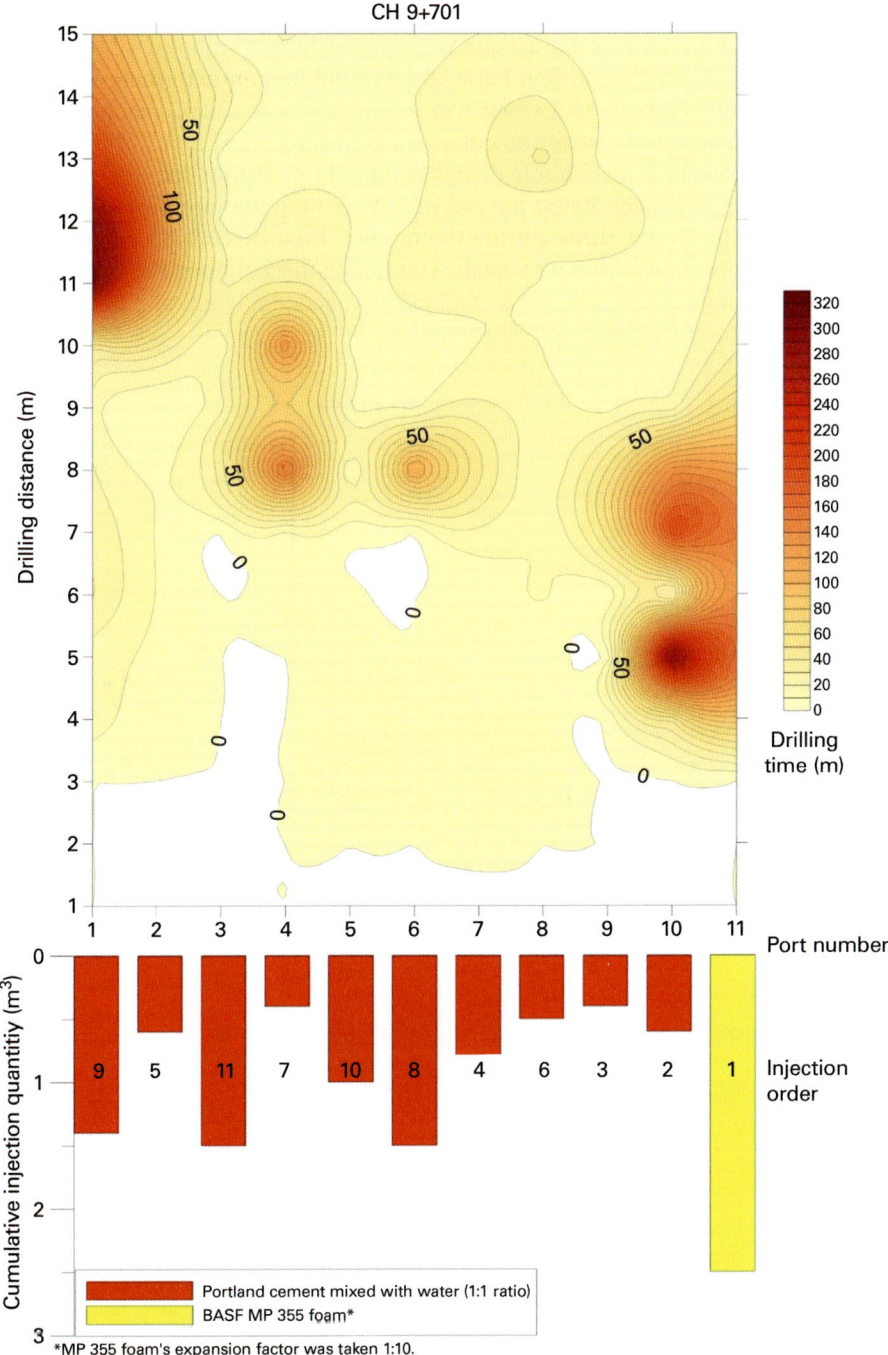

Figure 15.8 Port number, probe drilling rate and cumulative injection quantity at chainage 9+701 km, umbrella arch 3.

The second criterion used for UA was the increasing value of TBM thrust. Before the TBM was blocked it was noticed that torque was approaching a value of 42,000 kNm and due to the effect of East Anatolian Fault, the ground had an extremely fractured characteristic containing blocks of 1×2×0.7 m in size in most cases. Fractured rock masses were accumulated in front of the cutterhead, causing local face collapses and jamming of the cutterhead. Considerable delays in tunnel excavation due to the salvage of the blocked cutterhead necessitated the use of UAs to stop the jamming of the cutterhead and UA was used nine times during the project. Figures 15.9–15.11 represent three cases where umbrella arching was used. As noted in these figures the UA in front of the TBM was applied after TBM torque values approached a value of 37,000 kNm, and after using UA, torque decreased considerably.

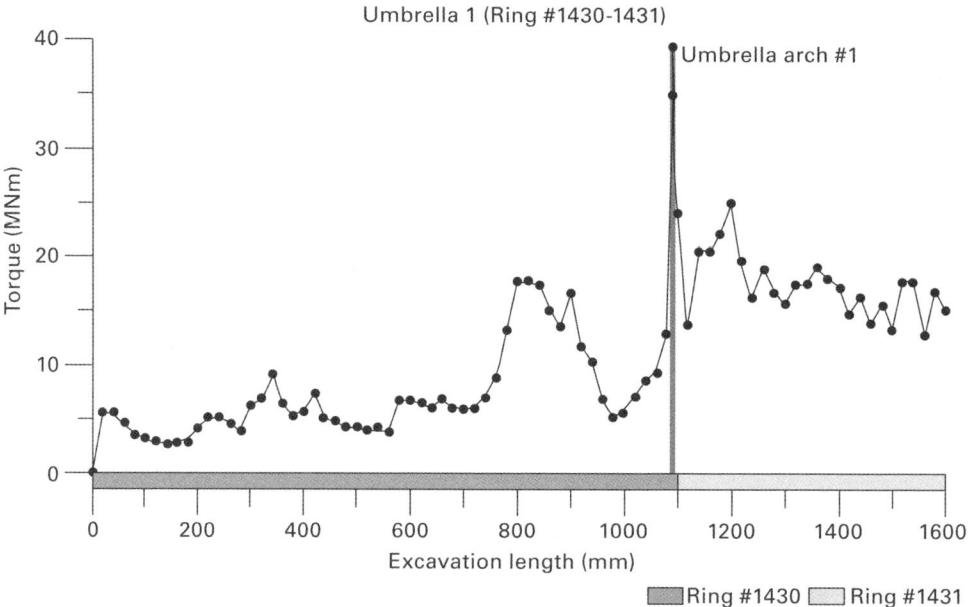

Figure 15.9 İncreasing of torque when excavating ring 1430, and application of umbrella arch 1.

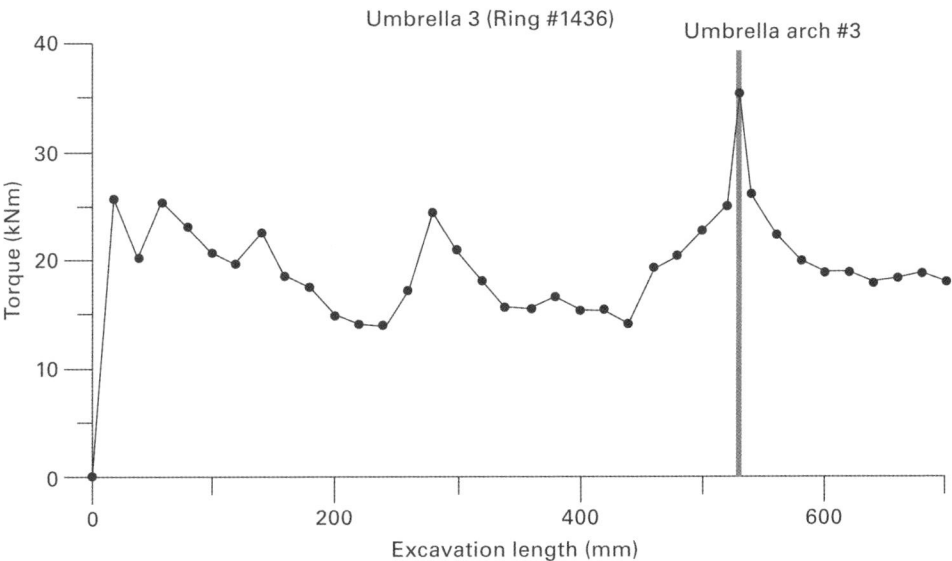

Figure 15.10 Increasing of torque when excavating ring 1436, and application of umbrella arch 3.

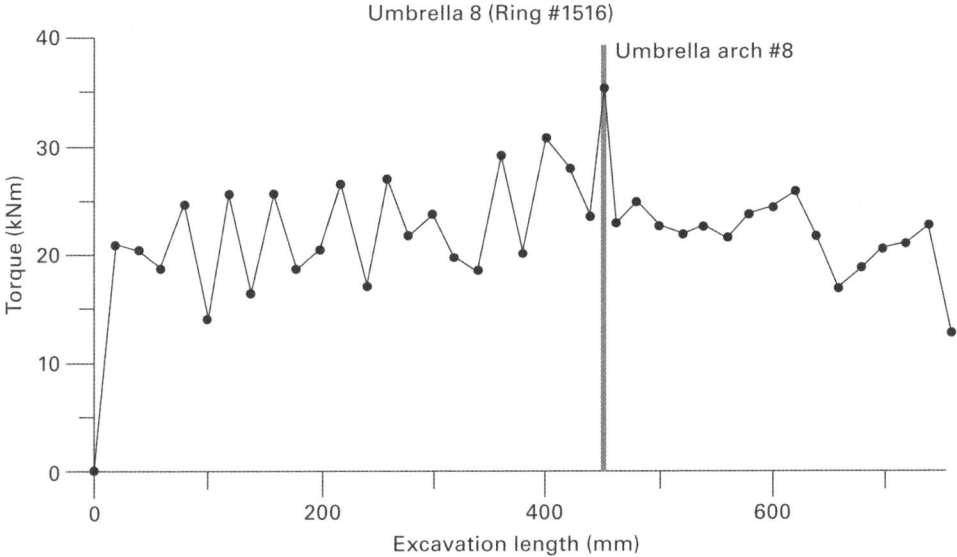

Figure 15.11 Increasing of torque when excavating ring 1516, and application of umbrella arch 3.

Drilling and injection time depends on the rock strength and the characteristics of the blocky ground, but on average time for using UA with two drill rigs was five days with two drills. Moreover, by reducing the drill number from 11 to 5, the total time was reduced to two days. A typical umbrella arch cycle is seen in Figure 15.12. It is worth mentioning that the last bypass tunnel that was excavated to release the cutterhead, took approximately 15 days to complete. In comparison with excavating a bypass tunnel, the umbrella arch method saves time significantly. It is also safer than bypass tunnels as the crew works under the shield.

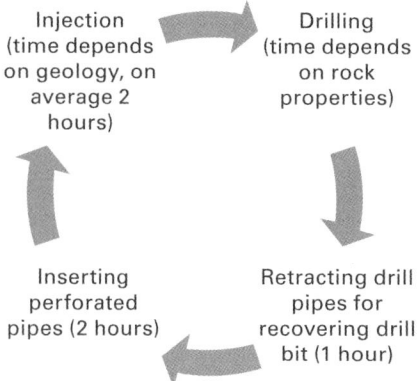

Figure 15.12 Umbrella arch cycle.

15.5   Conclusions

The umbrella arch is a very useful technique to prevent the jamming of the TBM cutterhead in blocky ground, as proved on the Kargi project, but it is important to note that some basic criteria should be established for using this technique. A map showing the contours of changing probe drilling rates and change of torque values were used as the two basic criteria in the Kargi project. The increase of TBM torque up to a certain limit can be used as a critical torque value for using UA. This was established as 42,000 kNm in the Kargi project.

The umbrella arch was used nine times in the Kargi project. As the crew gained experience, time for drilling and injection was reduced from five days to two. However, it should be said that the umbrella arch is a good alternative to excavating bypass tunnels for releasing a blocked cutterhead.

References

[1] Bilgin, N., Ates, U. (2016) Probe Drilling Ahead of Two TBMs in Difficult Ground Conditions in Turkey. *Rock Mechanics and Rock Engineering*, **49** (7), 2763–2772.

[2] Clark, J. (2015). *Extreme Excavation in Fault Zones and Squeezing Ground at the Kargi HEPP in Turkey. SEE Tunnel.* Promoting Tunneling in SEE Region"ITAWTC 2015 Congress and 41st General Assembly, May 22–28, 2015, Lacroma Valamar Congress Center, Dubrovnik, Croatia.

[3] Home, L. (2014a) *Case studies on mechanized rock tunneling.* In: Proceeding of World Tunnel Congress, Tunnels for Energy, 9–10 May, Brazil.

[4] Home, L. (2014b) *The Kargi Challenge, a study of TBM boring vs. conventional excavation.* In: third short course organized by Turkish Tunnelling Society, 28 August Istanbul, Turkey.

[5] Yurt, M., Ozturk, A., Arslan, Z., Nuhoglu, C., Oystein, L., Erdogan, E., Atlar, B., Palakci, Y., Bilgin, N. (2014*) Factors affecting the performance of a double shield TBM in a very complex geology in Kargi Turkey.* In: proceedings World Tunnel Congress 2014 – Tunnels for a better Life. Foz do Iguacu, Brazil.

[6] Ocak, I. (2008) Control of surface settlements with umbrella arch methods in second stage excavations of Istanbul Metro. *Tunn Undergr Sp Technol,* **23**, 674–681.

[7] Peila, D., Pelizza, S. (2003) *Ground reinforcing and steel pipe umbrella system in tunnelling.* In: Kolymbas D (ed) Rational tunneling: advances in geotechnical engineering and tunnelling. Logos Verlag Berlin, Innsbruck, Germany.

[8] Aksoy, C.O., Onargan, T. (2010) The role of umbrella arch and face bolt as deformation preventing support system in preventing building damages. *Tunn Undergr Sp Technol,* **25**, 553–559.

[9] Tuncdemir, H., Aksoy, C.O., Guclu, E., Ozer, S.C. (2012) *Umbrella arch and forepoling support methods: a comparison.* In: EUROCK, Stockholm, Sweden.

[10] Volkmann, G.M., Schubert, W. (2011) *Advantages and specifications for pipe umbrella support systems.* In: 14th Australasian Tunnelling Conference, Development of Underground Space; 8–10 March, Auckland; New Zealand.

[11] Oke, j., Vlachopoulos, N., Marinos, V. (2014) Umbrella Arch Nomenclature and Selection Methodology for Temporary Support Systems for the Design and Construction of Tunnels. *Geotech Geol Eng,* **32**, 97–130.

[12] Barla, G., Pelizza, S. (2000) *TBM tunnelling in difficult ground conditions.* GeoEng, Melbourne, Australia. pp. 1–21.

[13] Pcila, D., Pelizza, S. (2010) Ground probing and treatments in rock TBM tunnel to overcome limiting conditions. *Journal of Mining Sciences,* **45** (6), 602–619.

[14] Grandori, R. (2006) Abdalajis east railway tunnel (Spain) -Double Shield Universal TBM cope with extremely poor and squeezing formations. *Tunn Undergr Sp Technol* **21**. 268–280.

[15] http://www.therobbinscompany.com/en/case-study/the-niagara-tunnel-project/ (January 2016).

16   Index

A

Abrasive ring wear   228, 269
Abrasivity of soil   231
Alpide orogeny   3
Anatolide-Taurides   4
Ancient water wells   6, 91
Andesitic and diabasic dykes   6
Anti-clay agent   1, 192, 199, 200, 201,
    206 f.
Arabian Platform   4
Arcose   6, 44 f., 47, 50, 52, 97
Ayas Tunnel   XII, 126
Aydos Formation   10 f., 39, 241

B

Bakirkoy Formation   8, 15 f., 201
Baltalimani Formation   13 f., 159
Baltalimani Tunnel   XI, 1, 73 f., 90, 108
Basaksehir–Bagcilar metro tunnel   65
Basaksehir–Ikitelli metro tunnel   7
Belgrad Formation   7, 65
Beykoz Sewerage Tunnel   XIII f., 1 f., 84,
    143, 145, 147, 149 ff., 211, 261 f.
Blocky ground   VII, XIII, 2, 99, 129–132,
    134, 136 ff., 140, 167, 187, 295, 319,
    322, 327 f.
Bohme test   229
Bolu Tunnel   XII, 126 ff.
Boreability index   59, 61, 69
Buyukcekmece Wastewater Tunnel   XV,
    2, 262 ff.

C

Carboniferous rocks   6
CCS type discs cutters   VII, 1, 43, 57,
    109–112, 115, 211 ff., 222, 232 f., 238,
    255, 261, 265
Cekmece–Gungoren Formation   15, 16,
    201
Cenozoic formations   4, 68
Cherchar abrasivity index   28, 31, 33
Chisel cutter   VII, XII, 1, 109, 111 f., 115,
    225
Cohesionless soils   92

Complex geology   VII, XII, 1 f., 69 f., 73,
    75, 77, 86 f., 91, 93, 99, 106 f., 131, 140,
    154, 159 f., 163, 165, 175, 186, 293,
    295, 316, 318 f., 329
Cone penetrometer test   198
Conical cutter   109, 115
Consistency index   190, 201
Continent-continent collision   4
Creep behavior   167 f.
Cukurcesme Formation   15 f.
Cutter blockage   228, 251, 258 ff., 269
Cutter consumption   VII, XIV, 1 f., 90,
    222, 225, 232 f., 235, 237 ff., 241,
    243 ff., 247, 249, 250, 251, 253, 255,
    257, 259 ff., 263, 265, 269, 271 f.
Cutter cost   82, 255, 260
Cutter replacement   99, 130, 225, 243 f.,
    251, 255, 258, 267
Cutter wear and damage   3, 112, 171,
    226, 228 f., 264, 267, 269, 271 f.

D

Dogancay Energy Tunnel   XII, 1, 120 f.
Dolayoba Formation   20, 27, 32, 52, 138,
    146, 216
Double shield TBM   XI, 1, 73, 80 f., 83,
    88 ff., 109, 115, 128, 186, 207, 251,
    253 ff., 272, 286 f., 294, 318 f., 329
Dykes   VII, XIII, 2, 6, 8, 14, 54, 57, 63,
    68, 74, 81, 86, 94, 96, 98, 129, 131, 142,
    140, 143, 154, 163, 167, 216, 241, 261,
    293–296, 394, 305 ff., 318
Dynamic Young's modulus   57

E

East Anatolian Fault   VII, XII, 1 f., 117 f.,
    120, 122, 124, 126–129, 184, 336
Elastic modulus   27 f., 52, 63, 125, 154
EPB pressure   98
EPB-TBMs   VII, XII, XIV f., 1 f., 7 f., 55,
    58, 65, 73, 84, 87, 89, 91–94, 96 ff., 100,
    102 ff., 106 ff., 123, 159, 168, 171, 173,
    190, 192, 197, 201, 203, 207 ff., 216 f.,

TBM Excavation in Difficult Ground Conditions. Case Studies from Turkey. First Edition. Nuh Bilgin, Hanifi Copur, Cemal Balci.
© 2016 Ernst & Sohn GmbH & Co. KG. Published 2016 by Ernst & Sohn GmbH & Co. KG.

221 f., 226, 269, 275, 277 ff., 281, 283, 268, 291
Esenler–Otogar Metro Tunnel   65
Explosive gas   92

F
Fault zone   117, 143, 152 f., 159 f., 181, 184, 226, 253, 278, 309 ff., 315 f., 319
FER   198–201, 205, 207, 266,
FIR   192, 198–201, 205, 207, 266 f.
Foam   1, 138 f., 192, 197–201, 205, 207, 226, 267, 279, 320, 322, 324
Foaming agent   198, 200 f., 203, 206, 243
Full-scale linear rock cutting tests   40, 52, 69

G
Gaziantep Formation   195
Geology of Istanbul   2, 4 f., 73, 143, 163
Gerede Tunnel   118 f.
Gozdag Formation   11, 25, 34, 146, 241, 264
Graphitic schist   97, 123, 167, 169, 308, 312
Gripper type TBM   89, 93
Grizzly bars   84, 87 ff., 134, 140, 159 f., 206
Ground deformation   6, 91
GSI   16, 21–27, 29, 32–35, 100, 154
Gungoren Formation   7 f.
Gurpinar (Danisment) Formation   196

H
Herrenknecht   7, 58 f., 62, 64, 73 f., 86, 88, 120, 131 f., 140, 171, 173, 197, 216 f., 241, 262, 277, 287
High strength characteristics   12, 211, 305
Hitachi-Zosen   8
Hoek–Brown criteria   27
Hub breakage   228, 269
Hydropower projects   3

I
Ikitelli Formation   7, 15 f.
Injection ports   126, 178
Internal friction angle   27–36

Istanbul Water and Sewage Administration (ISKI)   196
Istinye Formation   6, 16, 62

J
Jamming of the cutterhead   VII, 179, 296, 326

K
Kaolinite   193
Karakaya formation   123, 125, 169, 171
Kargi Energy Tunnel   XV, 1, 127, 308 f., 311, 313, 315 f.
Karstic cavities   12
Kartal Formation   7, 12 f., 18, 22, 27, 29 f., 37 ff., 47, 52, 54 f., 57, 81, 86, 131, 134 f., 140, 144, 146, 149, 154, 159, 161, 216, 261, 295, 297
Kirklareli Formation   65 ff.
Kizlac Formation   122
Kosekoy–Bilecik high speed railway tunnel   167
Kozyatagi–Kadikoy metro tunnels   XIII, 2, 131, 153, 165
Kurtkoy Formation   10, 11, 19, 23 f., 27, 31 f., 37, 47, 52, 241
Kusdili Formation   150 f., 196, 262, 276

L
LCPC test   229, 231
Learning cost   2, 73, 88,
Learning curve   98
Learning period   101
Line spacing   41, 58, 66 f., 213, 247
Liquid limit   15, 104, 190, 198, 201
Lovat   7, 65
Lower Explosive Limit   273

M
Machine utilization time   XII f., 2, 80, 82, 89, 93, 98, 147 ff., 167, 175, 201, 222, 225, 242
Mahmutbey–Mecidiyekoy metro tunnels   XII, 2, 103, 106, 189
Markusa Formation   119
Marmaray Tunnel   XI, 2, 73, 84 f., 114
Menard modulus   29, 32–36
Mesozoic   4, 6, 68, 122

Meta-claystone   123, 169, 176

Meta-sandstone   97, 122 f., 169, 176, 213 ff.

Meta-siltstone   123, 169, 176, 300 f., 303, 312

Methane   VI, XV, 1, 201, 208, 273 ff., 278–282, 287 ff.

Methane Explosion   VII, XV, 1, 201, 273

Miocene   4, 8, 15 f.

Moda Collector Tunnel   XI, 80

Mohs hardness   229

N

Natural gas reservoir   1, 285, 289

Neolithic settlements   3

North Anatolian Fault   VII, XII, 1, 4, 117, 120, 126 f., 293, 308

NTNU abrasion test   229, 231

Nurdagi Railway Tunnel   XII, 1, 122, 127, 211

O

Opening ratio   VII, 2, 159, 164, 225

Ophiolite   3 f., 308

Ordovician   6, 13, 16, 122, 146

OSHA   274

Otogar Bagcilar Metro Tunnel   65

Otogar–Esenler Metro Tunnels   6, 68, 91

P

Palaeozoic   4, 6, 16, 68, 94, 144, 154, 216

Penetration rate   2, 61, 201, 207, 221 f., 226, 233 f., 239, 250, 253, 256, 266, 313, 316, 319

Permeability 29, 32, 168, 195

Plasticity index 15, 104, 190

Pliocene   15 f., 54, 81

Point load strength test   40

Poisson's ratio   57, 216

Polymer   199, 295, 322 f.

Pontides   3 f.

Porosity   27–32, 34 ff.

Pressure chamber of an EPB-TBM   XV, 275

p-wave velocity   57

Pyroclastic rocks   6

Q

Quartz arenite   6, 10, 241

Quartz content   84, 99, 226, 231, 234, 236, 301, 304

Quaternar   54, 81, 146, 196, 262, 276

R

Radial   41, 168, 225

Relieved cutting tests   50 f.

Ring breakage   228, 238, 243, 251, 258 ff., 269

Ring chipping   228, 233, 251, 258 ff., 267, 269,

Rippers   109

Risk classification   123

RMR   29, 32–35, 75, 121, 123, 125, 169, 179, 216, 253, 264, 285

Robbins   7, 53, 73, 80 f., 88, 117, 120, 146, 253 ff., 308, 320 f.

Rock density   27–35, 185

Rock discontinuities   XI, XIII, 52, 143

Rock rupture   VII, XIII, 2, 129–134

Rolling force   43, 47, 54, 57, 59, 63, 66, 94, 109 f., 213

Rotational speed   53, 58 f., 65, 67, 80, 86, 94, 121, 132, 147, 217, 220, 237, 239, 242, 253, 256

S

Schimazek abrasivity index   229

Schmidt hammer   XI, 37, 40, 52, 57, 229

Screw conveyor   88, 92, 159, 192, 217, 228, 242, 277–280, 282, 302

Segmental lining   81 f., 120, 240, 295

Selimpasa Tunnel   276

Shielded TBM   82, 119, 168, 288, 316

Sievers J value test (bit wear index)   229

Silvan Tunnel   1, 284

Slump testing   92

Slurry TBM   XI, XII, 1, 84 f., 88 f., 228

Smectite   119, 180, 185, 193

Soft ground   65, 91 ff., 104, 111, 222, 260, 269, 278

Soil conditioning experiments   197, 201

Specific energy   66 f., 95 f., 101 f., 106, 111, 114, 163, 196, 249

Squeezing of TBM   XIV, 2 f., 175, 184
Strike-slip faulting   4
Suleymaniye Formation   15 f.
Suruc Tunnel   1, 167, 192, 207
s-wave velocity   57
Swelling stresses   119

T
Tabor abrasivity index   229
Tarabya Sewerage Tunnel   62, 69
TBM breakdown   98 f.
TBM performance prediction model   129
Tectonic stresses   121, 123
Tertiary   4, 6, 55, 86, 131, 140
The clogging of a TBM   VII, 1, 189, 207
The stratigraphy of Istanbul   XI, 8
Trakya Formation   7 f., 14, 17, 21, 27 ff.,
   50 f., 74–77, 80, 88, 114, 154
Transition zones   XIII, 88, 93, 96, 143 f.,
   146, 148, 150, 293, 315 f.
Triassic   4, 6, 55, 86, 123, 131, 140, 169
Tuff   119, 167, 196, 276, 308
Tungsten carbide studded disc
   cutters   XII, 109, 114 f.

Tuzla Formation   13, 25 ff., 35, 216
Tuzla–Dragos Sewerage Tunnel   53, 69,
   80

U
Uluabat Energy Tunnel   XII, 1, 124, 127
Uludere volcanics   119
Unrelieved cutting tests   49
Upper Explosive Limit   273
Uskudar–Cekmekoy Metro Tunnels   2

V
V type disc cutters   VII, XIV, 1, 109–112,
   115, 211 ff., 222
Volcano-clastics   119

Y
Yamanli II HEPP   XIV, 2, 250, 252–260,
   269
Yenikapi–Istanbul Metro Tunnel   91

Z
Zeytinburnu   XIV, 1, 201, 206 f.